D0915945

DATE DUE

Modelling Transport

Second Edition

Modelling Transport

Second Edition

Juan de Dios Ortúzar
Department of Transport Engineering
Pontificia Universidad Católica de Chile
Santiago
Chile

Luis G. Willumsen
Steer Davies Gleave
London
UK

JOHN WILEY & SONS
Chichester • New York • Brisbane • Toronto • Singapore

Other Wiley Editorial Offices

John Wiley & Sons, Inc., 605 Third Avenue,
New York, NY 10158–0012, USA

Jacaranda Wiley Ltd, 33 Park Road, Milton,
Queensland 4064, Australia

John Wiley & Sons, (Canada) Ltd, 22 Worcester Road, Rexdale, Ontario M9W 1L1,
Canada

John Wiley & Sons (SEA) Pte Ltd, 37 Jalan Pemimpin 05–04, Block B, Union
Industrial Building, Singapore 2057

Library of Congress Cataloging-in-Publication Data

Ortúzar S., Juan de Dios.
    Modelling transport / Juan de Dios Ortúzar, Luis G. Willumen. —2nd ed.
        p.   cm.
    Includes bibliographical references and index.
    ISBN 0 471 94193 X
    1. Transportation—Mathematical models.   2. Choice of transportation—Mathematical models.   3.
Trip generation–Mathematical models.   I. Willumen, Luis G.   II. Title.
HE147.7.O77   1994
388'.01'5118—dc20
                                                                                      93-35746
                                                                                      CIP

British Library Cataloguing in Publication Data

A catalogue record for this book is available from the British Library

ISBN 0 471 94193 X

Typeset in 10/12pt Times by Keytec Typesetting Ltd, Bridport, Dorset
Printed and bound in Great Britain by Bookcraft (Bath) Ltd.

Contents

Preface

This book is a result of more than 15 years of collaboration, sometimes at a distance and sometimes working together in Britain and in Chile. Throughout these years we discussed many times what we thought were the strong and weak aspects of transport modelling and planning. We speculated, researched and tested in practice some new and some not so new ideas. We have agreed and disagreed on topics like the level of detail required for modelling or the value of disaggregate models in forecasting; we took advantage of a period when our views converged to put them in writing; here they are.

We wish to present the most important (in our view) tranport modelling techniques in a form accessible to students and practitioners alike. We attempt this giving particular emphasis to key topics in contemporary modelling and planning:

- the practical importance of theoretical consistency in transport modelling;
- the issues of data and specification errors in modelling, their relative importance and methods to handle them;
- the key role played by the decision-making context in the choice of the most appropriate modelling tool;
- the advantages of variable resolution modelling; a simplified background model coupled with a much more detailed one addressing the decision questions in hand;
- the need for a monitoring function relying on regular data collection and updating of forecasts and models so that courses of action can be adapted to a changing environment.

We have approached the subject from the point of view of a modelling exercise, discussing the role of theory, data, model specification in its widest sense, model estimation, validation and forecasting. Our aim in writing this book was to create both a text for a diploma or Master's course in transport and a reference volume for practitioners; however, the material is presented in such a way as to be useful for undergraduate courses in civil engineering, geography and town planning. The book is based on our lecture notes prepared and improved over several years of teaching at undergraduate and graduate levels; we have also used them to teach practitioners both through in-house training programmes and short skills-updating courses. We have extended and enhanced our lecture notes to cover additional material and to help the reader tackling the book without the support of a supervisor.

Chapters 3 to 9, 12 and 13 provide all the elements necessary to run a good 30 sessions course on transport demand modelling; in fact, such a course—with different emphasis in certain subjects—has been taught by us at undergraduate level in

Chile, and at postgraduate level in Britain, Portugal, Colombia and Spain; the addition of material from Chapters 10 and 11 would make it a transport modelling course. Chapters 4 to 6 and 10 to 12 provide the basic core for a course on equilibrium modelling in transport; a course on transport supply modelling would require more material, particularly relating to important aspects of public transport supply which we do not discuss in enough detail. Chapter 1 provides an introduction to transport planning issues and outlines our view on the relationship between planning and modelling. Chapter 2 is there mainly for the benefit of those wishing to brush up their analytical skills and to make the volume sufficiently self-contained.

During our professional life we have been fortunate to be able to combine teaching with research and consultancy practice. We have learnt from papers, research, experimentation and mistakes. We are happy to say the latter have not been too expensive in terms of inaccurate advice. This is not just luck; a conscientious analyst pays for mistakes by having to work harder and longer to sort out alternative ways of dealing with a difficult modelling task. We have learnt the importance of choosing appropriate techniques and technologies for each task in hand; the ability to tailor modelling approaches to decision problems is a key skill in our profession. Throughout the book we examine the practical constraints to transport modelling for planning and policy making in general, particularly in view of the limitations of current formal analytical techniques, and the nature and quality of the data likely to be available.

We have avoided the intricate mathematical detail of every model to concentrate instead on their basic principles, the identification of their strengths and limitations, and a discussion of their use. The level of theory supplied by this book is, we believe, sufficient to select and use the models in practice. We have tried to bridge the gap between the more theoretical publications and the too pragmatic 'recipe' books; we do not believe the profession would have been served well by a simplistic 'how to' book offering a blueprint to each modelling problem. There are no single solutions to transport modelling and planning. A recurring theme in the book is the dependence of modelling on context and theory. Our aim is to provide enough information and guidance so that readers can actually go and use each technique in the field; to this end we have strived to look into practical questions about the application of each methodology. Wherever the subject area is still under development we have striven to make extensive references to more theoretical papers and books which the interested reader can consult as necessary. In respect of other, more settled modelling approaches, we have kept the references to those essential for understanding the evolution of the topic or serving as entry points to further research.

We believe that nobody can aspire to become a qualified practitioner in any area without doing real work in a laboratory or in the field. Therefore, we have gone beyond the sole description of the techniques and have accompanied them with various application examples. These are there to illustrate some of the theoretical or practical issues related to particular models. We provide a few exercises at the end of key chapters; these can be solved with the help of a scientific pocket calculator and should assist the understanding of the models discussed.

Although the book is ambitious, in the sense that it covers quite a number of themes, it must be made clear from the outset that we do not intend (nor believe it possible) to be up-to-the-minute in every topic. The book is a good reflection of the state of the art but for leading-edge research the reader should use the references provided as signposts for further investigation.

The great majority of the final draft of this book we wrote together during a sabbatical visit by one of us to University College London in 1988–89. This was possible thanks to support provided by the UK Science and Engineering Research Council, The Royal Society, Fundación Andes (Chile), The British Council and The Chartered Institute of Transport. We thank them for their support as we acknowledge the funding provided for our research by many institutions and agencies over the past 20 years.

We have made an equal intellectual contribution to the contents of this book but in writing and researching material for it we have benefited from numerous discussions with friends and colleagues. Richard Allsop taught us a good deal about methodology and rigour. Huw Williams's ideas are behind many of the theoretical contributions in Chapter 7; Andrew Daly and Hugh Gunn helped to clarify many issues in Chapters 3, 7 and 8. Dirck Van Vliet's emphasis in explaining assignment and equilibrium in simple but rigorous terms inspired Chapters 10 and 11. Tony Fowkes made valuable comments on car ownership forecasting and stated-preference methods. Jim Steer provided a constant reference to practical issues and the need to develop improved approaches to address them.

Many parts of the book have benefited from a free, and sometimes very enthusiastic, exchange of ideas with our colleagues Sergio Jara and Jaime Gibson at the Universidad de Chile, Marc Gaudry at the Université de Montréal, Roger Mackett at University College London, Dennis Gilbert at Imperial College and Mike Bell at the University of Newcastle upon Tyne. Many others have contributed, without knowing, to our thoughts.

Our final thanks go to our graduate and undergraduate students in Britain, Chile, Colombia, Portugal and Spain; they are always sharp critics and provided the challenge to write this book.

For this second edition we have made an effort to strengthen the material on Stated-Preference techniques as this is a subject that has gained considerably in importance in the last three years. We have also improved a few other sections. These enhancements have benefited from ideas and comments from Rodrigo Garrido from the Pontificia Universidad Católica de Chile and David Pearmain at Steer Davies Gleave. We have also removed errors that had slipped into the first edition; a number of friends and readers have helped in identifying these and we would like thank them for this assistance. Frank Koppelman from Northwestern University, Mariëtte Kraan at University of Twente and Neil Chadwick from Steer Davies Gleave provided the most comprehensive lists of improvements whilst others like Michael Florian from Université de Montréal and Ben Heydecker from University College London also helped in this task.

We did not take on board all suggestions as we felt some required changing the style and approach of the text; we are sure future books in this field will continue to clarify issues or provide greater rigour to many of the topics discussed here. Despite this generous assistance, we are, as before, solely responsible for the errors that still remain in this second edition. We would genuinely value the opportunity to learn from them too.

Juan de Dios Ortúzar and **Luis G. Willumsen**

1 Introduction

1.1 TRANSPORT PLANNING AND MODELLING

1.1.1 Background

Transport problems and planning techniques suffered a revolution in the 1980s. We still encounter many of the same transport problems of the 1960s and 1970s: congestion, pollution, accidents, financial deficits and so on. However, it was possible to learn a good deal from a long period of weak transport planning, limited investment, emphasis on the short term and mistrust in strategic transport modelling and decision making. We have learnt, for example, that old problems do not fade away under the pressure of mild attempts to reduce them through better traffic management; old problems reappear with even greater vigour, pervading wider areas, and in their new forms they seem more complex and difficult to handle.

By the end of the 1980s, the developed world had entered a stage of greater confidence in technical solutions than during the previous twenty years. This was not the earlier confidence in technology as the magic solution to economic and social problems; we have also learnt that this is a mirage. However, electronics and computing have advanced so much as to make possible new conceptions of transport infrastructure (e.g. road transport informatics) and movement systems (e.g. automated driverless trains). Of particular interest to the subject of this book is the advent of low-cost and high-capacity computing; this has practically eliminated computing power as a bottleneck in transport modelling. The main limitations are now human and technical: contemporary transport planning require skilled professionals and, as we will argue below, theoretically sound modelling techniques with competent implementations in software.

A new contemporary dimension is the fact that most developing countries are suffering serious transport problems as well. These are no longer just the lack of roads to connect distant rural areas with markets. Indeed, the new transport problems bear some similarities with those prevalent in the industrialised world: congestion, pollution, and so on; however, they have a number of very distinctive features deserving a specific treatment: low incomes, fast urbanisation and change, high demand for public transport, scarcity of resources including capital, foreign currency, sound data and skilled personnel.

Important technical developments in transport modelling have taken place since the mid-1970s, in particular at major research centres; these developments have been improved and implemented by a small group of resourceful consultants. However, many of these innovations and applications have received limited attention outside the more academic journals. After these years of experimentation there is now a better recognition of the role of modelling in supporting transport planning. This book attempts a review of the best of current practice in transport modelling; in most areas it covers the 'state of the art' but we have selected those aspects which have already been implemented successfully in practice. The book does not represent the leading edge of research into modelling; it tries rather to provide a survival tool-kit for those interested in improving transport modelling and planning, a kind of bridge or entry-point to the more theoretical papers that will form the basis of transport modelling in the future.

1.1.2 Models and their Role

A *model* can be defined as a simplified representation of a part of the real world — the system of interest — which concentrates on certain elements considered important for its analysis from a particular point of view. Models are, therefore, problem and viewpoint specific. Such a broad definition allows us to incorporate both physical and abstract models. In the first category we find, for example, those used in architecture or in fluid mechanics which are basically aimed at design. In the latter, the range spans from the mental models all of us use in our daily interactions with the world, to formal and abstract (typically analytical) representations of some theory about the system of interest and how it works. Mental models play an important role in understanding and interpreting the real world and our analytical models. They are enhanced through discussions, training and, above all, experience. Mental models are, however, difficult to communicate and to discuss.

In this book we are concerned mainly with an important class of abstract models: mathematical models. These attempt to replicate the system of interest and its behaviour by means of mathematical equations based on certain theoretical statements about it. Although they are still simplified representations, these models may be very complex and often require large amounts of data to be used. However, they are invaluable in offering a 'common ground' for discussing policy and examining the inevitable compromises required in practice with a minimum of objectivity. Another important advantage of mathematical models is that during their formulation, calibration and use the planner can learn much, through experimentation, about the behaviour and internal workings of the system under scrutiny. In this way, we also enrich our mental models thus permitting more intelligent management of the transport system.

A model is only realistic from a particular perspective or point of view. It may be reasonable to use a knife and fork on a table to model the position of cars before a collision but not to represent their mechanical features, or their route choice patterns. The same is true of analytical models: their value is limited to a range of problems under specific conditions. The appropriateness of a model is, as discussed in the rest of this chapter, dependent on the context where it will be used. The ability to choose and adapt models for particular contexts is one of the most important elements in the complete planner's tool-kit.

This book is concerned with the contribution transport modelling can make to improved decision making and planning in the transport field. It is argued that the use of models is inevitable and that of formal models highly desirable. However, transport modelling is only one element in transport planning: administrative practices, an institutional framework, skilled professionals and good levels of communication with decision makers, the media and the public are some of the other requisites for an effective planning system. Moreover, transport modelling and decision making can be combined in different ways depending on local experience, traditions and expertise. However, before we discuss how to choose a modelling and planning approach it it worth outlining some of the main characteristics of transport systems and their associated problems. We will also discuss some very important modelling issues which will find application in other chapters of this book.

1.2 CHARACTERISTICS OF TRANSPORT PROBLEMS

Transport problems have become more widespread and severe than ever in both industrialised and developing countries alike. Fuel shortages are (temporarily) not a problem but the general increase in road traffic and transport demand has resulted in congestion, delays, accidents and environmental problems well beyond what has been considered acceptable so far. These problems have not been restricted to roads and car traffic alone. Economic growth seems to have generated levels of demand exceeding the capacity of most transport facilities. Long periods of under-investment in some modes and regions have resulted in fragile supply systems which seem to break down whenever something differs slightly from average conditions.

These problems are not likely to disappear in the near future. Sufficient time has passed with poor or no transportation planning to ensure that a major effort in improving most forms of transport, in urban and inter-urban contexts, is necessary. Given that resources are not unlimited, this effort will benefit from careful and considered decisions oriented towards maximising the advantages of new transport provision while minimising their money costs and undesirable side-effects.

1.2.1 Characteristics of Transport Demand

The demand for transport services is highly *qualitative* and *differentiated*. There is a whole range of specific demands for transport which are differentiated by time of day, day of week, journey purpose, type of cargo, importance of speed and frequency, and so on. A transport service without the attributes matching this differentiated demand may well be useless. This characteristic makes it more difficult to analyse and forecast the demand for transport services: tonne and passenger kilometres are extremely coarse units of performance hiding an immense range of requirements and services.

The demand for transport is *derived*, it is not an end in itself. With the possible exception of sight-seeing, people travel in order to satisfy a need (work, leisure, health) at their destination. This is even more true of goods movements. In order to understand the demand for transport, we must understand the way in which facilities to satisfy these human or industrial needs are distributed over space, in both urban and regional contexts. A good transport system widens the opportunities to satisfy these needs; a heavily congested or poorly connected system restricts options and limits economic and social development.

Transport demand takes place over *space*. This seems a trivial statement but it is the distribution of activities over space which makes for transport demand. There are a few transport problems that may be treated, albeit at a very aggregate level, without explicitly considering space. However, in the vast majority of cases, the explicit treatment of space is unavoidable and highly desirable. The most common approach to treat space is to divide study areas into zones and to code them, together with transport networks, in a form suitable for processing with the aid of computer programs. In some cases, study areas can be simplified assuming that the zones of interest form a corridor which can be collapsed into a linear form. However, different methods for treating distance and for allocating origins and destinations (and their attributes) over space are an essential element in transport analysis.

The spatiality of demand often leads to problems of lack of coordination which may strongly affect the equilibrium between transport supply and demand. For example, a taxi service may be demanded unsuccessfully in a part of a city while in other areas various taxis may be plying for passengers. On the other hand, the concentration of population and economic activity on well-defined corridors may lead to the economic justification of a high-quality mass transit system which would not be viable in a sparser area.

Finally, transport demand and supply have very strong *dynamic* elements. A good deal of the demand for transport is concentrated on a few hours of a day, in particular in urban areas where most of the congestion takes place during specific peak periods. This time-variable character of transport demand makes it more difficult — and interesting — to analyse and forecast. It may well be that a transport system could cope well with the *average* demand for travel in an area but that it breaks down during peak periods. A number of techniques exist to try to spread the peak and average the load on the system: flexible working hours, staggering working times, premium pricing, and so on. However, peak and off-peak variations in demand remain a central, and fascinating, problem in transport modelling and planning.

1.2.2 Characteristics of Transport Supply

The first distinctive characteristic of transport supply is that it is a *service* and not a good. Therefore, it is not possible to stock it, for example, to use it in times of higher demand. A transport service must be consumed when and where it is produced, otherwise its benefit is lost. For this reason it is very important to estimate demand with as much accuracy as possible in order to save resources by tailoring the supply of transport services to it.

Many of the characteristics of transport systems derive from their nature as a service. In very broad terms a transport system requires a number of fixed assets, the *infrastructure*, and a number of mobile units, the *vehicles*. It is the combination of these, together with a set of rules for their operation, that makes possible the movement of people and goods.

It is often the case that infrastructure and vehicles are not owned nor operated by the same group or company. This is certainly the case of most transport modes, with the notable exception of many rail systems. This separation between supplier of infrastructure and provider of the final transport service generates a rather complex set of interactions between government authorities (central or local), construction companies, developers, transport operators, travellers and shippers, and the general public. The latter plays several roles in the supply of transport services: it is usually the residents affected by a new scheme, or the unemployed in an area seeking improved accessibility to foster economic growth; it may even be car owners wishing to travel unhindered through somebody else's residential area.

The provision of transport infrastructure is particularly important from a supply point of view. Transport infrastructure is 'lumpy', one cannot provide half a runway or one-third of a railway station. In certain cases, there may be scope for providing a gradual build-up of infrastructure to match growing demand. For example, one can start providing an unpaved road, upgrade it later to one or two lanes with surface treatment; at a later stage a well-constructed single and dual carriageway road can be built, to culminate perhaps with motorway standards. In this way, the provision of infrastructure can be adjusted to demand and avoid unnecessary early investment in expensive facilities. This is more difficult in other areas such as airports, metro lines, and so on.

Investments in transport infrastructure are not only lumpy but also take a long time to be carried out. These are usually large projects. The construction of a major facility may take from 5 to 15 years from planning to full implementation. This is even more critical in urban areas where a good deal of disruption is also required to build them. This disruption involves additional costs to users and non-users alike.

Moreover, transport investment has an important political role. For example, politicians in developing countries often consider a road project a safe bet: it shows they care and is difficult to prove wrong or uneconomic by the popular press. In industrialised nations, transport projects usually carry the risk of alienating large numbers of residents affected by them or travellers suffering from congestion and delay in overcrowded facilities. Political judgement is essential in choices of this kind but when not supported by planning, analysis and research, these decisions result in responses to major problems and crises only; in the case of transport this is, inevitably, too late. Forethought and planning are essential; the emphasis of the 1980s on short term, tactical transport management measures reflected as much a mistrust of modelling as an avoidance of political problems and an inability to think strategically about future transport provision.

The separation of providers of infrastructure and suppliers of services introduces economic complexities too. For a start, it is not always clear that all travellers and shippers actually perceive the total costs incurred in providing the services they use. The charging for roadspace, for example, is seldom carried out directly and when it happens the charge does not include congestion costs or other external effects, perhaps the nearest approximation to this being toll roads and modern road-pricing schemes. The use of taxes on vehicles and fuels is only a rough approximation to charging for the provision of infrastructure.

But, why should this matter? Is it not the case that other goods and services like public parks, libraries and the police are often provided without a direct charge for them? What is wrong with providing free roadspace? According to elementary economic theory it does matter. In a perfect market a good allocation of resources to satisfy human needs is only achieved when the marginal costs of the goods equal their marginal utility. This is why it is often advocated that the price of goods and services, i.e. their perceived cost, should be set at their marginal cost. Of course real markets are not perfect and ability to pay is not a good indication of need; however, this general framework provides the basis for contrasting other ways of arranging pricing systems and their impact on resource allocation.

Transport is a very important element in the welfare of nations and the well-being of urban and rural dwellers. If those who make use of transport facilities do not perceive the resource implications of their choices, they are likely to generate an equilibrium between supply and demand that is inherently inefficient. Underpriced scarce resources will be squandered whilst other abundant but priced resources may not be used. The fact that overall some sectors of the economy (typically car owners) more than pay for the cost of the roadspace provided, is not a guarantee of more rational allocation of resources. Car owners probably see these annual taxes as fixed, *sunk*, costs which at most affect the decision of buying a car but not that of using it.

An additional element of distortion is provided by the number of concomitant- or *side-effects* associated with the production of transport services: accidents, pollution and environmental degradation in general. These effects are seldom *internalised*; the user of the transport service rarely perceives nor pays for the costs of cleaning the environment or looking after the injured in transport related accidents. Internalising these costs could also help to make better decisions and to improve the allocation of demand to alternative modes.

One of the most important features of transport supply is *congestion*. This is a term which is difficult to define as we all believe we know exactly what it means. However, most practitioners do know that what is considered congestion in Leeds or Lima is often accepted as uncongested flow in London or Lagos. Congestion arises when demand levels approach the capacity of a facility and the time required to use it (travel through it) increases well above the average under low demand conditions. In the case of transport infrastructure the inclusion of an additional vehicle generates supplementary delay to all other users as well, see for example Figure 1.1. Note that the contribution an additional car makes to the delay of all users is greater at high flows than at low flow levels.

This is the external effect of congestion, perceived by others but not by the driver originating it. This is a cost which schemes such as electronic road pricing attempt to internalise to help more reasoned decision making by the individual.

1.2.3 Equilibration of Supply and Demand

In general terms the role of transport planning is to ensure the satisfaction of a certain demand **D** for person and goods movements with different trip purposes, at different times of the day and the year, using various modes, given a transport

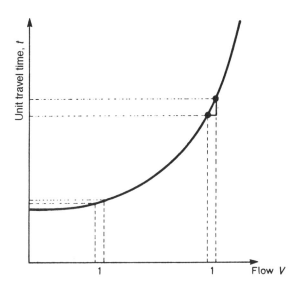

Figure 1.1 Congestion and its external effects

system with a certain operating capacity. The transport system itself can be seen as made up of:

- an infrastructure (e.g. a road network);
- a management system (i.e. a set of rules, for example driving on the right, and control strategies, for example at traffic signals);
- a set of transport modes and their operators.

Consider a set of volumes on a network $\mathbf{V}$, a corresponding set of speeds $\mathbf{S}$, and an operating capacity $\mathbf{Q}$, under a management system $\mathbf{M}$. In very general terms the speed on the network can be represented by:

$$S = f\{\mathbf{Q}, \mathbf{V}, \mathbf{M}\} \tag{1.1}$$

The speed can be taken as an initial proxy for a more general indicator of the *level of service* (LOS) provided by the transport system. In more general terms a LOS would be specified by a combination of speeds or travel times, waiting and walking times and price effects; we shall expand on these in subsequent chapters. The management system $\mathbf{M}$ may include traffic management schemes, area traffic control and regulations applying to each mode. The capacity $\mathbf{Q}$ would depend on the management system $\mathbf{M}$ and on the levels of investment $\mathbf{I}$ over the years, thus:

$$Q = f\{\mathbf{I}, \mathbf{M}\} \tag{1.2}$$

The management system may also be used to redistribute capacity among the infrastructure, producing $\mathbf{Q}'$, and/or giving priority to certain types of users over others, either on efficiency (public-transport users, cyclists), environmental (electric vehicles) or equity grounds (pedestrians).

As in the case of most goods and services, one would expect the level of demand **D** to be dependent on the level of service provided by the transport system and also on the allocation of activities **A** over space:

$$D = f\{S, A\} \tag{1.3}$$

Combining equations (1.1) and (1.3) for a fixed activity system one would find the set of equilibrium points between supply and demand for transport. But then again, the activity system itself would probably change as levels of service change over space and time. Therefore one would have two different sets of equilibrium points: short-term and long-term ones. The task of transport planning is to forecast and manage the evolution of these equilibrium points over time so that social welfare is maximised. This is, of course, not a simple task: modelling these equilibrium points should help to understand this evolution better and assist in the development and implementation of management strategies **M** and investment programmes **I**.

Sometimes very simple cause-effect relationships can be depicted graphically to help understanding the nature of some transport problems. A typical example is the car/public-transport vicious circle depicted in Figure 1.2.

Economic growth provides the first impetus to increase car ownership. More car owners means more people wanting to transfer from public transport to car; this in turn means fewer public-transport passengers, to which operators may respond by increasing the fares, reducing the frequency (level of service) or both. These measures make the use of the car even more attractive than before and induce more people to buy cars, thus accelerating the vicious circle. After a few cycles (years) car drivers are facing increased levels of congestion; buses are delayed, are becoming increasingly more expensive and running less frequently; the accumulation of sensible individual decisions results in a final state in which almost everybody is worse off than originally.

This simple representation can also help to identify what can be done to slow down or reverse this vicious circle. These ideas are summarised in Figure 1.3. Physical measures like bus lanes or other bus-priority schemes are particularly attractive as they also result in a more efficient allocation of roadspace. Public

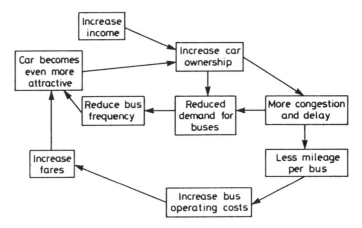

Figure 1.2   Car and public-transport vicious circle

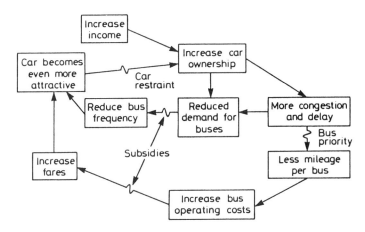

Figure 1.3   Breaking the car/public-transport vicious circle

transport subsidies have strong advocates and detractors; they may reduce the need for fare increases, at least in the short term, but tend to generate large deficits and to protect poor management from the consequences of their own inefficiency.

The type of model behind Figures 1.2 and 1.3 is sometimes called a *structural model*, as discussed in Chapter 12; these are simple but powerful constructs, in particular because they permit the discussion of key issues in a fairly parsimonious form. However, they are not exempt from dangers when applied to different contexts. Think, for example, of the vicious circle model in the context of developing countries. Population growth will maintain demand for public transport much longer than in industrialised countries. Indeed, some of the bus flows currently experienced in developing countries are extremely high, reaching 400 to 600 buses per hour one-way along some corridors. The context is also relevant when looking for solutions; it has been argued that one of the main objectives of introducing bus-priority schemes in developing countries is not to protect buses from car-generated congestion but to organise bus movements (Gibson *et al.* 1989). High bus volumes often implement a *de facto* priority, and interference between buses may become a greater source of delay than car-generated congestion. To be of value, the vicious circle model must be revised in this new context.

It should be clear that it is not possible to characterise all transport problems in a unique, universal form. Transport problems are context dependent and so should be the ways of tackling them. Models can offer a contribution in terms of making this identification of problems and selection of ways of addressing them more solidly based.

1.3 MODELLING AND DECISION MAKING

1.3.1 Decision-making Styles

Before choosing a modelling framework one needs to identify the general decision-making approach adopted in the country, government or decision unit. It must be recognised that there are several decision-making styles in practice and that not all of them use modelling as a basic building block. It is possible to characterise decision-making styles in many different ways; what follows is an adaptation of a scheme due to Nutt (1981).

1.3.1.1 Decisions Based on Master Plans

There is a long tradition of developing and applying this type of strategy in the area of transport and land-use planning. There is also a good deal of experience on the failure of such an approach. Decisions are based on interpretations of the master plan which provides the rules governing contingencies, performance expectations and what can or cannot be done. The master plan is normally prepared with a good deal of care and attention for the future, perhaps through the use of an expensive, one-off transport modelling exercise of the type implemented in the 1960s and 1970s.

Master plans may be reasonable when the environment is stable and therefore decision problems recur. They have the advantage of informing everybody what will be done. They have the major disadvantage that they seldom work: the economic, social and technological environments change faster than the plan can adapt. Moreover, as new information does not fit into a master plan it is seldom collected or used to improve decision making. It is unfortunate that many developing countries have tried to adopt this type of approach which is inappropriate to their changing environment.

1.3.1.2 Normative Decision Theory or Substantive Rationality

This is the rational decision-making approach implicit in most textbooks about transport planning. It is sometimes referred to as the 'systems approach' to planning. Here, quantification is essential. The decision problem is seen as one of choosing options from a complete set of alternatives and scenarios, with estimates on their probability of occurrence; the *utility* of each alternative is quantified in terms of benefits and costs and other criteria like environmental protection, safety, and so on.

In some cases it may be possible to cast a decision problem into a mathematical programming framework. This means that the objective function is well understood and specified, and that the same applies to the constraints defining a solution space. However, for most real problems some elements of the objective function or constraints may be difficult to quantify or to convert into common units of measurement, say money or time. It may also be difficult to include some of the probabilistic elements in each case, but a good deal about the problem is learnt in the process. Modelling is at the core of this approach. Some of the problems of applying normative decision theory are:

- the accusation of insensitivity to the aspirations of the public;
- its high costs; and
- the alienation of decision makers who may not understand nor accept the analytical treatment of the problem.

Moreover, this approach has often failed to deliver results on time and with acceptable accuracy; witness the backlash against large scale transport modelling in the 1980s.

1.3.1.3 Behavioural Decision Theory

This is an attempt to soften the edges of the normative decision-theory approach by recognising that often decision makers are not utility maximisers but simply *satisficers* (see the discussion in Chapter 7). The search for better solutions is often stopped once an acceptable one is found; this approach combines searching, learning and decision making but it is unlikely to generate solutions which are not marginal improvements on current practice.

Indeed, the approach is like a marginal analysis of the optimisation problem but starting from a non-optimal situation; minor but acceptable improvements are explored and chosen with the hope of moving towards a better scale of the environment. Modelling plays a more restricted role here and may well be of the simpler kind discussed in Chapter 12. Marginal demand and supply models fit very well with this approach.

1.3.1.4 Group Decision Making

This is a common approach followed in many areas where committees rule. Decision making becomes a learning process inside a group with decision authority and a specific remit. Individuals contribute their expertise and knowledge and the group tries to apply these to the decision problem. Qualitative and quantitative information, as well as forecasts, are combined in this approach but not in any systematic way. The views of persuasive or powerful members of the group may predominate beyond their intrinsic value.

Participation in a decision group of this kind often fosters the acceptance of decisions and this is an important element in any planning context. Sometimes *steering groups* are set up to guide and advise on the implementation of major modelling exercises. They potentially provide good advice on the inclusion or otherwise of issues in the modelling task and promote and acceptability of the resulting plan.

1.3.1.5 Adaptive Decision Making

This general approach is a more flexible version of group decision making. It recognises the interaction between pressure groups, none holding complete decision-making power. Each group sees the problem in a different way and therefore negotiation and compromise are required to reach a decision. The approach is advocated wherever the problem contains many ill-defined variables and interactions where no normative or behavioural theory exists to suggest cause-and-effect relationships. The approach is common in legislative decision making and diplomacy, as well as in many decision groups, in particular where survival becomes the dominant objective.

Within this approach transport modelling plays only a minor instrumental role. A study and its recommendations are used as arguments in negotiations and pressure groups, almost regardless of their intrinsic value. 'State-of-the-art' techniques may be used, not because they are more sensitive or accurate but because this claim renders greater value to the study results and hence power to its sponsors.

1.3.1.6 Mixed-mode Decision-making Strategies

Finally, it is often possible to combine many of the above approaches into a flexible strategy. This is quite common in transport studies. This type of strategy acknowledges that the manner in which decisions are made could be as important as the course of action selected. The mixed-mode approach uses analysis, persuasion, bargaining and political strategies in different *arenas* and under different *goals*. The latter are sometimes considered fixed and known but as bargaining proceeds it may become necessary to enlarge them in order to include the concerns of conflicting parties.

This is a realistic approach which accepts that for important problems the goals and arenas often shift as part of the decision-making process. Consider for example plans to introduce major road improvements or a new motorway network into an urban area; or the task of choosing the best alignment for a rail link from the Channel Tunnel to London; or the expansion of airport capacity serving a conurbation. Modelling often plays an important role in this approach; emphasis should be placed on flexibility and capacity for adaptation, the inclusion of new variables and the quick analysis of innovative policies and designs.

1.3.2 Choosing Modelling Approaches

This book assumes that the decision style adopted involves the use of models, for example substantive rationality or mixed-mode strategies, but it does not advocate a single (i.e. a normative) decision-making approach. The acceptability of modelling, or a particular modelling approach, within a decision style is very important. Models which end up being ignored by decision makers not only represent wasted resources and effort, but result in frustrated analysts and planners. It is further proposed that there are several features of transport problems and models which must be taken into account when specifying an analytical approach:

1. The decision-making context. This involves the adoption of a particular *perspective* and a choice of a *scope* or coverage of the system of interest. The choice of perspective defines the type of decisions that will be considered: strategic issues or schemes, tactical (transport management) schemes, or even specific operational problems. The choice of scope involves specifying the level of analysis: is it just transport or does it involve activity location too? In terms of the transport system, are we interested in just demand or also on the supply side at different levels: system or suppliers' performance, cost minimisation issues within suppliers, and so on? The question of how many options need to be considered to satisfy different interest groups or to develop a single best scheme is also crucial. The decision-making context, therefore, will also help define requirements on the models to be used, the variables to be included in the model, or considered given or exogenous.

2. Accuracy required. This stems from the first point and it is heavily influenced by the next two. However, it is often the case that the accuracy required is just that necessary to discriminate between a good scheme and a less good one. In some cases the best scheme may be quite obvious, thus requiring less accurate modelling. Remember, however, that common sense has been blamed for some very poor transport decisions in the past.

3. The availability of suitable data, their stability and the difficulties involved in forecasting their future values. In some cases very little data may be available; in others, there may be reasons to suspect the information, or to have less confidence in future forecasts for key planning variables as the system is not sufficiently stable. In many cases the data available will be the key factor in deciding the modelling approach.

4. The state of the art in modelling for particular type of interventions in the transport system. This in turn can be subdivided into:

- behavioural richness;
- mathematical and computer tractability;
- availability of good solution algorithms.

It has to be borne in mind that in practice all models assume that some variables are exogenous to it. Moreover, many other variables are omitted from the modelling framework on the grounds of not being relevant to the task in hand, too difficult to forecast or expected to change little and not influence the system of interest. An explicit consideration of what has been left out of the model may help to decide on its appropriateness for a given problem.

5. Resources available for the study. These include money, data, computer hardware and software, technical skills, and so on. Two types of resource are, however, worth highlighting here: time and level of communication with decision makers and the public. *Time* is probably the most crucial one: if little time is available to make a choice between schemes, shortcuts will be needed to provide timely advice. Decision makers are prone to setting up absurdly short timescales for the assessment of projects which will take years to process through multiple decision instances, years to implement and many more years to be confirmed as right or wrong. On the other hand, a good level of communication with decision makers and the public will alleviate this problem: fewer unrealistic expectations about our ability to accurately model transport schemes will arise, and a better understanding of the advantages and limitations of modelling will moderate the extremes of blind acceptance or total rejection of study recommendations.

6. Data processing requirements. This aspect used to be interpreted as something like 'how big a computer do you need?'. The answer to that question today is 'not very big', as a good microcomputer will do the trick in most cases. The real bottleneck in data processing is the human ability to collect, code, input the data, run the programs and interpret the output. The greater the level of detail, the more difficult all these human tasks will be. The use of computer-assisted data collection and graphics for input–output of programs reduces the burden somewhat. However, more progress is needed in these areas before this bottleneck is cleared.

7. Levels of training and skills of the analysts. Training costs are usually quite high; so much so that it is sometimes better to use an existing model that is well understood, than to embark on acquiring and learning to use a slightly more advanced one. This looks, of course, like a recipe for stifling innovation and progress; however, it should always be possible to spend some time building up strengths in new advanced techniques without rejecting the experience gained with earlier models.

Florian *et al.* (1988) specify in greater detail decision-making contexts using a two-dimensional framework based on two key propositions. The first proposition states that an (analytical) transport planning problem is specified by the choice of a scope and a perspective.

Proposition 1

To frame a transportation planning problem is simultaneously:

- *to identify the appropriate level of analysis for the situation in hand, and*
- *to adopt a perspective that determines what is endogenous and exogenous to the situation.*

The second proposition states that to solve an analytical transport planning problem it is necessary to select a model and a set of suitable solution methods.

Proposition 2

To resolve a transportation planning problem is to choose the precise model-specific functional relationships and computational algorithms that produce an acceptable answer for a particular frame.

In this context a frame is a combination of a suitable level of analysis and perspective. These two propositions are 'fleshed out' through two dimensions: the *levels* dimension and the *perspectives* dimension. The levels dimension includes six groups of *procedures*, where a procedure centres on one or more models and their specific solution algorithms. These are:

- *activity location* procedures **L**;
- *demand procedures* **D**;
- *transport system performance* procedures **P**, which produce as output levels of service, expenditure and practical capacities, and depend on demand levels and on transport supply conditions;
- *supply actions* procedures **S**, which determine the actions taken by suppliers of transport services and infrastructure; these depend on their objectives (profit

maximisation, social welfare), institutional environment, their costs and estimates of future states of the system;

- *cost minimisation* procedures **CM**;
- *production* procedures **PR**.

The last two have more to do with the microeconomic issues affecting the suppliers in their choice of input combinations to minimise costs.

The perspectives dimension considers the six level procedures **L, D, P, S, CM, PR** and three perspectives: a *strategic* perspective **STR**, a *tactical* perspective **TAC** and an *operational* perspective **OPE**. These are, of course, related to the planning horizons and the levels of investment; however, in this context they must be seen as generic concepts dealing with the capacity:

- to visualise the levels **L, D, P, S, CM, PR** in their true and relative importance;
- to choose, at any level, what is to be regarded as fixed and what as variable.

Figure 1.4 summarises the way in which different perspectives and levels usually interact. The largest and most aggregate is, of course, the strategic level; analysis and choice at this level have major system-wide and long-term impacts, and usually involve resource acquisition and network design. Tactical issues have a narrower perspective and concern questions like making the best use of existing facilities and infrastructure. The narrowest perspective, the operational one, is concerned with the

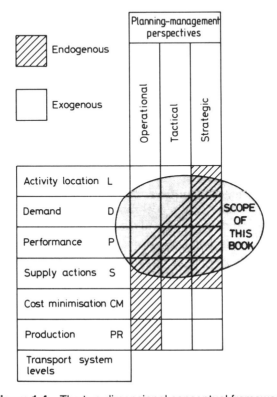

Figure 1.4   The two-dimensional conceptual framework

short-term problems of suppliers of transport services which fall outside the scope of this book; nevertheless, the actual decisions on, for example, levels of service or vehicle size, are important exogenous input to some of the models discussed in this book, and this is depicted in Figure 1.4.

This is, of course, a rather abstract and idealised way of visualising planning problems. However, it helps to clarify the choices the analyst must face in developing a transport modelling approach. In this book we are mainly concerned with strategic and tactical issues at the demand and performance procedure levels. Nevertheless, some of the models discussed here sometimes find application outside these levels and perspectives.

1.4 ISSUES IN TRANSPORT MODELLING

We have already identified the interactions between transport problems, decision-making styles and modelling approaches. We need to discuss now some of the critical modelling issues which are relevant to the choice of model. These issues cover some general points like the roles of theory and data, model specification and calibration. But perhaps the most critical choices are those between the use of aggregate or disaggregate approaches, cross-section or time-series models, and revealed or stated preference techniques.

1.4.1 General Modelling Issues

Wilson (1974) provides an interesting list of questions to be answered by any would-be modeller; they range from broad issues such as the *purpose* behind the model-building exercise, to detailed aspects such as *what techniques* are available for building the model. We will discuss some of these below, together with other modelling issues which are particularly relevant to the development of this book.

1.4.1.1 The Roles of Theory and Data

Many people tend to associate the word 'theory' with endless series of formulae and algebraic manipulations. In the urban transport modelling field this association has been largely correct: it is difficult to understand and replicate the complex interactions between human beings which are an inevitable feature of transport systems.

Some theoretical developments attempting to overcome these difficulties have resulted in models lacking adequate data and/or computational software for their practical implementation. This has led to the view, held strongly by some practitioners, that the gap between theory and practice is continually widening; this is something we have tried to redress in this book.

An important consideration on judging the contribution of a new theory is whether it places any meaningful restrictions on, for example, the form of a demand function. There is at least one documented case of a 'practical' transport planning study, lasting several years and costing several million dollars, which relied on 'pragmatic' demand models with a faulty structure (i.e. some of its elasticities had a wrong sign; see Williams and Senior 1977). Although this could have been diagnosed *ex ante* by the pragmatic practitioners, had they not despised theory, it was only discovered *post hoc* by theoreticians.

Unfortunately (or perhaps fortunately, a pragmatist would say), it is sometimes possible to derive similar functional forms from different theoretical perspectives (this, the *equifinality issue*, is considered in more detail in Chapter 8). The interpretation of the model output, however, is heavily dependent on the theoretical framework adopted. For example, the same functional form of the gravity model can be derived from analogy with physics, from entropy maximisation and from maximum utility formalisms. The interpretation of the output, however, may depend on the theory adopted. If one is just interested in flows on links it may not matter which theoretical framework underpins the analytical model function. However, if an evaluation measure is required, the situation changes, as only an economically based theory of human behaviour will be helpful in this task. In other cases, phrases like: 'the gravitational pull of this destination will increase', or 'this is the most probable arrangement of trips' or 'the most likely trip matrix consistent with our information about the system' will be used; these provide no help in devising evaluation measures but assist in the interpretation of the nature of the solution found. The theoretical framework will also lend some credence to the ability of the model to forecast future behaviour. In this sense it is interesting to reflect on the influence practice and theory may have on each other. For example, it has been noted that models or analytical forms used in practice have had traditionally a guiding influence on the assumptions employed in the development of subsequent theoretical frameworks. It is also well known that widely implemented forms, like the gravity-logit model we will discuss in Chapters 6 and 7, have been the subject of strong *post hoc* rationalisation:

> theoretical advances are especially welcome when they fortify existing practice which might be deemed to lack a particularly convincing rationale (Williams and Ortúzar, 1982b).

The two classical styles of approach to the development of theory are known as *deductive* (building a model and testing its predictions against observations) and *inductive* (starting with data and attempting to infer general laws). The deductive approach has been found more productive in the pure sciences and the inductive approach has been preferred in the analytical social sciences. It is interesting to note that data are central to both; in fact, it is well known that data availability usually leaves little room for negotiation and compromise in the trade-off between modelling *relevance* and modelling *complexity*. Indeed, in very many cases the nature of the data restricts the choice of model to a single option.

The question of data is closely connected with issues such as the type of variables to be represented in the model and this is, of course, closely linked again to questions about theory. Models predict a number of dependent (or endogenous) variables given other independent (or explanatory) variables. To test a model we would normally need data about each variable. Of particular interest are the *policy variables*, which are those assumed to be under the control of the decision maker, e.g. those the analyst may vary in order to test the value of alternative policies or schemes.

Another important issue in this context is that of aggregation:

- How many population strata or types of people do we need to achieve a good representation and understanding of a problem?
- In how much detail do we need to measure certain variables to replicate a given phenomenon?
- Space is crucial in transport; at what level of detail do we need to code the origin and destination of travellers to model their trip making behaviour?

1.4.1.2 Model Specification

In its widest and more interesting sense this issue considers the following themes.

Model Structure   Is it possible to replicate the system to be modelled with a simple structure which assumes, for example, that all alternatives are independent? Or is it necessary to build more complex models which proceed, for example, to calculate probabilities of choice conditional on previous selections? Contemporary models, such as those discussed in Chapters 7 to 9, usually have parameters which represent aspects of model structure and the extensions to methodology achieved by the mid-1980s have allowed the estimation of more and more general model forms. However, as Daly (1982b) has remarked, although it might be supposed that ultimately all issues concerned with model form could be resolved by empirical testing, such resolution is neither possible nor appropriate.

Functional Form   Is it possible to use linear forms or does the problem require postulating more complex non-linear functions? The latter may represent the system of interest more accurately, but certainly will be more demanding in terms of resources and techniques for model calibration and use. Although theoretical considerations may play a big role in settling this question, it is also possible to examine it in an inductive fashion by means of 'laboratory simulations', for example in stated intentions/preferences experiments.

Variable Specification   This is the more usual meaning attached to the specification issue; which variables to use and how (which form) they should enter a given model. For example, if income is assumed to influence individual choice, should the variable enter the model as such or deflating a cost variable? Methods to advance on this question range from the deductive ('constructive') use of theory, to the inductive statistical analysis of the data using transformations.

1.4.1.3 Model Calibration, Validation and Use

A model can be simply represented as a mathematical function of variables X and parameters θ, such as:

$$Y = f(\mathbf{X}, \boldsymbol{\theta}) \tag{1.4}$$

It is interesting to mention that the twin concepts of *model calibration* and *model estimation* have taken traditionally a different meaning in the transport field. Calibrating a model requires choosing its parameters, assumed to have a non-null value, in order to optimise one or more *goodness-of-fit* measures which are a function of the observed data. This procedure has been associated with the physicists and engineers responsible for the first generation of transport models who did not worry unduly about the statistical properties of these indices, e.g. how large any calibration errors could be.

Estimation involves finding the values of the parameters which make the observed data more likely under the model specification; in this case one or more parameters can be judged *non-significant* and left out of the model. Estimation also considers the possibility of examining empirically certain specification issues; for example, structural and/or functional form parameters may be estimated. This procedure has tended to be associated with the engineers and econometricians responsible for the second generation of models, who placed much importance on the statistical testing possibilities offered by their methods. However, in essence both procedures are the same because the way to decide which parameter values are better is by examining certain previously defined goodness-of-fit measures. The difference is that these measures generally have well-known statistical properties which in turn allow confidence limits to be built around the estimated values and model predictions.

Because the large majority of transport models have been built on the basis of *cross-sectional* data, there has been a tendency to interpret model *validation* exclusively in terms of the goodness-of-fit achieved between observed behaviour and base-year predictions. Although this is a *necessary*, it is by no means a *sufficient* condition for model validation; this has been demonstrated by a number of cases which have been able to compare model predictions with observed results in *before-and-after* studies (see the discussion in Williams and Ortúzar, 1982a). Validation requires comparing the model predictions with information *not used* during the process of model estimation. This obviously puts a more stringent test on the model and requires further information or more resources.

One of the first tasks a modeller faces is to decide which variables are going to be predicted by the model and which are possibly required as inputs to it. Some will not be included at all, either because the modeller lacks control over them or simply because the theory behind the model ignores them (see Figure 1.5). This implies immediately a certain degree of error and uncertainty (we will come back to this problem in Chapter 3) which of course gets compounded by other errors which are also inherent to modelling; for example, sampling errors and, more important, errors due to the unavoidable simplifications of reality the model demands in order to be practical (see Figure 1.5).

Thus, the main use of models in practice is for *conditional forecasting*: the model will produce estimates of the dependent variables given a set of independent variables. In fact, typical forecasts are conditional in two ways (Wilson 1974):

- in relation to the values assigned to the policy variables in the plan, the impact of which is being tested with the model;
- in relation to the assumed values of other variables.

A model is normally used to test a range of alternative plans for a range of

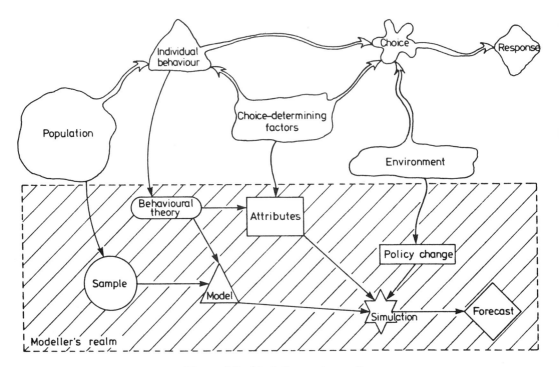

Figure 1.5   Modelling and sampling

possible assumptions about the future value of the other variables (e.g. low- and high-income scenarios). This means that it might be 'run' many times in the context of examining a particular problem. For this reason it may be of crucial importance that its specification allows for quick turn-around time in a computer; this is not an easy task in the case of a full-scale transportation model which involves complex processes of equilibration between supply and demand, as we will discuss in Chapter 11.

1.4.2 Aggregate and Disaggregate Modelling

The level of aggregation selected for the measurement of data is an important issue in the general design of a transportation planning study. Although a greater level of detail—leading to a higher degree of accuracy—should improve the quality of a forecasting model, the costs of data collection and analysis, and of most other aspects of the modelling exercise, will probably increase.

Of central interest is the aggregation of exogenous data, that is, information about items other than the behaviour of travellers which is assumed endogenous (i.e. the model attempts to replicate it). For example, throughout the years it has been a cause for concern whether a given data item represents an average over a group of travellers rather than being collected specifically for a single individual. When the model at base aims at representing the behaviour of more than one individual (e.g. a population segment like car owners living in a zone), such as in the case of the *aggregate* or *first-generation* models we will examine in Chapters 5 and 6, a certain degree of aggregation of the exogenous data is inevitable. But when the model at base attempts to represent the behaviour of individuals, such as in the case of the *disaggregate* or *second-generation* models we will study in Chapters 7 to 9, it is conceivable that exogenous information can be obtained and used separately for each traveller. An important issue is then whether, as often, it might be preferable on cost or other grounds to use less detailed data (see Daly and Ortúzar 1990).

Forecasting future demand is a crucial element of the majority of transport planning studies. Being able to predict the likely usage of new facilities is an essential precursor to rational decision making about the advantages or otherwise of providing such facilities. It may also be important to have an idea about the sensitivities of demand to important variables under the control of the analyst (e.g. the price charged for its use). In most cases the forecasts and sensitivity estimates must be provided at the aggregate level, that is, they must represent the behaviour of an entire population of interest. Therefore, the analyst using disaggregate models must find a sound method for aggregating model results to provide these indicators.

First-generation models were used almost without exception in transportation studies up to the late 1970s; they became familiar, demanded relatively few skills on the part of the analyst (but required arcane computer knowledge) and have the property of offering a 'recipe' for the complete modelling process, from data collection through the provision of forecasts at the level of links in a network. The output of these models, perhaps because they were generated by obscure computer programs, were often considered more accurate than intended, for example predicting turning movement flows 15 years in the future. First-generation models have been severely (and sometimes justifiably) criticised for their inflexibility, inaccuracy and cost. Unfortunately, many second-generation approaches which have adopted sophisticated treatments of the choices and constraints faced by individual travellers have failed to take the process through to the production of forecasts, sometimes because they require data which cannot reasonably be forecast.

Disaggregate models, which became increasingly popular during the 1980s, offer substantial advantages over the traditional methods while remaining practical in many application studies. However, one important problem in practice is that they demand from the analyst quite a high level of statistical and econometric skills for their use (in particular for the interpretation of results), certainly much higher than in the case of aggregate models. Moreover, the differences between first- and second-generation model systems have often been overstated. For example, the disaggregate models were first marketed as a radical departure from classical methods, a 'revolution' in the field, while eventually it became clear that an 'evolutionary' view was more adequate (see Williams and Ortúzar 1982b). In fact, in many cases there is complete equivalence between the forms of the forecasting models (Daly 1982a). The essential difference lies in the treatment of the description of behaviour, particularly during the model development process; in many instances the disaggregate approach is clearly superior to the grouping of behaviour zonally and by predefined segments.

Attempts to clarify the issue of whether disaggregate or aggregate approaches were to be preferred, and in what circumstances, have basically concluded that there is no such thing as a definitive approach appropriate to all situations (see Daly and Ortúzar 1990). These attempts have also established the need for guidelines to help the despairing practitioner to choose the most appropriate model tools to apply in a particular context. We have striven to answer that call in this book.

1.4.3 Cross-section and Time Series

The vast majority of transport planning studies up to the late 1980s relied on information about trip patterns revealed by a cross-section of individuals at a single point in time. Indeed, the traditional use of the cross-sectional approach transcended the differences between the two generations of models discussed above.

A fundamental assumption of the cross-sectional approach is that a measure of the response to incremental change may simply be found by computing the derivatives of a demand function with respect to the policy variables in question. This makes explicit the assumption that a realistic *stimulus-response* relation may be derived from model parameters estimated from observations at one point in time. This would be reasonable if there were always enough people changing their choices, say of mode or destination, in *both* directions and without habit or time-lag effects.

However, the cross-sectional assumption has two potentially serious drawbacks. First, a given cross-sectional data set may correspond to a particular 'history' of changes in the values of certain key variables influencing choice. For example, changes in mode or location in time may have been triggered by a series of different stimuli (petrol prices, life-cycle effects, etc.) and the extent to which a system is considered to be in *disequilibrium* (because of, say, inertia) will depend on these. The trouble is that it can be shown (see Chapter 7) that the response of groups with exactly the same current characteristics, but having undergone a different path of changes, may be very different indeed. Second, data collected at only one point in time will usually fail to discriminate between alternative model formulations, even between some arising from totally different theoretical postulates. It is always possible to find 'best-fit' parameters from base-year data even if the model suffers severe mis-specification problems; the trouble is, of course, that these do not guarantee good response properties for a future situation. As we saw in section 1.4.1, a good base-year fit is not a sufficient condition for model validation.

Thus, in general it is not possible to discriminate between the large variety of possible sources of dispersion within a cross-sectional data set (i.e. preference dispersion, habit effects, constraints, and so on). Real progress in understanding and assessing the effectiveness of forecasting models, however, can only be made if information is available on response over time. From a theoretical point of view, it is also desirable that appropriate frameworks for analysis are designed which allow the eventual refutation of hypotheses relating to response. Until this is achieved, a general problem of potential misrepresentation will continue to cast doubts on the validity of cross-sectional studies.

The discussion above has led many people to believe that, where possible, longitudinal or time-series data should be used to construct more dependable forecasting models. This type of data incorporates information on response by design. Thus, in principle, it may offer the means to directly test and even perhaps reject hypotheses relating to response.

Longitudinal data can take the form of *panels* or more simply *before-and-after* information. Unfortunately, models built on this type of data have severe technical problems of their own; in fact, up to the end of the 1980s progress in this area had been very limited. We will discuss some of the issues involved in the collection and use of this type of information in Chapters 3 and 7.

1.4.4 Revealed and Stated Preferences

The development of good and robust models is quite difficult if the analyst cannot set up experiments to observe the behaviour of the system under a wide range of conditions. Experimentation of this kind is not practical nor viable in transport and the analyst is restricted, like an astronomer, to make observations on events and choices she does not control. Up to the mid-1980s it was almost axiomatic that modelling transport demand should be based on information about observed choices and decisions, i.e. *revealed-preferences* data. Within this approach, project evaluation requires expressing policies in terms of changes in attributes which 'map onto' those considered to influence current behaviour. However, this has practical limitations basically associated with survey costs and the difficulty of distinguishing the effects of attributes which are not easy to observe, e.g. those related to notions such as quality or convenience. Another practical embarrassment has been traditionally the 'new option' problem, whereby it is required to forecast the likely usage of a facility not available at present and perhaps even radically different to all existing ones.

Stated-preferences/intentions techniques, borrowed from the field of market research, were put forward by the end of the 1970s as offering a way of experimenting with transport-related choices, thus solving some of the problems outlined above. Stated-preferences techniques base demand estimates on an analysis of the response to *hypothetical choices*; these, of course, can cover a wider range of attributes and conditions than the real system. However, these techniques were severely discredited in their beginning because it was not known how to discount for the overenthusiasm of certain respondents, e.g. not even half of the individuals stating they would take a given course of action actually did so when the opportunity eventually arose.

It took a whole decade for the situation to change, but by the end of the 1980s stated-preferences methods were perceived by many to offer a real chance to solve the above-mentioned difficulties. Moreover, it has been found that, in appropriate cases, revealed- and stated-preferences data and methods may be employed in complementary senses with the strengths of both approaches recognised and combined. In particular, they are considered to offer an invaluable tool for assisting the modelling of completely new alternatives. We will examine data-collection aspects of stated-preferences methods in Chapter 3 and modelling issues in Chapter 8.

1.5 THE STRUCTURE OF THE CLASSIC TRANSPORT MODEL

Years of experimentation and development have resulted in a general structure which has been called the classic transport mode. This structure is, in effect, a result from practice in the 1960s but has remained more or less unaltered despite major improvements in modelling techniques during the 1970s and 1980s.

The general form of the model is depicted in Figure 1.6. The approach starts by considering a zoning and network system, and the collection and coding of planning, calibration and validation data. These data would include base-year levels for population of different types in each zone of the study area as well as levels of economic activity including employment, shopping space, educational and recreational facilities. These data are then used to estimate a model of the total number of trips generated and attracted by each zone of the study area (*trip generation*). The next step is the allocation of these trips to particular destinations, in other words their *distribution* over space, thus producing a trip matrix. The following stage normally involves modelling the choice of mode and this results in *modal split*, i.e. the allocation of trips in the matrix to different modes. Finally, the last stage in the classic model requires the *assignment* of the trips by each mode to their corresponding networks: typically private and public transport.

The classic model is presented as a sequence of four sub-models: trip generation, distribution, modal split and assignment. It is generally recognised that travel decisions are not actually taken in this type of sequence; a contemporary view is that the 'location' of each sub-model depends on the form of the utility function assumed

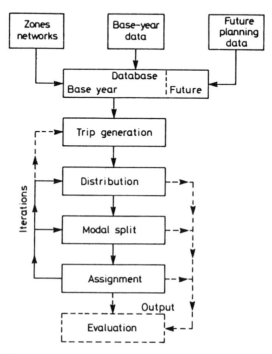

Figure 1.6   The classic four-stage transport model

to govern all these travel choices (see Williams 1977). Moreover, the four-stage model is seen as concentrating attention on only a limited range of travellers' responses. Current thinking requires an analysis of a wider range of responses to transport problems and schemes. For example, when faced with increased congestion a trip maker can respond with a range of simple changes to:

- the **route** followed to avoid congestion or take advantage of new links; this includes choice of parking place or combination of services in the case of public transport;
- the **mode** used to get to the destination;
- the **time** of departure to avoid the most congested part of the peak;
- the **destination** of the trip to a less congested area;
- the **frequency** of journeys by undertaking the trip at another day, perhaps combining it with other activities.

Furthermore, other more complex responses take place in the longer term, for example changes in jobs, residential location, choice of shopping areas and so on; all of these will respond, at least partially, to changes in the accessibility provided by the transport system.

Despite these comments, the four-stage sequential model provides a point of reference to contrast alternative methods. For example, some contemporary approaches attempt to treat simultaneously the choices of trip frequency (trips per week), destination and mode of travel thus collapsing trip generation, distribution and mode choice in one single model. Other approaches emphasise the role of household activities and the travel choices they entail; concepts like sojourns, circuits, and time and money budgets are used in this context to model travel decisions and constraints. These modelling strategies are more difficult to cast in terms of the four main decisions or sub-models above. They have played more of a research role so far and their operational use is some time away yet. However, the improved understanding of travel behaviour these activity based models provide is likely to enhance more conventional modelling approaches in the future.

The trip generation–distribution–modal split and assignment sequence is the most common but not the only possible one. Some past studies have put modal split before trip distribution and immediately after (or with) trip generation. This permits a greater emphasis on decision variables depending on the trip generation unit, perhaps the household. However, forcing modal split before the destination is known, makes it difficult to include the attributes of the journey and modes in the model. This detracts policy relevance from the modal-split model. Perhaps a better approach is to perform distribution and mode choice simultaneously, as discussed in Chapter 6. Note also that the classic model makes trip generation inelastic, that is independent of the level of service provided in the transport system. This is probably unrealistic but only recently techniques have been developed which can take systematic account of these effects.

Once the model has been calibrated and validated for base-year conditions it must be applied to one or more planning horizons. In order to do this it is necessary to develop *scenarios* and plans describing the relevant characteristics of the transport system and planning variables under alternative futures. The preparation of realistic and consistent scenarios is not a simple task as it is very easy to fall into the trap of constructing futures which are not financially viable nor realistic in the context of the likely evolution of land use and activities in the study area. Despite these difficulties, scenario writing is still more of an art than a technique and requires a good deal of engineering expertise combined with sound political judgement; unfortunately these are scarce resources seldom found together in planning teams.

Having prepared realistic scenarios and plans for testing, the same sequence of models is run again to simulate their performance. A comparison is then made between the costs and benefits, however measured, of different schemes under different scenarios; the idea is to choose the most attractive programme of investment and transport policies which satisfies the demand for movement in the study area.

An important issue in the classic four-stage model is the consistent use of variables affecting demand. For example, at the end of the traffic assignment stage new flow levels, and therefore new travel times will be obtained. These are unlikely to be the same travel times assumed when the distribution and mode choice models were run, at least when the models are used in the forecasting mode. This seems to call for the re-run of the distribution and modal-split models based now on the new travel times. The subsequent application of the assignment model may well result in a new set of travel times; it will be seen that in general the naive recycling of the model does not lead to a stable set of distribution, modal split and assignment models with consistent travel times. This problem will be treated in some detail in Chapter 11; its particular relevance is in the risk of choosing the wrong plan depending on how many cycles one is prepared to undertake.

1.6 CONTINUOUS TRANSPORT PLANNING

Transport planning models on their own do not solve transport problems. To be useful they must be utilised within a decision process adapted to the chosen decision-making style. The classic transport model was originally developed for an idealised normative decision-making approach. Its role in transport planning can be presented as contributing to the key steps in a 'rational' decision-making framework as in Figure 1.7:

1. **Formulation of the problem**. A problem can be defined as a mismatch between expectations and perceived reality. The formal definition of a transport problem requires reference to objectives, standards and constraints. The first reflect the values implicit in the decision-making process, a definition of an ideal but achievable future state. Standards are provided in order to compare, at any one time, whether minimum performance is being achieved at different levels of interest. For example, the fact that many signalised junctions in a city operate at more than 90% degree of saturation can be taken to indicate an overloaded network. Constraints can be of many types, financial, temporal, geographical, technical or simply certain areas or types of building that should not be threatened by new proposals.

2. **Collection of data** about the present state of the system of interest in order to support the development of the analytical model. Of course, data collection is not independent from model development, as the latter defines which types of data are needed: data collection and model development are closely interrelated.

3. **Construction of an analytical model** of the system of interest. The tool-set provided in this book can be used to build transport models including demand and system performance procedures from a tactical and strategic perspective. In general, one would select the simplest modelling approach which makes possible a choice between schemes on a sound basis. The construction of an analytical model involves specifying it, estimating or calibrating its parameters and validating its performance with data not used during calibration.

4. **Generation of solutions** for testing. This can be achieved in a number of ways, from tapping the experience and creativity of local transport planners and interested parties, to the construction of a large-scale design model, perhaps using optimisation techniques. This involves supply- and cost-minimisation procedures falling outside the scope of this book.

5. In order to test the solutions or schemes proposed in the previous step it is necessary to **forecast the future values of the planning variables** which are used as inputs to the model. This requires the preparation of consistent quantified descriptions, or scenarios, about the future of the area of interest, normally using forecasts from other sectors and planning units. We will come back to this issue in Chapter 13.

6. **Testing the model and solution**. The performance of the model is tested under different scenarios to confirm its reasonableness; the model is also used to simulate different solutions and estimate their performance in terms of a range of suitable indicators. These must be consistent with the identification of objectives and problem definition above.

7. **Evaluation of solutions** and recommendation of a plan/strategy/policy. This involves operational, economic, financial and social assessment of alternative courses of action on the basis of the indicators produced by the models. A combination of skills is required here, from economic analysis to political judgement.

8. **Implementation of the solution** and search for another problem to tackle; this requires recycling through this framework starting again at point (1).

Although based on the idea of a normative decision theory approach, this framework could also be used within behavioural decision-theory styles, to formulate master plans or to provide ammunition in the bargaining involved in adaptive decision making. It implicitly assumes that the problem can be fully specified, the constraints and decision space can be defined and the objective function identified, even if not necessarily completely quantified.

However, one of the main arguments of this book is that real transport systems do not obey the restrictions above: objective functions and constraints are often difficult to define. With hindsight these definitions often turn out to be blinkered: by

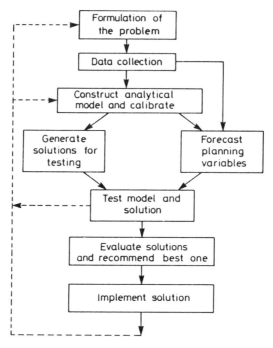

Figure 1.7   A framework for rational decision making with models

narrowing a transport problem we may gain the illusion of being able to solve it; however, transport problems have the habit of 'biting back', of reappearing in different places and under different guises; new features and perspectives are added as our understanding of the transport system progresses; changes in the external factors and planning variables throw our detailed transport plans off course. A strong but fixed normative decision-making framework may be suitable for simpler, well-defined and constrained problems but it hardly helps to deal with richer, more complex, many-featured and multi-dimensional transport issues.

How can we improve this general approach to cope with an ever-changing world? It seems essential to recognise that the future is much more tenuous than our forecasting models would lead us to believe. If this is the case, master plans need revising at regular intervals and other decision-making strategies need supporting with the inclusion of fresh information regularly collected to check progress and correct course where necessary. Adaptive or mixed-mode decision-making styles seem more flexible and appropriate to the characteristics of transport problems. They recognise the need to continually redefine problems, arenas and goals as we understand them better, identify new solution strategies, respond to political and technological changes and enhance our modelling capabilities through training, research and experience.

The introduction of a monitoring function is an important addition to the scheme in Figure 1.7. A monitoring system is not restricted to regular data collection; it should also facilitate all other stages in the decision-making framework, as highlighted in Figure 1.8. There are two key roles for a monitoring system. First, it should provide data to identify departures from the estimated behaviour of the

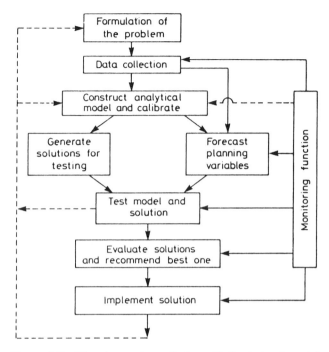

Figure 1.8   Planning and monitoring with the help of models

transport system and of exogenous key variables such as population and economic growth. Second, the data collected should be valuable in further validating and enhancing the modelling approach followed in preparing the plans.

A good monitoring system should also facilitate learning by the planning team and provide ideas on how to improve and modify models. In this sense, major disruptions to the transport system, like public-transport strikes, short-term fuel shortages or major roadworks which may temporarily change the network structure and its characteristics, should provide a major source of information on transport behaviour to contrast with model predictions. These unplanned experiments should enable analysts to test and enhance their models. A monitoring system fits very well with the idea of a regular or continuous planning approach in transport. If the monitoring system is not in place, it should be established as part of any transportation study.

Monitoring the performance of a transport system and plans is such an important function that it deserves to influence the choice of transport models used to support planning and policy making. The use of models which can be re-run and updated using low-cost and easy-to-collect data, seems particularly appropriate to this task. As we shall see in subsequent chapters, these simpler models cannot provide all the behavioural richness of other more detailed approaches. However, there is scope for combining the two techniques, applying the tool with the highest resolution to the critical parts of the problem and using coarser tools that are easier to update to monitor progress and identify where and when a new detailed modelling effort is needed. We have made an attempt to identify the scope for trade-offs of this kind in the remainder of this book.

The adoption of a monitoring function enables the implementation of a continuous planning process. This is in contrast to the conventional approach of spending considerable resources over a period of one or two years to undertake a large-scale transport study. This burst of activity may be followed by a much longer period of limited effort in planning and updating of plans. Soon the reports and master plans become obsolete or simply forgotten, and nobody capable of running the models again is left in the planning unit. Some years later a new major planning and modelling effort is embarked upon and the cycle is repeated. This style of planning with the help of models in fits and starts is wasteful of resources, does not encourage learning and adaptation as a planning skill, and alienates analysts from real problems. This approach is particularly painful in developing countries: they do not have resources to waste and the rapid change experienced there speeds up plan and data obsolescence. The use of models that are simpler and easier to update is advocated in Chapter 12 to help the implementation of a sound but low-cost monitoring function.

1.7 THEORETICAL BASIS VERSUS EXPEDIENCE

One of the recurring themes of transport modelling practice is the distance, and some would say mistrust, between theoreticians and practitioners. The practitioner would often refer to the need to choose between a theoretically sound but difficult to implement set of models, and a more pragmatic modelling approach reflecting the limitations of the data, time and resources available for a study. The implication is that the 'pragmatic' method can deliver the answers needed in the time period available for the study, even if shortcuts must be taken.

The authors have nothing against pragmatic approaches provided they deliver the answers needed to make sound decisions. There is no point in using sophisticated and expensive (but presumably theoretically sound) models for the sake of winning some credit in the academic fraternity. However, there are several reasons to prefer a model based on a sound theoretical background:

1. To guarantee stable results. The recommendations from a study should not depend on how many iterations of a model were run. Prescriptions like 'always start from free flow costs' or 'iterate twice only' are not good enough reasons to assume stable results: next time somebody will suggest running a couple more iterations or a different, and quite justifiable, starting point; this should not be able to change the recommendations for or against a particular scheme.
2. To guarantee consistency. One should be careful about using a particular model of travellers' choice in one part of a model system and a different one in another. Pragmatic models sometimes fail to pass this test. Model consistency is necessary to pass the test of 'reasonableness' and public scrutiny.
3. To give confidence in forecasting. It is almost always possible to fit a model to an existing situation. However, there are plenty of examples of well-fitting models that make no sense, perhaps because they are based on correlated variables. Variables which are correlated today may not be so tomorrow; for example, a

strong correlation between banana production and car ownership in a particular country may disappear once oil is discovered there. Therefore models should be backed by some theory of travel behaviour so that one can interpret them consistently and have some confidence that they will remain valid in the future.

4. To understand model properties and develop improved algorithms for their solution. When one is able to cast a problem in mathematical programming or maximum likelihood terms, to mention two popular approaches to model generation, one has a wealth of technical tools to assist in the development of good solution algorithms. These have been developed over the years by researchers working in many areas besides transport.

5. To understand better what can be assumed constant and what must be accepted as variable for a particular decision context and level of analysis. The identification of exogenous and endogenous variables and those which may be assumed to remain constant is a key issue in modelling economics. For example, for some short-term tactical studies the assumption of a fixed trip matrix may be reasonable as in many traffic management schemes. However, even in the short term, if the policies to be tested involve significant price changes or changes to accessibility, this assumption no longer holds valid.

On the other hand practitioners have often abandoned the effort to use theoretically better models; some of the reasons for this are as follows:

1. They are too complex. This implies that heuristic approaches, rules of thumb, and *ad hoc* procedures are easier to understand and therefore preferable. This is a reasonable point; we do not advocate the use of models as 'black boxes'; quite the contrary. Model output needs interpretation and this is only possible if a reasonable understanding of the basis for such a model is available. Without ignoring the important role of academic literature in advancing the state of the art, there is a case for more publications explaining the basis of models without recourse to difficult notation or obscure (to the practitioner) concepts. Most models are not that complex, even if some of the statistics and computer implementations needed may be quite sophisticated. Good publications bridging the gap between the practitioner and the academic are an urgent need.

2. Theoretical models require data which are not available and are expensive to collect. This is often not entirely correct; many advanced models make much better use of small-sample data than some of the most pragmatic approaches. Improvements in data-collection methods have also reduced these costs and improved the accuracy of the data.

3. It is better to work with 'real' matrices than with models of trip making behaviour. This is equivalent to saying that it is better to work with fixed trip matrices, even if they have to be grossed up for the planning horizon. We will see that sampling and other data-collection errors cast doubts on the accuracy of such 'real' matrices; moreover, they cannot possibly respond to most policies (e.g. improvements in accessibility, new services, price changes) nor be reasonable for oversaturated do-minimum future conditions. Use of observations alone may lead to 'blinkered' decision making, to a false sense of accuracy and to underestimating the scope for change.

4. Theoretical models cannot be calibrated to the level of detail needed to analyse some schemes. There may be some truth in this statement, at least in some cases where the limitations of the data and time available make it necessary to compromise in detail if one wishes to use a better model. However, it may be preferable to err in this way than to work with the illusion of sufficient detail but undermined by potentially pathological (predictions of the wrong sign or direction) or insensitive results from *ad hoc* procedures.

5. It is better to use the same model (or software) for most problems because this ensures consistency in the evaluation methods. This is, in principle, correct provided the model remains appropriate to these problems. It has the advantage of consistent approach, ease of use and interpretation, and reduced training costs. However, this strategy breaks down when the problems are not of the same nature. Assumptions of fixed trip matrices, or insensitivity to mode choice or pricing policies, may be reasonable in some cases but fail to be acceptable in others. The use of the same model with the same assumptions may be appropriate in one case and completely misleading in another.

The importance of these criteria depends, of course, on the decision context and the levels of analysis involved in the study. What we argue in this book is for the use of the appropriate level of resolution to the problem in hand. Our own preference is for striving to use good, sound models as far as possible even if some level of detail has to be sacrificed. One has to find the best balance between theoretical consistency and expedience in each particular case and decision-making context. We have striven to provide material to assist in this choice.

2 Mathematical Prerequisites

2.1 INTRODUCTION

This book is aimed at practitioners and students in transport modelling and planning. Some of these may have a sound mathematical background; they may skip this chapter without loss of continuity. Other readers may have a weaker mathematical background or may simply welcome the opportunity to refresh ideas and notation. This chapter is addressed to these readers. It aims only to outline the most important mathematical prerequisites needed to benefit from this book.

Most of the mathematical prerequisites, however, are not that demanding; the reader can get by with little more than school algebra and some calculus. We introduce first the idea of functions and some specialised notation together with the idea of plotting functions in Cartesian (orthogonal) coordinates. After introducing the concept of series we treat the very important topic of matrix algebra; this is particularly important in transport as we often deal with trip and other matrices. Elements of calculus come next, including differentiation and integration. Logarithmic and exponential functions deserve some special attention as we will find them often in transport models. Finding maxima and minima of functions plays an important role in model development and the generation of solution algorithms. Finally, a few statistical concepts are introduced in the last section of this chapter. Statistics play a key part in contemporary transport modelling techniques and this section provides only an elementary entry point to the subject. A few other statistical concepts and techniques will be introduced in subsequent chapters as needed.

There are several books available as reference works for the more informed reader and as first textbooks for readers needing to brush up their mathematical background. These include those by Morley (1972), Stone (1966), and Wilson and Kirby (1980). We see the future of transport modelling practice moving steadily away from expedience through shortcuts and 'fudge factors', and increasingly adopting models with sounder theoretical backing. This trend results from the need to provide consistent advice to decision makers; this advice should not depend on an arbitrarily chosen number of iterations or starting point, nor on models likely to produce pathological results when used to forecast completely new options. This increased rigour will rely on better mathematical and statistical representations of problems and therefore requires further reading in these areas.

2.2 ALGEBRA AND FUNCTIONS

2.2.1 Introduction

Elementary algebra consists of forming expressions using the four basic operations of ordinary mathematics on letters which stand for numbers. It is useful to distinguish between *variables* (generally denoted by letters such as x, y and z), which represent measured quantities, and *constants* or *parameters* (generally denoted by letters such as a, b, c, ..., k, m, n, ..., or by letters from the Greek alphabet). The value of a constant is supposed to remain invariant for the particular situation examined.

Variables, and constants, are related through equations such as:

$$y = a + bx \qquad (2.1)$$

and if we were interested in x, we could 'solve' (2.1) for x, obtaining:

$$x = (y - a)/b \qquad (2.2)$$

The variables x and y in (2.1) and (2.2) are related by the '=' sign. However, in algebra we may also have *inequalities* of the following four types:

$<$ which means *less than*
$\leqslant$ which means *less than or equal to*
$>$ which means *greater than*, and
$\geqslant$ which means *greater than or equal to*

and which are used to constrain variables, for example:

$$x + 2y \leqslant 5 \qquad (2.3)$$

This expression, unlike an equation, cannot be 'solved' for x or y, but note that both variables can take only a restricted range of values. For example, if we restrict them further to be positive integers, it can easily be seen that x cannot be greater than 3 and y cannot be greater than 2.

It is possible to manipulate inequalities in much the same way as equations, thus:

- we can add or subtract the same quantity to/from each side;
- we can also multiply or divide each side by the same quantity, but if the number which is being multiplied or divided is negative, the inequality is reversed.

For example, if we subtract 5 on both sides of (2.3) we get

$$x + 2y - 5 \leqslant 0$$

which is certainly the same constraint. However, if we multiply it by -2 we obtain:

$$-2x - 4y \geqslant -10$$

which can be checked by the reader to provide the same constraint as (2.3).

The use of different letters to denote each variable is only convenient up to a certain point. Soon it becomes necessary to use indices (e.g. subscripts or super-scripts) to define additional variables, as in x_1, x_2, x_3, ..., x_n, which we can conveniently summarise as x_i, $i = 1, 2, ..., n$; it does not matter if we use another letter for the index if it has the same numerical range. For example, we could have defined also x_k, $k = 1, 2, ..., n$.

The use of indices facilitates a very convenient notation for summations and products:

$$\sum_{i=1}^{n} x_i = x_1 + x_2 + x_3 + ... + x_n \tag{2.4}$$

or

$$\prod_{j=1}^{m} y_j = y_1 y_2 y_3 ... y_m \tag{2.5}$$

In certain cases a single index is not enough and two or more may be used. For example we could define the following six variables, T_{11}, T_{12}, T_{21}, T_{22}, T_{31}, T_{32} as T_{ij}, $i = 1, 2, 3$, and $j = 1, 2$. With two-subscript variables we can have double summations or double products, as in:

$$\sum_{i=1}^{3} \sum_{j=1}^{2} T_{ij} = \sum_{i=1}^{3} (T_{i1} + T_{i2}) = T_{11} + T_{12} + T_{21} + T_{22} + T_{31} + T_{32} \tag{2.6}$$

2.2.2 Functions and Graphs

We have already referred to variables as being related by equations and inequalities; in general these can be called functional relations. A particular function is some specific kind of relationship between two or more variables. For example, the power function:

$$y = \phi x^n \tag{2.7}$$

yields values of the *dependent* variable y, given values of the parameters ϕ and n, and of the *independent* variable x; a function requires that for every value of x in some range, a corresponding value of y is specified. Often we do not wish to refer to a particular function, but only to state that y is 'some function of x' or vice versa; this can be written as:

$$y = f(x) \tag{2.8}$$

A large range of functions exists and readers should familiarise themselves with these as they arise. It is usually convenient to plot functions graphically on a Cartesian co-ordinate system (see Figure 2.1).

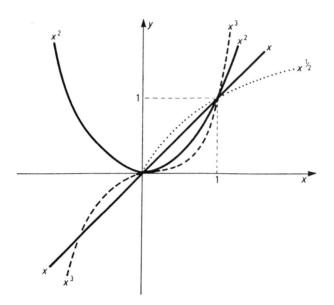

Figure 2.1   Plot of various power functions

A dependent variable may be a function of several independent variables, for example:

$$y = f(x_1, x_2, \ldots, x_n) \tag{2.9}$$

but this would require $n + 1$ dimensions to represent it (n for the independent variables and 1 for y). Cartesian coordinates can be used in three or more dimensions, in the case of three dimensions the orientation of the third axis is out of this side of the paper in Figure 2.1. More than three dimensions cannot be easily visualised physically but are dealt with algebraically in just the same way as one, two and three dimensions. For example, in the case of $n = 2$ the function can be represented by a surface over the relevant part of the (x_1, x_2) plane.

Generally, any equation for an unknown quantity x can by put in the form $f(x) = 0$; for example, the linear equation:

$$ax = b$$

is equivalent to

$$ax - b = 0$$

where $f(x) = ax - b$. Solving the equation is therefore equivalent to finding the points on the curve $y = f(x)$ which meet the x axis. These points are called *real solutions* or *zeros* of $f(x)$; for example, x_1 and x_2 in Figure 2.2.

We are sometimes interested in what happens to the value of a function $f(x)$, as x

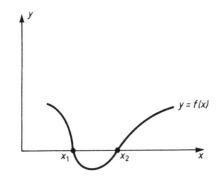

Figure 2.2   Real solutions of a general function

increases indefinitely $(x \to \infty)$; it can easily be seen that the possibilities are only the following:

- tend to infinity (e.g. when $f(x) = x^2$)
- tend to minus infinity (e.g. when $f(x) = -x$)
- oscillate infinitely (e.g. when $f(x) = (-1)^x x^2$)
- tend to a finite limit (e.g. $f(x) = 1 + 1/x$).

For more complex functions some ingenuity may be required to find out it they tend to a finite limit when $x \to \infty$.

We may also be interested in finding the *limit* when x approaches a finite value. For example, if $f(x) = 1/(x + 3)$, it can easily be seen that the limit when $x \to 0$ is $1/3$. If for some value α we have that $f(x) \to \infty$ as $x \to \alpha$, the curve $y = f(x)$ is said to have an asymptote $x = \alpha$ (see Figure 2.3).

One of the most important functions is the *straight line*, shown in Figure 2.4 and whose general equation is (2.1). It can easily be seen that b is the value of y when

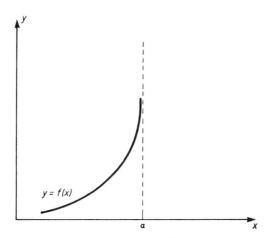

Figure 2.3   General function with asymptote at α

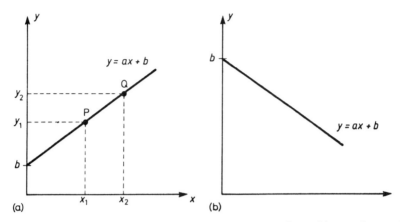

Figure 2.4   Two straight lines $y = ax + b$: (a) positive gradient, (b) negative gradient

$x = 0$; this is usually called the *intercept* on the y axis. The constant a is called the *gradient* and it can be shown to be given by:

$$a = \frac{y_2 - y_1}{x_2 - x_1} \tag{2.10}$$

where (x_1, y_1) and (x_2, y_2) are any two points on the line (see Figure 2.4a). Although a straight line has by definition a constant gradient, this can be either positive or negative as shown in the figure.

Unless two straight lines are parallel they will intersect at one point; this can be represented either graphically (as in Figure 2.5) or algebraically as a system of equations as follows:

$$y = x + 2 \tag{2.11a}$$

$$y = -x + 4 \tag{2.11b}$$

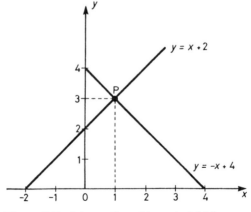

Figure 2.5   Intersection of two straight lines

Solving for x in (2.11b) and replacing this value (i.e. $-y + 4$) in (2.11a) we get that the solution is point P with coordinates $(x = 1, y = 3)$.

2.2.3 Sums of Series

A series is simply defined as a sequence of numbers u_n, $n = 1, 2, \ldots, N$. In many cases it may be interesting to find its sum given by:

$$S_N = u_1 + u_2 + \ldots + u_N = \sum_n u_n \tag{2.12}$$

In some cases, such as the *arithmetic progression* given by:

$$u_n = u_{n-1} + d \tag{2.13}$$

it can be shown that the series has a sum to N terms. For example, if the first term is b the sum can be shown to be:

$$S_N = Nb + N(N - 1)d/2 \tag{2.14}$$

The *geometric progression* (2.15), formed by multiplying successive terms by a constant factor r, also has an expression for the sum of N terms. If b is again the first term, the sum can be shown to be given by (2.16).

$$u_n = ru_{n-1} \tag{2.15}$$

$$S_N = \frac{b(1 - r^n)}{1 - r} \tag{2.16}$$

In other cases the series may have a simple expression for its sum, such as in:

$u_n = n$, where the sum is given by $S_N = N(N + 1)/2$, or

$u_n = x^n$, where it is given by $S_N = x(1 - x^N)/(1 - x)$ for x different from 1;

but still *diverge* (i.e. S_N keeps increasing indefinitely when N tends to infinity). That happens to $u_n = n$; it also happens to $u_n = x^n$ if $x > 1$ above; however, the latter converges to $S_N = x/(1 - x)$ for the range $0 < x < 1$.

2.3 MATRIX ALGEBRA

2.3.1 Introduction

Any variable with two subscripts can be called a *matrix*. We will denote matrices by the notation $\mathbf{B} = \{B_{ij}\}$, where the variables B_{ij}, $i = 1, 2, \ldots, N$; $j = 1, 2, \ldots, M$ are the elements of $\mathbf{B}$. This can be written as follows:

$$\mathbf{B} = \begin{pmatrix} B_{11} & B_{12} & B_{13} & \cdots & B_{1M} \\ B_{21} & B_{22} & B_{23} & \cdots & B_{2M} \\ \vdots & & & & \\ B_{N1} & B_{N2} & B_{N3} & \cdots & B_{NM} \end{pmatrix}$$

(2.17)

As can be seen, the matrix has N rows and M columns; for this reason it is known as a $N \times M$ matrix. A *vector* is an important special case, being a one-dimensional array or a $N \times 1$ matrix. In these cases the second index is redundant, so we write:

$$\mathbf{V} = \{V_i\} = \begin{pmatrix} V_1 \\ V_2 \\ V_3 \\ \vdots \\ V_N \end{pmatrix}$$

(2.18)

Formally, a non-indexed variable, or even a constant, can be thought of as a 1×1 matrix and it is known as a *scalar*.

If we interchange rows and columns we obtain an $M \times N$ matrix known as the *transpose* $\mathbf{B}^T$ of $\mathbf{B}$, which is given by:

$$\mathbf{B}^T = \begin{pmatrix} B_{11} & B_{21} & B_{31} & \cdots & B_{N1} \\ B_{12} & B_{22} & B_{32} & \cdots & B_{N2} \\ \vdots & & & & \\ B_{1M} & B_{2M} & B_{3M} & \cdots & B_{NM} \end{pmatrix}$$

(2.19)

Similarly, the transpose of an $N \times 1$ vector (also known as a *column* vector) is a *row* vector:

$$\mathbf{V}^T = [V_1 V_2 V_3 \cdots V_N]$$

(2.20)

A *square* matrix $\mathbf{S}$ is one where $N = M$; a square matrix such that $\mathbf{S} = \mathbf{S}^T$ is called symmetric. A *diagonal* matrix $\mathbf{D} = \{D_{ij}\}$ is one where $D_{ij} = 0$ unless $i = j$. The *unit* matrix is a square diagonal matrix with each diagonal element equal to 1, that is:

$$\mathbf{I} = \begin{pmatrix} 1 & 0 & 0 & \cdots & 0 \\ 0 & 1 & 0 & \cdots & 0 \\ 0 & 0 & 1 & \cdots & 0 \\ \vdots & & & & \\ 0 & 0 & 0 & \cdots & 1 \end{pmatrix}$$

(2.21)

2.3.2 Basic Operations of Matrix Algebra

We will define the operations between two matrices $\mathbf{A}$ and $\mathbf{B}$ by setting a new matrix $\mathbf{C}$ which will represent the combination required. First matrix *addition*:

$$C = A + B = B + A \tag{2.22}$$

is defined by $C_{ij} = A_{ij} + B_{ij}$ and requires that both matrices being combined are of the same size, say both $N \times M$ matrices; then C is also an $N \times M$ matrix. This is also a requirement for matrix subtraction:

$$C = A - B \tag{2.23}$$

similarly defined as $C_{ij} = A_{ij} - B_{ij}$. An operation which is unique to matrix algebra is *multiplication by a scalar*:

$$C = kA \tag{2.24}$$

defined by $C_{ij} = kA_{ij}$, where obviously the new 'grossed up' matrix has the same size as the old one.

Matrix *multiplication* is more complex, as:

$$C = AB \tag{2.25}$$

is defined by $C_{ij} = \Sigma_k A_{ik} B_{kj}$, where A is an $N \times M$ matrix and B is any $M \times L$ matrix (i.e. the number of columns in A must equal the number of rows in B but there are no other restrictions). In this case C is an $N \times L$ matrix.

It is easy to see that in general AB is not equal to BA, i.e. the operation is non-commutative, as opposed to elementary algebra. However, this is not the case with the unit matrix I; in fact, it can easily be checked that:

$$IA = AI = A \tag{2.26}$$

Thus, although it is possible to define the product of any number of matrices, order must always be preserved. In fact we refer to *pre-multiplication* of A by B to form the product BA, and to *post-multiplication* to form AB.

To define *division* it is convenient to use the concept of inverse of a matrix. Unfortunately this only exists for square matrices and then not always. If the inverse exists, it is denoted as B^{-1} and is the matrix that satisfies:

$$B^{-1}B = BB^{-1} = I \tag{2.27}$$

In this case B is said to be *non-singular*. We will not give a procedure for the calculation of the elements of the inverse matrix as it is fairly complicated. It is sufficient to know that under suitable conditions it exists. Division is then just pre- or post-multiplication by B^{-1}.

In this book matrices and vectors are mostly used to provide a shorthand notation for such things as sets of simultaneous equations and for obtaining their solution in terms of the inverse matrix.

2.4 ELEMENTS OF CALCULUS

The two main branches of calculus are differentiation and integration; their basic nature can be intuitively identified by reference to the function $y = f(x)$ depicted in Figure 2.6. Consider the points P and Q and the straight line (*chord*) connecting them. Differentiation is concerned with the calculation of the gradient of a curve at a point. To do this, it is useful to consider Q approaching P; in the limit the chord PQ becomes the *tangent* to the curve at $P = Q$ (i.e. when their horizontal 'distance' h is 0) and by definition its gradient is equal to that of the curve.

Integration, on the other hand, is concerned with calculating the area under a curve, say the shaded area in Figure 2.6; as we will see below, these two operations are closely related.

2.4.1 Differentiation

Using (2.10) the gradient of the chord PQ in Figure 2.6 can be written as:

$$\delta(x) = [f(x_0 + h) - f(x_0)]/h$$

If the limit of $\delta(x)$ when $h \to 0$ exists and is the same whether h tends to zero from above or below, it is called the *derivative* of y, or of $f(x)$, with respect to x at x_0 and it is often written as $f'(x_0)$ or $dy/dx|_{x_0}$. The process of finding the derivative is called differentiation.

If $f(x)$ is given as an expression in x, it is usually not difficult to find $f'(x)$ as a function of x using the results in Table 2.1, plus others we will give below.

Since derivatives are themselves functions of x, we can also define second- and higher-order derivatives (i.e. $f_0''(x)$ or d^2y/dx^2 and so on). For example, if we differentiate the first derivative of $y = x^b$ in Table 2.1, we get:

$$\frac{d^2y}{dx^2} = b(b - 1)x^{b-2} \tag{2.28}$$

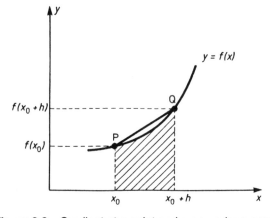

Figure 2.6   Gradient at a point and area under a curve

Table 2.1 Common derivatives

Function $f(x)$		Derivative $f'(x)$
k	(k constant)	0
x^b	(b constant, $x > 0$)	bx^{b-1}
$ku(x)$	(k constant)	$ku'(x)$
$u(x) + v(x)$		$u'(x) + v'(x)$
$u(x)v(x)$		$u'(x)v(x) + u(x)v'(x)$
$u[v(x)]$		$u'[v(x)]v'(x)$

2.4.2 Integration

This is the reverse of differentiation; if we know the gradient of some curve at every point then the equation of the curve itself is known as the *integral*. For example, if $g = g(x)$ is the gradient, the equation of the curve is written

$$y = \int_x g(x)\,dx$$

and this result is always arbitrary up to an additive constant; for example, if $g = bx^{b-1}$ we know from Table 2.1 that the *indefinite* integral of $g(x)$ is given by:

$$y = G(x) = \int_x bx^{b-1}\,dx = x^b + C \tag{2.29}$$

where C is an arbitrary constant of integration (i.e. the derivative of $x^b + C$ is bx^{b-1} no matter the value of C). The most practical elementary use of integration is to obtain the *area under a curve* as the *definite* integral, as shown in Figure 2.7a.

$$\text{Area abcd} = [F(x)]_a^b = F(b) - F(a) = \int_a^b y\,dx = \int_a^b f(x)\,dx \tag{2.30}$$

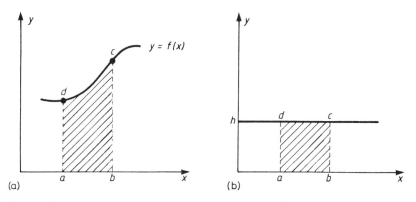

Figure 2.7 Areas under a curve: (a) general case, (b) line parallel to *x* axis

For example, if we take the simple case of a straight line parallel to the x axis at height h and want to integrate between the values a and b (see Figure 2.7b), we get:

$$y = f(x) = h$$

and

$$F(x) = hx + C$$

then

$$\text{Area} = F(b) - F(a) = h(b - a)$$

which is indeed the area of the shaded rectangle in the figure.

Table 2.1 can be used in reverse to help to find indefinite integrals. In particular, if

$$\int u(x)\,dx = U(x) + C_1 \quad \text{and} \quad \int v(x)\,dx = V(x) + C_2$$

then

$$\int u[v(x)]v'(x)\,dx = U[v(x)] + C_3$$

and

$$\int U(x)v(x)\,dx = U(x)V(x) - \int u(x)V(x)\,dx$$

Of course not all functions, even some that are deceptively simple in appearance, have indefinite integrals which are similarly simple expressions. However, for those that do not it is still possible to evaluate definite integrals numerically.

2.4.3 The Logarithmic and Exponential Functions

Among the functions we have considered so far, the simplest one with no indefinite integral is the inverse function $f(x) = 1/x$, depicted in Figure 2.8.

The integral of this function has been defined as the *natural logarithm* of x, or $\log_e(x)$, where e is Nepper's constant. Its value of approximately 2.7183 corresponds to the point on the x axis of Figure 2.8 such that the shaded area is 1, i.e. $\log_e(e) = 1$. As in this book we will only use natural logarithms, we will drop the base e from our notation.

In common with any other logarithm, $\log(x)$ has the following properties:

$$\log(1) = 0;$$
$$\text{As } t \to \infty, \log(t) \to \infty;$$
$$\text{As } t \to 0, \log(t) \to -\infty;$$
$$\log(uv) = \log(u) + \log(v).$$

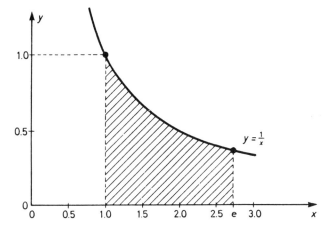

Figure 2.8   Inverse function and Nepper's constant

Another useful related function is the *exponential function* $\exp(x)$ or e^x for short, defined as the number w such that $\log(w) = x$. Then, as expected of a power function, we have:

$$e^{(x+y)} = e^x e^y;$$

moreover,

$$e^{\log(x)} = x$$

Both functions $\log(x)$ and $\exp(x)$ are easy to differentiate; by definition:

$$\frac{d}{dx}\log(x) = \frac{1}{x} \tag{2.31}$$

and it is not difficult to show that:

$$\frac{d}{dx}(e^x) = e^x \tag{2.32}$$

Thus the exponential is the function which remains unaltered under differentiation.

2.4.4 Finding Maximum and Minimum Values of Functions

This is one important use of differentiation. Consider Figure 2.9 for example; the function shown has a *maximum* at x_1 and a *minimum* at x_2. Both are characterised by the gradient of the curve being zero at those points, so the first step in finding them is to solve the equation $f'(x) = 0$.

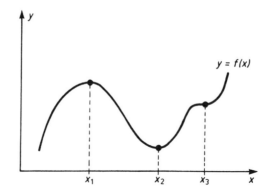

Figure 2.9   Maximum, minimum and point of inflexion

It is important to note, however, that not all zeros are maxima or minima; an example of one that is not (called a *point of inflexion*) is x_3 in Figure 2.9. To find out more precisely what a zero gradient stands for, it is necessary to evaluate $f''(x)$ at each zero of $f'(x)$. Thus, for a maximum we require:

$$f''(x) < 0 \qquad (2.33)$$

For a minimum we need:

$$f''(x) > 0 \qquad (2.34)$$

and for a point of inflexion,

$$f''(x) = 0 \qquad (2.35)$$

These cases are illustrated in Figure 2.10, which suggests a good mnemonic. Consider the function as a cup of water; it if is facing downwards as in the case of the maximum, the liquid will drop (i.e. a minus sign). Conversely if its is facing upwards (e.g. a minimum) the liquid will stay (i.e. a plus sign).

In order to develop a theory directed toward characterising global, rather than local, minimum (or maximum) points mathematicians have found it necessary to

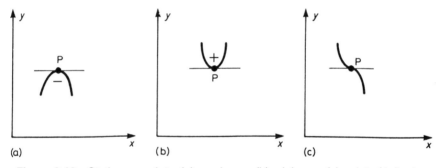

Figure 2.10   Stationary points: (a) maximum, (b) minimum, (c) point of inflexion

introduce the complementary notions of *convexity* and *concavity*. These result not only in a more powerful (although more restrictive) theory, but also provide an interesting geometric interpretation of the second-order conditions (2.33) to (2.35).

Figure 2.11 presents some examples of convex and non-convex functions. Geometrically, a function is convex if the line joining two points on its graph lies nowhere below the graph, as shown in Figure 2.11a; in two dimensions, a convex function would have a bowl-shaped graph. Similarly and simply, a function g is said to be concave if the function $f = -g$ is convex. A nice property of convex functions is that the sum of two such functions is also convex.

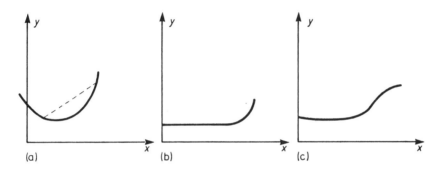

Figure 2.11   Convex and nonconvex functions: (a) convex, (b) convex, (c) nonconvex

2.4.5 Functions of More Than One Variable

It is useful to consider the application of differential and integral calculus to this kind of function. Suppose that we have:

$$y = f(x_1, x_2, \ldots, x_n) \tag{2.36}$$

Then the derivative of y with respect to one of these variables may be calculated assuming the other variables remain constant during the operation. This is known as a *partial* derivative and is written $\partial y/\partial x_i$. For example, if:

$$y = 2x_1 + x_2^3 x_3$$

then

$$\frac{\partial y}{\partial x_1} = 2$$

$$\frac{\partial y}{\partial x_2} = 3x_2^2 x_3$$

$$\frac{\partial y}{\partial x_3} = x_2^3$$

It can be shown that maxima and minima of a function such as (2.36) can be found by setting all the partial derivatives to zero:

$$\frac{\partial y}{\partial x_i} = 0, \qquad i = 1, 2, \ldots, n \qquad (2.37)$$

which gives a set of simultaneous equations to solve. A particularly interesting case is that of the restricted maximum or minimum. Assume we wish to maximise (2.36) subject to the following restrictions:

$$r_1(x_1, x_2, \ldots, x_n) = b_1$$
$$r_2(x_1, x_2, \ldots, x_n) = b_2$$
$$\vdots \qquad (2.38)$$
$$r_K(x_1, x_2, \ldots, x_n) = b_K$$

This can be done by defining *Lagrangian multipliers* $\lambda_1, \lambda_2, \ldots, \lambda_K$ for each of the equations (2.38) in turn, and maximising

$$L = f(x_1, x_2, \ldots, x_n) + \sum_k \lambda_k [r_k(x_1, \ldots, x_n) - b_k] \qquad (2.39)$$

as a function of $x_1, x_2, \ldots, x_n$ and $\lambda_1, \lambda_2, \ldots, \lambda_K$. Thus, we solve:

$$\frac{\partial L}{\partial x_i} = 0, \qquad i = 1, 2, \ldots, n \qquad (2.40)$$

and

$$\frac{\partial L}{\partial \lambda_k} = 0, \qquad k = 1, 2, \ldots, K \qquad (2.41)$$

The equations (2.41) are simply the restrictions (2.38) in another form; the device of introducing the multipliers as additional variables enables the restricted maximum to be found.

2.4.6 Multiple Integration

In the case of integration, multiple integrals can be defined. For example, given (2.36) we might have:

$$V = \int\int \ldots \int f(x_1, x_2, \ldots, x_n) \, dx_1 \, dx_2, \ldots \, dx_n \qquad (2.42)$$

with n integral signs. In order to get an intuitive feeling of its meaning it is useful to consider the two-dimensional case. The function

$$S = f(x_1, x_2) \qquad (2.43)$$

may be considered as defining a surface in a three-dimensional Cartesian system. Therefore,

$$V = \iint f(x_1, x_2) \, dx_1 \, dx_2 \tag{2.44}$$

measures a volume under this surface, in a similar way to the single variable measuring an area under a curve.

2.4.7 Elasticities

The elasticity of a dependent variable y with respect to another variable x_i in a function such as (2.9), is given by the expression:

$$E(y, x_i) = \frac{\partial y}{\partial x_i} \frac{x_i}{y} \tag{2.45}$$

and can be interpreted as the percentage change in the dependent variable with respect to a given percentage change in the relevant independent variable.

In econometrics we will be often interested in the elasticities of a given demand function with respect to changes in the values of some explanatory variables or *attributes*. We will generally distinguish between *direct-* and *cross-elasticities*; the first relate to attributes of the service or good under consideration and the second to attributes of competing options or goods. For example, it is often stated that the elasticity of public transport demand to fares is around -0.33; this means that if we increase fares by 1% we should expect patronage to decrease by approximately 0.3%.

2.4.8 Series Expansions

It is sometimes necessary to estimate the values of a function $f(x)$ in the neighbour-hood of a particular value x_0 of x, in terms of the values of the function and its derivatives at this value. For suitable functions this can be done by means of *Taylor's series* expansion; first we require to define the concept of a *factorial* number ($n!$) which applies to non-negative integers:

$$n! = n(n-1)(n-2) \ldots 3\cdot2\cdot1$$
$$0! = 1 \tag{2.46}$$

A Taylor's series expansion is defined as:

$$f(x_0 + h) = f(x_0) + hf'(x_0) + (h^2/2!)f''(x_0) + (h^3/3!)f'''(x_0) + \ldots \tag{2.47}$$

and it is most useful when h is small enough for the higher-order terms to become rapidly smaller so that a good approximation is obtained by stopping the summation after just a few terms—even just after two terms.

The special case when $x_0 = 0$ is known as *Maclaurin's series*, which upon setting h to x in the left-hand side of (2.47) yields:

$$f(x) = f(0) + hf'(0) + (h^2/2!)f''(0) + (h^3/3!)f'''(0) + \ldots \tag{2.48}$$

This provides a method of expressing certain functions as power series, for example:

$$e^x = 1 + x + x^2/2! + x^3/3! + \ldots$$

which allows us very easily to see why expression (2.32) holds.

2.5 ELEMENTARY MATHEMATICAL STATISTICS

In this section we provide only a basic review of the more fundamental statistical concepts. In the rest of the book we take for granted that the reader is not only aware of the most important distributions (e.g. binomial, normal, Student, chi-squared and Fisher) but also has some knowledge about basic statistical inference (e.g. estimators, confidence intervals and tests of hypotheses). As there are very good textbooks about this subject, the reader is strongly advised to consult them for further reference. In particular we recommend Wonnacott and Wonnacott (1977) and Chapter 7 of Wilson and Kirby (1980).

Certain specialised subjects, such as basic sampling theory, linear regression analysis and maximum likelihood estimation, are presented at greater length in the relevant chapters (i.e. 3, 4 and 8 respectively).

2.5.1 Probabilities

The most intuitive definition of the probability that a certain result will occur (e.g. obtaining a six by rolling a dice) is given by the limit of its *relative frequency*, that is:

$$P(e_i) = p_i = \lim_{n \to \infty} \frac{n_i}{n} \tag{2.49}$$

where e_i is the desired result, n is the number of times the experiment is repeated and n_i the number of times e_i occurs. Expression (2.49) allows to deduce certain basic properties of probabilities:

$$0 \leq p_i \leq 1 \tag{2.50}$$

as n_i can take both the values 0 and n, and

$$\sum_i p_i = 1 \tag{2.51}$$

as $n_1 + n_2 + \ldots = n$. An alternative view of the expected probability of the result can be expressed in terms of a fair bet. If a person regards as fair a bet in which they win \$35 if e_i happens and loses \$$x$ if it does not, then their estimate of p_i is $x/(x + 35)$. This is so because they have solved the following equation which makes their expected gains or losses equal to zero, i.e. a fair bet:

$$35p_i - x(1 - p_i) = 0$$

On many occasions the probabilities of certain experiments are not simple to calculate. It is convenient to define an event (E) as a subset of the set of possible results of an experiment, $E = \{e_1, \ldots, e_i\}$. The probability of an event is the sum of the probabilities of the results it is composed of,

$$P(E) = \sum_i p_i, \qquad e_i \in E$$

Example 2.1: The event E: {to obtain at least two heads in three throws of a coin} includes (the first) four results out of the eight possible ones: (H, H, H), (H, H, T), (H, T, H), (T, H, H), (T, T, H), (T, H, T), (H, T, T) and (T, T, T). As each result has a probability of $1/8$ (if the probabilities of getting heads and tails are equal), the probability of the event is $1/2$.

For *combinations* of events (i.e. two heads but such that not all throws give the same result) it becomes necessary to work with the concepts of *union* ($\cup$) and *intersection* ($\cap$) of set theory as presented in Figure 2.12. The rectangle in the figure represents the event space and A and B are events within it.
In general, it is true that

$$P(A \cup B) = P(A) + P(B) - P(A \cap B) \tag{2.52}$$

and if A and B are *mutually exclusive*,

$$P(A \cup B) = P(A) + P(B) \tag{2.53}$$

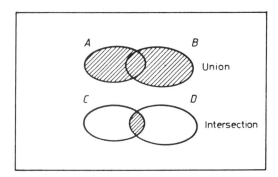

Figure 2.12   Venn diagram for events and probabilities

The *conditional probability* $P(A/B)$, of A happening given that B is true, is:

$$P(A/B) = P(A \cap B)/P(B) \tag{2.54}$$

An event F is statistically independent of another event E, *if and only if* (iff) $P(F|E)$ is equal to $P(F)$. Therefore, for independent events we have:

$$P(E \cap F) = P(E)P(F) \tag{2.55}$$

which we applied intuitively when estimating event probability in Example 2.1.

2.5.2 Random Variables

These can be defined as those which take values following a certain probability *distribution* (see Figure 2.13).

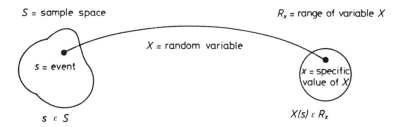

Figure 2.13   Random variable mapping from sample space

Example 2.2: The experiment 'spinning a coin twice', can yield only the following results (sample space): $S = \{HH, HT, TH, TT\}$. If we define the random variable X = number of heads, it is easy to see that it can only take the following three values: $X(HH) = 2$, $X(HT) = X(TH) = 1$ and $X(TT) = 0$. Therefore, an advantage of the random variable concept becomes immediately apparent: the set of results (sample space) is reduced to a smaller, more convenient, numerical set (the range of the variable). The probabilities of X are as follows:

$$P(X = 1) = P(HT \cup TH) = P(HT) + P(TH) = 1/2$$
$$P(X = 2) = P(X = 0) = 1/4.$$

Random variables may be *discrete* or *continuous*. In the former case they can take values from a finite set with probabilities belonging to a set $P(X)$ which satisfy (2.51) and $p(x_i) \geq 0$. In the latter case it is necessary to define a probability *density* function $f(x)$, such that:

$$\int_x f(x)\,dx = 1$$
$$f(x) \geq 0, \quad \forall x \tag{2.56}$$

2.5.3 Descriptive Statistics

When dealing with statistical data, summary information may be provided conveniently by specifying certain key features rather than the whole of a distribution. For example, the distribution of a random variable might be described with reference to its mean value and the dispersion around it. These descriptive statistics can be used to make simple comparisons between distributions without going into full details. More interestingly, certain standard distributions can be completely specified by just a few descriptive statistics.

The most usual descriptive statistic attempts to indicate the 'middle' of the distribution; three measures of *central tendency* are commonly defined to do this:

1 The *mean* (or *expectation*) $E(X)$, given by:

$$E(X) = \sum_i x_i p_i(x_i), \qquad \text{discrete case}$$

(2.57)

$$E(X) = \int_a^b x f(x)\, dx, \qquad \text{continuous case}$$

where $f(x)$ is defined for the range $(a \leqslant x \leqslant b)$. The mean is normally found by direct application of the *expectation operator* to the random variable X. It can be applied also to functions of random variables. The operator has the important property of *linearity*, whereby for any random variables X and Y, and constants a, b and c we have:

$$E(a + bX + cY) = a + bE(X) + cE(Y)$$

(2.58)

2 The *mode* X^* is the value of X which maximises $p_i(x_i)$.
3 The *median* $X_{0.5}$ is the value of X below which lies half of the distribution, that is:

$$P(X_{0.5}) = \sum_{x=1}^{X_{0.5}} P(X) = 0.5 \qquad \text{discrete case}$$

(2.59)

$$P(x < X_{0.5}) = \int_a^{X_{0.5}} f(x)\, dx = 0.5 \quad \text{continuous case}$$

Neither the median nor the mode can be found by direct calculation, but need the solution of a problem.

Another important feature of a distribution is its spread or width; statistics commonly used for this purpose are:

1 The *variance*, which is the expectation of the square of the deviation from the mean:

$$\text{Var}(X) = E\{[X - E(X)]^2\}$$

(2.60)

but it is normally calculated from the following result:

$$\begin{aligned} \mathrm{Var}(X) &= E\{X^2 - 2XE(X) + [E(X)]^2\} \\ &= E(X^2) - 2[E(X)]^2 + [E(X)]^2 \\ &= E(X^2) - [E(X)]^2 \end{aligned} \tag{2.61}$$

Unlike expectation, the variable is not a linear operator:

- $\mathrm{Var}(a + bX) = b^2\,\mathrm{Var}(X)$, i.e. adding a constant does not affect the spread of the distribution.
- $\mathrm{Var}(aX + bY) = a^2\,\mathrm{Var}(X) + b^2\,\mathrm{Var}(Y) + 2ab\,\mathrm{Cov}(X, Y)$, where the *covariance* of X and Y is given by:

$$\mathrm{Cov}(X, Y) = E(XY) - E(X)E(Y) \tag{2.62}$$

Thus, the covariance of two mutually independent random variables is 0.

2 The *standard deviation* $\mathrm{se}(x)$, which is the square root of the variance. This, in contrast with the variance, has the same dimensions as the random variable X and the measures of central tendency.

3 The *coefficient of variation* CV, which is the ratio of the standard deviation to the mean, and is a dimensionless measure of spread.

3 Data and Space

This chapter is devoted to issues in data collection and their representation for use in transport modelling. We will consider four subjects which are a prerequisite for the rest of the book. Firstly, we will provide a brief introduction to statistical sampling theory, which will complement in part the elementary concepts discussed in section 2.5. Interested readers are advised that there is a complete book on the subject (Stopher and Meyburg 1979) which should be consulted for more details. In section 3.2 we will discuss the nature and importance of errors which can arise both during model estimation and when forecasting with the aid of models; the interesting question of data accuracy versus model complexity and cost is also addressed.

In section 3.3 we will consider the various types of surveys used in applied transport planning; we will be particularly interested in problems such as the correction, expansion and validation of survey data. We will also discuss issues involved in the collection of both stated preferences and longitudinal (e.g. panel) data. Finally, section 3.4 considers the important practical problems of network representation and zoning design; this is where the 'spatial capabilities' of the model are actually decided. Poor network representations or too coarse zoning systems may invalidate the results of even the most theoretically appealing model.

3.1 BASIC SAMPLING THEORY

3.1.1 Statistical Considerations

Statistics may be defined as the science concerned with gathering, analysing and interpreting data in order to obtain the maximum quantity of useful information. It may also be described as one of the disciplines concerned with decision making under uncertainty; its goal would be in this case to help determine the level of uncertainty associated with measured data in order to support better decisions.

Data usually consist of a sample of observations taken from a certain population of interest which is not economically (or perhaps even technically) feasible to observe in its entirety. These observations are made about one or more attributes of each member of the population (say income). Inferences can be made then about the mean value of these attributes, often called parameters of the population. Sample design aims at ensuring that the data to be examined provide the greatest amount of useful information about the population of interest at the lowest possible cost; the problem remains of how to use the data (i.e. expand the values in the sample) in order to make correct inferences about this population. Thus two difficulties exist:

- how to ensure a *representative* sample; and
- how to extract valid conclusions from a sample satisfying the above condition.

Neither of these would constitute a problem if there was no variability in the population. To solve the second difficulty, a well-established procedure exists which does not present major problems if certain conditions and assumptions hold. The identification of a representative sample, however, may be a more delicate task in certain cases, as we shall see below.

3.1.1.1 Basic Definitions

Sample   The sample is defined as a collection of units which has been especially selected to represent a larger population with certain attributes of interest (i.e. height, choices, options). Three aspects of this definition have particular importance: first, which population the sample seeks to represent; second, how large the sample should be; and third, what is meant by 'especially selected'.

Population of Interest   This is the complete group about which information is sought; in many cases its definition stems directly from the study objectives. The population of interest is composed of individual elements; however, the sample is usually selected on the basis of sampling units which may not be equivalent to these individual elements as aggregation of the latter is often deemed necessary. For example, a frequently used sampling unit is the household while the elements of interest are individuals residing in it.

Sampling Method   Most of the acceptable methods are based on a form of random sampling. The key issue in these cases is that the selection of each unit is carried out independently, with each unit having the same probability of being included in the sample. The more interesting methods are:

- *Simple random sampling*, which is not only the simplest method but constitutes the basis of all the rest. It consists in first associating an identifier (number) to each unit in the population and then selecting these numbers at random to obtain the sample; the problem is that far too large samples may be required to ensure sufficient data about minority options of particular interest. For example, it may well be that sampling households at random in a developing country would provide little information on multiple car ownership.
- *Stratified random sampling*, where *a priori* information is first used to subdivide the population into homogeneous strata (with respect to the stratifying variable) and then simple random sampling is conducted inside each stratum using the same sampling rate. The method allows the correct proportions of each stratum in the sample to be obtained; thus it may be important in those cases where there are relatively small subgroups in the population as they could lack representation in a simple random sample.

It is also possible to stratify with respect to more than one variable, thus creating an *n*-dimensional matrix of group cells. However, care must be taken with the number of cells created as it increases geometrically with the number of strata; large figures may imply that the average number of sampling units per cell is too small. Nevertheless, even stratified sampling does not help when data are needed about options with a low probability of choice in the population; in these cases a third method called *choice-based sampling*, actually a subset of the previous one, is required. The method consists in stratifying the population based on the result of the choice process under consideration. This method is fairly common in transport studies, as we will see in section 3.3. A major advantage is that data may be produced at a much lower cost than with the other sampling methods; its main drawback is that the sample thus formed may not be random and therefore the risk of bias in the expanded values is greater.

Sampling Error and Sampling Bias   These are the two types of error that might occur when taking a sample; combined, they contribute to the measurement error of the data. The first is simply due to the fact that we are dealing with a sample and not with the total population, i.e. it will always be present due to random effects. This type of error does not affect the expected values of the means of the estimated parameters; it only affects the variability around them, thus determining the degree of confidence that may be associated to the means; it is basically a function of sample size and of the inherent variability of the parameter under investigation.

The sampling bias, on the other hand, is caused by mistakes made either when defining the population of interest, or when selecting the sampling method, the data collection technique or any other part of the process. It differs from the sampling error in two important respects:

- it can affect not only the variability around the mean of the estimated parameters but the values themselves; therefore it implies a more severe distortion of the survey results;
- while the sampling error may not be avoided (it can only be reduced by increasing sample size), the sampling bias may be virtually eliminated by taking extra care during the various stages of sampling design and data collection.

Sample Size   Unfortunately, there are no straightforward and objective answers to the calculation of sample size in every situation. This happens, in spite of the fact that sample size calculations are based on precise statistical formulae, because many of their inputs are relatively subjective and uncertain; therefore they must be produced by the analyst after careful consideration of the problem in hand.

Determining sample size is a problem of trade-offs, as:

- a much too large sample may imply a data-collection and analysis process which is too expensive given the study objective and its required degree of accuracy; but
- a far too small sample may imply results which are subject to an unacceptably high degree of variability reducing the value of the whole exercise.

Somewhere between these two extremes lies the most efficient (in cost terms) sample size given the study objective. In what follows it will be assumed that this consists in estimating certain population parameters by means of statistics calculated from sample data; as any sample statistics are subject to sampling error, it is also necessary to include an estimate of the accuracy that may be associated to its value.

3.1.1.2 Sample Size to Estimate Population Parameters

This depends on three main factors: variability of the parameters in the population under study, degree of accuracy required for each, and population size. Without doubt the first two are the most important; this may appear surprising at first sight because, to many, it seems intuitively necessary to take bigger samples in bigger populations in order to maintain the accuracy of the estimates. However, as will be shown below, the size of the population does not significantly affect sample size except in the case of very small populations.

The Central Limit Theorem, which is at the heart of the sample size estimation problem, postulates that the estimates of the mean of a sample tend to become distributed Normal as the sample size (n) increases. This holds for any population distribution if n is greater than or equal to 30; the theorem holds even in the case of smaller samples, if the original population has a Normal-like distribution.

Consider a population of size N and a specific property which is distributed with mean μ and variance σ^2. The Central Limit Theorem states that the distribution of the mean ($\bar{x}$) of successive samples is distributed Normal with mean μ and standard deviation se $(\bar{x})$, known as the standard error of the mean, and given by:

$$se(\bar{x}) = \sqrt{(N - n)\sigma^2/[n(N - 1)]} \tag{3.1}$$

If only one sample is considered, the best estimate of μ is $\bar{x}$ and the best estimate of σ^2 is S^2 (the sample variance); in this case the standard error of the mean can be estimated as:

$$se(\bar{x}) = \sqrt{(N - n)S^2/nN} \tag{3.2}$$

and, as mentioned above, it is a function of three factors: the parameter variability (S^2), the sample size (n) and the size of the population (N). However, for large populations and small sample sizes (the most frequent case) the factor $(N - n)/N$ is very close to 1 and equation (3.2) reduces to:

$$se(\bar{x}) = \frac{S}{\sqrt{n}} \tag{3.3}$$

Thus, for example, quadrupling sample size will only halve the standard error, i.e. it is a typical case of diminishing returns of scale. The required sample size may be estimated solving equation (3.2) for n; it is usually simpler to do it in two stages, first calculating n from equation (3.3) such that:

$$n' = \frac{S^2}{se(\bar{x})^2} \tag{3.4}$$

then correcting for finite population size, if necessary, by:

$$n = \frac{n'}{1 + \dfrac{n'}{N}}$$ (3.5)

Although the above procedure appears to be both objective and relatively trivial it has two important problems that impair its application: estimating the sample variance S^2 and choosing an acceptable standard error for the mean. The first one is obvious: S^2 can only be calculated once the sample has been taken, so it has to be estimated from other sources. The second one is related with the desired degree of confidence to be associated with the use of the sample mean as an estimate of the population mean; normal practice does not specify a single standard error value, but an interval around the mean for a given confidence level. Thus, two judgements are needed to calculate an acceptable standard error:

- First a confidence level for the interval must be chosen; this expresses how frequently the analyst is prepared to make a mistake by accepting the sample mean as a measure of the true mean (e.g. the typical 95% level implies a willingness to err in 5% of cases).
- Second, it is necessary to specify the limits of the confidence interval around the mean, either in absolute or relative terms; as the interval is expressed as a proportion of the mean in the latter case, an estimate of this is required to calculate the absolute values of the interval. A useful option considers expressing sample size as a function of the expected coefficient of variation ($CV = \sigma/\mu$) of the data.

For example, if a Normal distribution is assumed and a 95% confidence level is specified, this means that a maximum value of 1.96 se $(\bar{x})$ would be accepted for the confidence interval (i.e $\mu \pm 1.96\sigma$ contains 95% of the Normal probability distribution); if a 10% error is specified we would get a $(\mu \pm 0.1\mu)$ interval and it may be seen that:

$$\text{se}(\bar{x}) = 0.1\mu/1.96 = 0.051\mu$$

and replacing this value in (3.4) we get:

$$n' = (S/0.051\mu)^2 = 384CV^2$$ (3.6)

Note that if the interval is specified as $(\mu \pm 0.05\mu)$, i.e. with half the error, n' would increase fourfold to $1536\ CV^2$.

To complete this point it is important to emphasise that the above exercise is relatively subjective; thus, more important parameters may be assigned smaller confidence intervals and/or higher levels of confidence. However, each of these actions will result in smaller acceptable standard errors and, consequently, bigger samples and costs. If multiple parameters need to be estimated the sample may be chosen based on that requiring a larger sample size.

3.1.1.3 Obtaining the Sample

The last stage of the sampling process is the extraction of the sample itself. In some cases the procedure may be easily automated, either on site or at the desk (in which case care must be taken that it is actually followed on the field), but it must always be conducted with reference to a random process. Although the only truly random processes are those of a physical nature (i.e. roll of a dice or flip of a coin), they are generally too time consuming to be useful in sample selection. For this reason pseudo-random processes, capable of generating easily and quickly a set of suitable random-like numbers, are usually employed in sampling.

Example 3.1: Consider a certain area whose population may be classified in groups according to: automobile ownership (with and without a car); and household size (up to four and more than four residents).

Let us assume that *m* observations are required by cell in order to guarantee a 95% confidence level in the estimation of, say, trip rates; assume also that the population can be considered to have approximately the following distribution (i.e. from historic data).

Car ownership	Household size	% of population
With car	Four or less	9
	More than four	16
Without car	Four or less	25
	More than four	50

There are two possible ways to proceed:

1. Achieve a sample with *m* observations by cell by means of a random sample. In this case it is necessary to select a sample size which guarantees this for each cell, including that with the smallest proportion of the population. Therefore, the sample size would be:

$$n = 100m/9 = 11.1m$$

2. Alternatively, one can undertake first a preliminary random survey of $11.1m$ households where only cell membership is asked for; this low-cost survey can be used to obtain the addresses of *m* households even in the smallest group. Subsequently, as only *m* observations are needed by cell, it would suffice to randomly select a (stratified) sample of $3m$ households from the other groups to be interviewed in detail (together with the *m* already detected for the most restrictive cell).

   As can be seen, a much higher sample is obtained in the first case; its cost (approximately three times more interviews) must be weighed against the cost of the preliminary survey.

3.1.2 Conceptualisation of the Sampling Problem

In this part we will assume that the final objective of taking the sample is to calibrate a choice model for the whole population. Following Lerman and Manski (1976) we will denote by P and f population and sample characteristics respectively. We will also assume that each sampled observation may be described on the basis of the following two variables:

i = observed choice of the sample individual (e.g. took a bus);

$\mathbf{X}$ = vector of characteristics (attributes) of the individual (age, sex, income, car ownership) and of the alternatives in his choice set (walking, waiting and travel times, cost)

We will finally assume that the underlying choice process in the population may be represented by a model with parameters $\boldsymbol{\theta}$; in this case, the joint distribution of i and $\mathbf{X}$ is given by:

$$P(i, \mathbf{X}/\boldsymbol{\theta})$$

and the probability of choosing alternative i among a set of options with attributes $\mathbf{X}$ is:

$$P(i/\mathbf{X}, \boldsymbol{\theta})$$

Depending on the form in which each observation is extracted, the sample will have its own joint distribution of i's and $\mathbf{X}$'s which we will denote by $f(i, \mathbf{X}/\boldsymbol{\theta})$. On the basis of this notation the sampling problem may be formalised as follows (Lerman and Manski, 1979).

3.1.2.1 Random Sample

In this case the distribution of i and $\mathbf{X}$ in the sample and population should be identical, that is:

$$f(i, \mathbf{X}/\boldsymbol{\theta}) = P(i, \mathbf{X}/\boldsymbol{\theta}) \tag{3.7}$$

3.1.2.2 Stratified Sample

In this case the sample is not random with respect to certain independent variables of the choice model (e.g. a sample with 50% low-income households and 50% high-income households is stratified if and only if a random sample is taken inside each stratum). The sampling process is defined by a function $f(\mathbf{X})$, giving the probability of finding an observation with characteristics $\mathbf{X}$; in the population this probability is of course $P(\mathbf{X})$. The distribution of i and $\mathbf{X}$ in the sample is thus given by:

$$f(i, \mathbf{X}/\boldsymbol{\theta}) = f(\mathbf{X}) P(i/\mathbf{X}, \boldsymbol{\theta}) \tag{3.8}$$

It is simple to show that a random sample is just a special case of stratified sample where $f(\mathbf{X}) = P(\mathbf{X})$, because:

$$f(i, \mathbf{X}/\boldsymbol{\theta}) = P(\mathbf{X})P(i/\mathbf{X}, \boldsymbol{\theta}) = P(i, \mathbf{X}/\boldsymbol{\theta}) \qquad (3.9)$$

3.1.2.3 Choice-based Sample

In this case the sampling procedure is defined by a function $f(i)$, giving the probability of finding an observation that chooses option i. Now the distribution of i and $\mathbf{X}$ in the sample is given by:

$$f(i, \mathbf{X}/\boldsymbol{\theta}) = f(i)P(\mathbf{X}/i, \boldsymbol{\theta}) \qquad (3.10)$$

We had not defined this latter probability, but we may obviate it on the basis of a Bayes theorem stating:

$$P(\mathbf{X}/i, \boldsymbol{\theta}) = P(i/\mathbf{X}, \boldsymbol{\theta})P(\mathbf{X})/P(i/\boldsymbol{\theta}) \qquad (3.11)$$

The expression in the denominator, which has not been defined either, may be obtained assuming discrete $\mathbf{X}$ from:

$$P(i/\boldsymbol{\theta}) = \sum_{\mathbf{X}} P(i/\mathbf{X}, \boldsymbol{\theta})P(\mathbf{X}) \qquad (3.12)$$

Therefore the final expression for the joint probability of i and $\mathbf{X}$ for a choice-based sample is clearly more complex:

$$f(i, \mathbf{X}/\boldsymbol{\theta}) = f(i)P(i/\mathbf{X}, \boldsymbol{\theta})P(\mathbf{X})/\sum_{\mathbf{X}} P(i/\mathbf{X}, \boldsymbol{\theta})P(\mathbf{X}) \qquad (3.13)$$

and it serves to illustrate not only that choice-based sampling is intuitively more problematic than the others, but also that it has higher bias potential in what really concerns us: choice.

Thus, each sampling method yields a different distribution of choices and characteristics in the sample, and there are no *a priori* reasons to expect that a single parameter estimation method would be applicable in all cases.

Example 3.2: Assume that for the purposes of a transport study the population of a certain area has been classified according to two income categories, and that there are only two means of transport available (car and bus) for the journey to work. Let us also assume that the population distribution is given by:

	Low income	High income	Total
Bus user	0.45	0.15	0.60
Car user	0.20	0.20	0.40
Total	0.65	0.35	1.00

1. Random sample. If a random sample is taken, it is clear that the same population distribution would be obtained.
2. Stratified sample. Consider a sample with 75% low income (LI) and 25% high income (HI) travellers. From the previous table it is possible to calculate the probability of a low-income traveller using bus, as:

$$P(\text{Bus}/\text{LI}) = \frac{P(\text{LI and Bus})}{P(\text{LI and Bus}) + P(\text{LI and Car})} = \frac{0.45}{0.45 + 0.20} = 0.692$$

Now, given the fact that the stratified sample has 75% of individuals with low income, the probability of finding a bus user with low income in the sample is: $0.75 \times 0.692 = 0.519$. Proceeding analogously the following table of probabilities for the stratified sample may be built:

	Low income	High income	Total
Bus user	0.519	0.107	0.626
Car user	0.231	0.143	0.374
Total	0.750	0.250	1.000

3. Choice-based sample. Let us assume now that we take a sample of 75% bus users and 25% car users. In this case the probability of a bus user having low income may be calulated as:

$$P(\text{LI}/\text{Bus}) = \frac{P(\text{LI and Bus})}{P(\text{LI and Bus}) + P(\text{HI and Bus})} = \frac{0.45}{0.45 + 0.15} = 0.75$$

Therefore, the probability of finding a low-income traveller choosing bus in the sample is 0.75 times 0.75, or 0.563. Proceeding analogously, the following table of probabilities for the choice-based sample may be built:

	Low income	High income	Total
Bus user	0.563	0.187	0.750
Car user	0.125	0.125	0.250
Total	0.688	0.312	1.000

As was obviously expected, each sampling method produces in general a different distribution in the sample. The importance of the above example will increase when we consider what is involved in the estimation of models using the various samples. For this it is necessary to acquire an intuitive understanding of what calibration programs do; they simply search for the 'best' values of the model coefficients associated to a set of explanatory variables; in this case best consists in replicating the observed choices more accurately.

For the population as a whole the probability of actually observing a given data set may be found, conceptually, simply by calculating the probabilities of choosing the observed option by different types of traveller (with given attributes and choice sets). For example, in the first table in Example 3.2 (simple random sample) the probability that a high-income traveller selects car is given by the ratio between probability of him having high income and using car, and the probability of him having high income, that is:

$$\frac{0.20}{0.15 + 0.20} = 0.572$$

If we consider the second table (stratified sample), the above probability is now given by:

$$\frac{0.143}{0.107 + 0.143} = 0.572$$

This is no coincidence; in fact it was one of the most important findings of an interesting piece of research by Lerman *et al.* (1976) in the USA. In practice it means that standard software may be used to estimate models with data obtained from a stratified sample.

It is also important to note that this is not the case for choice-based samples. To prove this consider calculating the same probability but using information from the third table:

$$\frac{0.125}{0.187 + 0.125} = 0.400$$

As can be seen, the result is completely different. To end this theme it is interesting to mention that Lerman *et al.* (1976) did also propose a method to use data from choice-based samples in model estimation avoiding bias at the expense only of requiring knowledge of the actual market shares. This involves weighting the observations by factors calculated as:

$$\frac{\text{Prob (select the option in a random sample)}}{\text{Prob (select the option in a choice based sample)}}$$

Thus, in our example the weighting factor for bus-based observation should be:

$$\frac{0.45 + 0.15}{0.563 + 0.187} = 0.8$$

and for car users:

$$\frac{0.20 + 0.20}{0.125 + 0.125} = 1.6$$

Note that it is necessary to have data about choices on each alternative, i.e. it would not be possible to calibrate a model for car and bus, based on data for the latter mode only.

3.1.3 Practical Considerations in Sampling

3.1.3.1 The Implementation Problem

Stratified (and choice-based) sampling requires random sampling inside each stratum; to do so it is first necessary to isolate the relevant group and this may turn out to be not easy in practice. Consider for example a case where the population of interest consists of all potential travellers in a city. Thus if we stratify by area of residence, it may be relatively simple to isolate the subpopulation of residents inside the city (e.g. using data from a previous survey); the problem is that it is extremely difficult to isolate and take a sample of the rest, i.e. those living outside the city.

An additional problem is that in certain cases even if it is possible to isolate all subpopulations and conforming strata, it may still be difficult to ensure a random sample inside each stratum. For example, if we are interested in taking a mode choice-based sample of travellers in a city we will need to interview bus users and for this it is first necessary to decide which routes will be included in the sample. The problem is that certain routes might have, say, higher than average proportions of students and/or old age pensioners, and this would introduce bias.

3.1.3.2 Finding the Size of Each Subpopulation

This is a key element in determining how many people will be surveyed. Given a certain stratification, there are several methods available to find out the size of each subpopulation, such as:

1. Direct measurement. This is possible in certain cases. Consider a mode choice-based sample of journey-to-work trips; the number of bus and metro tickets sold, plus traffic counts during the peak hour in an urban corridor, may yield an adequate measure (although imperfect as not all trips during peak are to work) of the number of people choosing each mode. If we have a geographical (i.e. zonal) stratification, on the other hand, the last census may be used to estimate the number of inhabitants in each zone.
2. Estimation from a random sample. If a random sample is taken, the proportion of observations corresponding to each stratum is a consistent estimator of the fraction of the total corresponding to each subpopulation. It is important to note that the cost of this method is low as the only information sought is that necessary to establish the stratum to which the interviewee belongs.
3. Solution of a system of simultaneous equations. Assume we are interested in stratifying by chosen mode and that we have data about certain population characteristics (e.g. mean income and car ownership). Taking a small on-mode sample we can obtain modal average values of these variables and postulate a system of equations which has the subpopulation fractions as unknowns.

Finally, the 'failure rate' of different types of surveys must be considered when designing sampling frameworks. The sample size discussed above correspond to the number of successful and valid responses to the data-collection effort. Some survey procedures are known to generate low valid response rates (e.g. some postal surveys), but they may still be used because of their low cost.

Example 3.3: Assume the following information is available:

> Average income of population (I): 33 600 $/year
> Average car ownership (CO): 0.44 cars/household

Assume also that small on-mode surveys yield the following:

Mode	I ($/year)	CO (cars/household)
Car	78 000	1.15
Bus	14 400	0.05
Metro	38 400	0.85

If F_i denotes the subpopulation fraction of the total, the following system of simultaneous equations holds:

$$33\,600 = 78\,000 F_1 + 14\,400 F_2 + 38\,400 F_3$$

$$0.44 = 1.15 F_1 + 0.05 F_2 + 0.85 F_3$$

$$1 = F_1 + F_2 + F_3$$

the solution of which is:

$$F_1 = 0.2451$$

$$F_2 = 0.6044$$

$$F_3 = 0.1505$$

This means that if the total population of the area was 180 000 inhabitants, there would be approximately 44 100 car users, 108 800 bus users and 27 100 metro users.

3.2 ERRORS IN MODELLING AND FORECASTING

The statistical procedures normally used in (travel demand) modelling assume not only that the correct functional specification of the model is known *a priori*, but also that the data used to estimate the model parameters have no errors. In practice, however, these conditions are often violated; furthermore, even if they were satisfied, model forecasts are usually subject to errors due to inaccuracies in the values assumed for the explanatory variables in the design year.

The ultimate goal of modelling is often forecasting (i.e. the number of people choosing given options); an important problem model designers face is to find which combination of model complexity and data accuracy fits best the required forecasting precision and study budget. To this end, it is important to distinguish between different types of errors, in particular:

- those that could cause even correct models to yield incorrect forecasts, e.g. errors in the prediction of the explanatory variables, transference and aggregation errors; and
- those that actually cause incorrect models to be estimated, e.g. measurement, sampling and specification errors.

In the next section consideration is given first to the types of errors that may arise with the broad effects they may cause; then the trade-off between model complexity and data accuracy is examined with particular emphasis on the role of simplified models in certain contexts.

3.2.1 Different Types of Error

Consider the following list of errors that may arise during the processes of building, calibrating and forecasting with models.

3.2.1.1 Measurement Errors

These occur due to the inaccuracies inherent in the process of actually measuring the data in the base year, such as: questions badly registered by the interviewee, answers badly interpreted by the interviewer, network measurement errors, coding and digitising errors, and so on. These errors tend to be higher in less developed countries but they can always be reduced by improving the data-collection effort (e.g. by appropriate use of computerised interview support) or simply by allocating more resources to data quality control; however, both of these cost money.

Measurement error, as defined here, should be distinguished from the difficulty of defining the variables that ought to be measured. The complexity that may arise in this area is indicated in Figure 3.1; ideally, modelling should be based on the information perceived by individual travellers but whilst reported data may give some insight into perception, its use raises the difficult question of how to forecast what users are going to perceive in the future. So it appears inevitable that models will be endowed with perception errors which tend to be greater for non-chosen alternatives due to the existence of *self-selectivity* bias (i.e the attributes of the chosen option are perceived as better and those of the rejected option as worse than they are, such as to reinforce the rationality of the choice made).

3.2.1.2 Sampling Errors

These arise because the models must be estimated using finite data sets. Sampling errors were approximately inversely proportional to the square root of the number of observations (i.e. to halve them it is necessary to quadruple sample size); thus, reducing them may be costly. Daganzo (1980) has examined the problem of defining optimal sampling strategies in the sense of refining estimation accuracy.

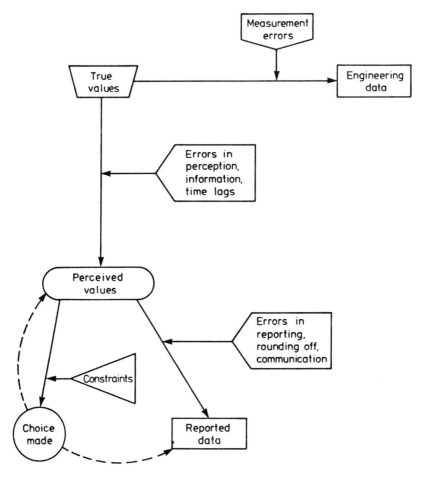

Figure 3.1   Attribute measurement and choice

3.2.1.3 Computational Errors

These arise because models are generally based on iterative procedures for which the exact solution, if it exists, has not been found for reasons of computational costs. These errors are typically small in comparison with other errors, except for cases such as assignment to congested networks and problems of equilibration between supply and demand in complete model systems, where they can be very large (see ESTRAUS 1989).

3.2.1.4 Specification Errors

These arise either because the phenomenon being modelled is not well understood or because it needs to be simplified for whatever reason. Important subclasses of this type of error are the following:

- Inclusion of an irrelevant variable (i.e. one which does not affect the modelled choice process). This error will not bias the model (or its forecasts) if the parameters appear in linear form, but it will tend to increase sampling error; in a non-linear model, however, bias may be caused (see Tardiff 1979).
- Omission of a relevant variable; perhaps the most common specification error. Interestingly, models incorporating a random error term (such as many of those we will examine in Chapters 4 and 7) are designed to accommodate this error; however, problems can arise when the excluded variable is correlated with variables in the model or when its distribution in the relevant population is different from its distribution in the sample used for model estimation (see Horowitz 1981b).
- Exclusion of *taste variations* on the part of the individuals; this will always produce biased models, as shown in Chapter 8. Unfortunately this is the case in most practical models of choice; a notable exception is the generally unyielding multinomial probit model which we will discuss also in Chapter 8 (Daganzo 1979).
- Other specification errors, in particular the use of model forms which are not appropriate, such as linear functions to represent non-linear effects, *compensatory* models to represent behaviour that might be *non-compensatory* (see the discussion in Chapter 8), or the omission of effects such as *habit* or *inertia*. A full discussion of these forms of error is given by Williams and Ortúzar (1982a).

All specification errors can be reduced in principle simply by increasing model complexity; however, the total costs of doing this are not easy to estimate as they relate to model operation, but may induce other types of errors which might be costly or impossible to eliminate (e.g. when forecasting more variables and at a higher level of disaggregation). Moreover, removal of some specification errors may require extensive behavioural research and it must simply be conceded that such errors may be present in all feasible models.

3.2.1.5 Transfer Errors

These occur when a model developed in one context (time and/or place) is applied in a different one. Although adjustments may be made to compensate for the transfer, ultimately the fact must be faced that behaviour might just be different in different contexts. In the case of spatial transfers, the errors can be reduced or eliminated by partial or complete re-estimation of the model to the new context (although the latter would imply discarding the substantial cost savings obtainable from transfer). However, in the case of temporal transfer (i.e. forecasting), this re-estimation is not possible and any potential errors must just be accepted.

3.2.1.6 Aggregation Errors

These arise basically out of the need to make forecasts for groups of people while modelling often needs to be done at the level of the individual in order to capture behaviour better. The following are important subclasses of aggregation error:

- Data aggregation. In most practical studies the data used to define the choice situation of individual travellers is aggregated in some form or another; even when the traveller is asked to report the characteristics of his available options, they can only have based their choice on the expected values of these characteristics. When network models are used there is aggregation over routes, departure times and even zones; this means that the values thus obtained for the explanatory variables are, at best, averages for groups of travellers rather than exact values for any particular individual. Models estimated with aggregate data will suffer from some form of specification error (see Daly and Ortúzar 1990). Reducing this type of aggregation error implies making measurements under many more sets of circumstances: more zones, more departure times, more routes, more socio-economic categories; this costs time and money and increases model complexity.
- Aggregation of alternatives. Again due to practical considerations it may just not be feasible to attempt to consider the whole range of options available to each traveller; even in relatively simpler cases such as the choice of mode, aggregation is present as the large variety of services encompassing a bus option, say (e.g. one-man operated single decker, two-man operated double decker, mini-buses, express services), are seldom treated as separate choices.
- Model aggregation. This can cause severe difficulties to the analyst except in the case of linear models where it is a trivial problem. Aggregate quantities such as flow on links are a basic modelling result in transportation planning, but methods to obtain them are subject to aggregation errors which are often impossible to eliminate. We will examine this problem in some detail in Chapters 4 and 9.

3.2.2 The Model Complexity/Data Accuracy Trade-off

Given the difficulties discussed above, it is reasonable to consider the dual problem of how to optimise the return of investing in increasing data accuracy, given a fixed study budget and a certain level of model complexity, to achieve a reasonable level of precision in forecasts. In order to tackle this problem we must understand first how errors in the input variables influence the accuracy of the model we use.

Consider the observed variables x with the associated errors e_x (i.e. standard deviation); to find the output error derived from the propagation of input errors in a function such as:

$$z = f(x_1, x_2, \ldots, x_n)$$

the following formula may be used:

$$e_z^2 = \sum_i \left(\frac{\partial f}{\partial x_i}\right)^2 e_{x_i}^2 + \sum_i \sum_{j \neq i} \frac{\partial f}{\partial x_i} \frac{\partial f}{\partial x_j} e_{x_i} e_{x_j} r_{ij} \tag{3.14}$$

where r_{ij} is the coefficient of correlation between x_i and x_j; the formula is exact for linear functions and a reasonable approximation in other cases. Alonso (1968) uses it to derive some simple rules to be followed during model building in order to prevent large output errors; for example, an obvious one is to avoid using correlated variables, thus reducing the second term of the right-hand side of equation (3.14) to zero.

If we take the partial derivative of e_z with respect to e_{x_i} and ignore the correlation term, we get:

$$\frac{\partial e_z}{\partial e_{x_i}} = \left(\frac{\partial f}{\partial x_i}\right)^2 \frac{e_{x_i}}{e_z} \tag{3.15}$$

Using these marginal improvement rates and an estimation of the marginal costs of enhancing data accuracy it is possible, in principle, to determine an optimum improvement budget; in practice this problem is not easy though, not least because the law of diminishing returns (i.e. each further percentage reduction in the error of a variable will tend to cost proportionately more) might operate, leading to a complex iterative procedure. However, equation (3.15) serves to deduce two logical rules:

- concentrate the improvement effort on those variables with a large error; and
- concentrate the effort on the most relevant variables, i.e. those with the largest value of $(\partial f / \partial x_i)$ as they have the largest effect on the dependent variable.

Example 3.4: Consider the model $z = xy + w$, and the following measurement of the independent variables:

$$x = 100 \pm 10; \quad y = 50 \pm 5; \quad w = 200 \pm 50$$

Assume also that the marginal cost of improving each measurement is the following:

$$\text{Marginal cost of improving } x \text{ (to } 100 \pm 9) \quad = \$5.00$$
$$\text{Marginal cost of improving } y \text{ (to } 50 \pm 4) \quad \ = \$6.00$$
$$\text{Marginal cost of improving } w \text{ (to } 200 \pm 49) = \$0.02$$

Applying equation (3.14) we get:

$$e_z^2 = y^2 e_x^2 + x^2 e_y^2 + e_w^2 = 502\,500$$

then $e_z = 708.87$; proceeding analogously, values of improved e_z in the cases of improving x, y or w may be found to be 674.54, 642.26 and 708.08 respectively. Also from (3.15) we get:

$$\frac{\partial e_z}{\partial e_x} = \frac{10y^2}{708.87} = 35.2; \qquad \frac{\partial e_z}{\partial e_y} = 70.5; \qquad \frac{\partial e_z}{\partial e_w} = 0.0705$$

These last three values are the marginal improvement rates corresponding to each variable. To work out the cost of the marginal improvements in e_z we must divide the marginal costs of improving each variable by their respective marginal rates of improvement. Therefore we get the following marginal costs of improving e_z arising from the various variable improvements:

$$\text{Marginal improvement in } x = 5/35.2 \qquad = \$0.142$$

$$\text{Marginal improvement in } y = 6/70.5 \qquad = \$0.085$$

$$\text{Marginal improvement in } w = 0.02/0.0705 = \$0.284$$

Therefore it would be decided to improve the measurement accuracy of variable y if the marginal reduction in e_z is worth at least $0.085.

Let us now define complexity as an increase in the number of variables of a model and/or an increase in the number of algebraic operations with the variables. It is obvious that in order to reduce specification error (e_s) complexity must be increased; however, it is also clear that as there are more variables to be measured and/or greater problems for their measurement, data measurement error (e_m) will probably increase as well.

If total modelling error is defined as $E = \sqrt{(e_s^2 + e_m^2)}$, it is easy to see that the minimum of E does not necessarily lie at the point of maximum complexity (i.e. maximum realism). Figure 3.2 shows not only that this is intuitively true, but also that as measurement error increases, the optimum value can only be attained at decreasing levels of model complexity.

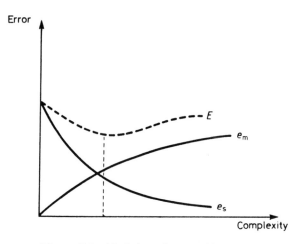

Figure 3.2   Variation of error with complexity

Example 3.5: Consider the case of having to make a choice between an extremely simple model, which is known to produce a total error of 30% in forecasts, and a new model which has a perfect specification (i.e. $e_s = 0$) given by:

$$z = x_1 x_2 x_3 x_4 x_5$$

where the x_i are independent variables measured with a 10% error (i.e. $e_m = 0.1x_i$). To decide which model is more convenient we will apply equation (3.14):

$$e_z^2 = 0.01[x_1^2(x_2 x_3 x_4 x_5)^2 + x_2^2(x_1 x_3 x_4 x_5)^2 + \ldots + x_5^2(x_1 x_2 x_3 x_4)^2]$$

$$e_z^2 = 0.05[x_1 x_2 x_3 x_4 x_5]^2 = 0.05z^2$$

that is, $e_z = 0.22z$ or a 22% error, in which case we would select the second model.

It is interesting to mention that repeating the exercise in a situation where measurement errors are higher produces different results. The interested reader can check that if it is assumed that the x_i variables can only be measured with 20% error, the total error of the second model comes out as 44.5% (i.e we would now select the first model even if its total error increased up to 44%).

Figure 3.3 serves to illustrate this point, which may be summarized as follows: if the data are not of a very good quality it might be safer to predict with simpler and more robust models. However, to learn about and understand the phenomenon, a better-specified model will always be preferable. Moreover, most models will be used in a forecasting mode where the values of the planning variables x_i will not be observed but forecast. We know that some planning variables are easier to forecast than others and that disaggregation makes predicting their future values an even less certain task. Therefore, in choosing a model for forecasting purposes preference should be given to those using planning variables which can be forecast, in turn, with greater confidence.

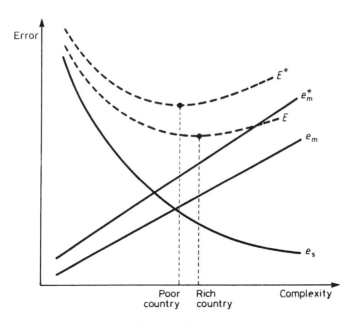

Figure 3.3   Influence of the measurement error

3.3 DATA-COLLECTION METHODS

3.3.1 Practical Considerations

Practical limitations have a strong influence in determining the most appropriate type of survey for a given situation. The design of a survey is not a simple matter and requires considerable skill and experience. For basic information on recruiting, training, questionnaire design, supervision and quality control, the reader is referred to the still good book by Moser and Kalton (1971). Information about survey procedures with a particular transport planning flavour may also be found in Stopher and Meyburg (1979). In what follows we briefly discuss some of the most typical practical constraints in transport studies.

3.3.1.1 Length of the Study

This obviously has great importance because it determines indirectly how much time and effort it is possible to devote to the data-collection stage. It is very important to achieve a balanced study (in terms of its various stages) avoiding the all too frequent problem of eventually finding that the largest part of the study budget (and time) was spent in data collection, analysis and validation (see Boyce *et al*. 1970).

3.3.1.2 Study Horizon

There are two types of situation worth considering in this respect:

- If the design year is too close, as in a tactical transport study, there will not be much time to conduct the study; this will probably imply the need to use a particular type of analysis tool, perhaps requiring data of a special kind.
- In strategic transport studies, on the other hand, the usual study horizon is 20 or more years into the future. Although in principle this allows time to employ almost any type of analytical tool (with their associated surveys), it also means that errors in forecasting will only be known in 20 or more years time. This calls for flexibility and adaptation if a successful process of monitoring and re-evaluation is to be achieved.

3.3.1.3 Limits of the Study Area

Here it is important to ignore formal political boundaries (i.e. of county or district) and concentrate on the whole area of interest. It is also necessary to distinguish between this and the study area as defined in the project brief; the former is normally larger as we would expect the latter to develop in a period of, say 20 years. The definition of the area of interest depends again on the type of policies examined and decisions to be made; we will come back to this issue below.

3.3.1.4 Study Resources

It is necessary to know, as clearly and in as much detail as possible, how many personnel and of what level will be available for the study; it is also important to know what type of computing facilities will be available and what restrictions to their use will exist. In general, the time available and study resources should be commensurate with the importance of the decisions to be taken as a result. The greater the cost of a wrong decision, the more resources should be devoted to get it right.

There are many other possible restrictions, ranging from physical (i.e. sheer size and topography of the locality) to social and environmental (e.g. known reluctance of the population to answer certain types of questions), which need to be taken into account and will influence sample design.

A general practical consideration is that travellers are often reluctant to answer 'yet another' questionnaire. Responding to questions takes time and may sometimes be seen as a violation of privacy. This may result in either flatly refusing to answer or in the provision of simplistic but credible responses, which is actually worse. In many countries it is necessary to obtain permission from the authorities before embarking on any traffic survey involving disruptions to travellers.

3.3.2 Types of Surveys

During the 1960s and up to the mid-1970s a large number of household origin–destination (O–D) surveys, using a simple random sample technique, were undertaken in urban areas of industrialised countries and also in many important cities in developing countries. This large effort was very expensive and demanded enormous quantities of time (a problem with collecting too much information is that a lot of time and money must also be spent analysing it); in fact, as we have commented already, the data-collection effort has traditionally absorbed a vital part of the resources available to conduct these large studies leaving, in many cases, little time and money for the crucial tasks of preparing and evaluating plans.

What follows considers the requirements of a strategic transport study with an analysis horizon of, say, 20 years. In such a case, data may be needed not only about trips but also about land use, employment and activities in general.

3.3.2.1 Typical Information Needs

These are as follows:

1. Infrastructure and existing services inventories (e.g. public and private transport networks, traffic signals), for model calibration, especially assignment models.
2. Land use inventory; residential zones (housing density), commercial and industrial zones (by type of establishment), parking spaces, etc; these are particularly useful for trip-generation models.
3. O–D travel surveys (at households, cordons and screen lines) and associated traffic counts; flow, speed and travel time measurements (to build speed–flow curves, not an easy task), for model calibration, especially trip distribution.
4. Socioeconomic information (income, car ownership, family size and structure, etc.), for trip generation and modal split

As we already commented, it is first necessary to define the area of interest of the study. Its external boundary is known as the *external cordon*; once this is defined, the area is divided into zones (we will look at some basic zoning rules in section 3.4) in order to have a clear and spatially disaggregated idea of the origin and destination of trips; also, to be able to spatially quantify some variables such as population and employment. The area outside the external cordon is also divided into zones but at a lesser level of detail (larger zones); inside the study area there can be also other *internal* cordons, as well as *screen lines* (i.e. an artificial divide following a natural or artificial boundary with few crossings, such as a river or a railway line), whose purposes will be discussed below. There are no hard rules for deciding the location of the external cordon and hence which areas will be considered external to the study; this decision is a function of the scope and levels of decision adopted for the study, i.e. it is a very 'contextual' problem.

3.3.2.2 O–D Surveys

Although the most expensive and difficult type of O–D survey is household-based, it is also that which offers, in general, the possibility of obtaining more useful data. However, on many occasions interest will not be centred on gathering data for the complete model system, but only for parts of it: the most typical case is that of mode choice and assignment in short-term studies.

An interesting method, particularly suitable for corridor-based journey-to-work studies and which has proved very efficient in practice, is the use of workplace interviews (see Dunphy 1979; Ortúzar and Donoso 1983). These involve the local authority asking a sample of institutions (employers) in, for example, the Central Business District (CBD) permission to interview a sample of their employees; in certain cases it has been found efficient to ask for the sample to be distributed by residence of the employee (e.g. those living in a certain corridor). It must be noted, however, that contrary to the case of random household survey, the data obtained in this case are choice based in terms of destination; nevertheless it is random with respect to mode.

In what follows we will be mainly referring to household surveys, although most aspects of the general discussion and indeed to the design of measurement instrument are equally applicable to any other type of O–D survey.

General Considerations It is widely recognised that both the procedures and measurement instruments used to collect information on site have a direct and profound influence on the results derived from any data-collection effort. This is why it has been recommended to include the measurement procedure as yet another element to be considered explicitly in the design of any project requiring empirical data for its development. Wermuth (1981), for example, has even proposed a categorisation of all the stages comprising a measurement procedure. In this part we will refer to only two of these categories: the *development* and the *use* of measurement instruments designed to measure activity patterns outside the household.

We have already mentioned that the empirical measurement of travel behaviour is one of the main inputs to the decision-making process in urban transport planning; in fact, it provides the basis for the formulation and estimation of models to explain and predict future travel activities. For this reason, methodological deficiencies at this stage will have direct repercussions in all subsequent stages of the transport planning process.

Frequent criticisms about household or workplace O–D surveys include:

- the surveys only measure average rather than actual travel behaviour of individuals;
- only part of the individual's movements can be investigated;
- information (for example about travel times) is often poorly estimated by the interviewee.

In fact, it has been found that variable measurements derived from traditional O–D surveys—for example related to times, distances and costs of travel—have proved inadequate when compared with values measured objectively for the same variables. That is, the reported characteristics have tended to differ substantially from reality in spite of the fact that the individuals responding to the survey experience the actual values of these level-of-service variables twice per day. It has also been concluded that the bias has a systematic nature and is apparently related with user attitudes with respect to each mode; for example in the case of public transport, access, waiting and transfer times (which are apparently very bothersome) tend to be severely overestimated. It is interesting to note that from a conceptual point of view these results would indicate that the subjective perception of level-of-service variables constitutes an important determinant in modal choice (see the discussion in Ortúzar *et al*. 1983).

A methodological analysis of these criticisms leads to two conclusions. In the first place, travel behaviour information should not be sought in general terms (i.e. average values) but with reference to a concrete temporal point of reference. In the second place, it is not recommended to examine the various activities in isolation, but rather to take the complete activity pattern as the basis for analysis; for example, it can be shown that asking for starting and ending times of a trip yields more accurate results than asking for its total duration (Ampt 1981). This critical analysis has brought two important consequences:

- the substantial improvement of the measurement procedure associated with household and other O–D surveys, which has allowed most of the deficiencies mentioned above to be overcome (Brög and Ampt 1982);
- the development of alternative or complementary data-collection methods such as the travel diary, which we will briefly discuss below.

Survey Date   Determining an appropriate date to conduct an O–D survey is strongly dependent on its objectives; normally, however, its main goal will be to obtain travel behaviour data about the inhabitants of a study area during a typical working day. In this case experience indicates that the best times of the year to collect the data are spring and autumn; the summer is normally discarded because it includes holidays, and the mid-winter period is not considered because the climatic conditions might translate into abnormal travel behaviour, both from the point of view of the trips made and the means of transport used.

Days and Times to Conduct the Survey   If the main objective is to obtain data for a typical working day, this often rules out collecting data about Mondays and Fridays; the former sometimes present a higher rate of labour and study absenteeism. The latter, being prior to the weekend, usually register more trips than other working days. In order to ensure a good recollection of events the survey should ask for information about the previous day; for this reason it is usually carried out during Wednesdays, Thursdays and Fridays.

With respect to survey times, past experience in many countries has shown that the best results for household surveys are obtained at the period of the day where the probability of finding the people at their homes is highest, usually between 18:00 and 21:00 hours. For workplace surveys, on the other hand, the best times are, of course, the normal working hours. This helps to see why this type of survey might be so interesting if feasible; the survey period is extended considerably, allowing for much better use of interviewers' time.

It is important to experiment with the times, circumstances and interviewing methods on site, in order to determine the most appropriate set before the actual conducting of the survey. Small pilot surveys serve this role, together with the actual testing of the questionnaire design and data-processing techniques. Completing the collection of all the necessary data on schedule may become a much more difficult assignment if this preliminary task is not done. Experience indicates that it may be necessary to visit each household up to three times in order to complete the questionnaire adequately; this is, of course, another important potential advantage of workplace interviews, which are usually completed in one visit.

Survey Period   Ideally all the selected sample should be interrogated on one day in order to obtain a real *snapshot* of what happened on the previous day. However, given the high number of people to be interviewed, it has become standard practice to conduct the survey during a period of several days; this makes it possible to use a relatively small group of interviewers (who actually become highly experienced in the job) who can be better trained and supervised. This is equivalent to assuming that the sum of responses for several working days is a good representation of the answers that would be obtained in a single day, which seems a reasonable hypothesis in most cases.

Questionnaire Design   Normally the order in which the questions are asked seeks to minimise resistance on the part of the interviewee; for this reason, whenever possible, 'difficult' questions (e.g. relating to income) are formulated at the end of the interview. In terms of its formal aspect, the questionnaire, and the interview itself, should try to satisfy the following objectives:

- the questions should be simple and direct;
- the number of open questions should be minimised;
- the information about travel must be elicited with reference to the activities which originated the trips;
- in the case of household surveys, each member older than 12 years old should be personally interviewed; the rest (e.g. those between 5 and 12) may be considered indirectly, letting the head of the household, or any other member, answer for them.

In general, household O–D surveys have three distinct sections: personal characteristics and identification, trip data, and household characteristics; we will briefly review the information sought in each of them.

- Personal characteristics and identification. This part includes questions designed to classify the household members (older than 5) according to the following aspects: relation to the head of the household (e.g. wife, son), sex, age, possession of a driving licence, educational level, and activity. In order to reduce the possibility of subjective classification, it is important to define a complete set of activities. Non-household surveys are usually concerned only with the person being interviewed; however, the relevant questions are the same or very similar.
- Trip data. This part of the survey aims at detecting and characterising all trips made by the household members identified in the first part. A trip is normally defined as any movement greater than 300 metres from an origin to a destination with a given purpose. Trips are characterised on the basis of variables such as: origin and destination (normally expressed by their nearest road junction or full post code in the UK), trip purpose, trip start and ending times, mode used, walking distance (including transfers), public-transport line and transfer station or bus stop (if applicable).
- Household characteristics. This section includes a set of questions designed to obtain socioeconomic information about the household. Relevant questions are: characteristics of the house, identification of household vehicles (including a code to identify their usual user), house ownership and income.

Sample Size   Household O–D surveys have been traditionally taken on the basis of very large random samples. Table 3.1 shows the values which have been postulated as recommended practice for more than 20 years (see Bruton 1985) but are rarely used in practice.

The problems created by these enormous sample sizes have been compounded by the fact that many agencies, particularly in developing countries, believe them to be essential and thus require up to 20% bigger samples to be contacted in the survey to allow for eventual validation losses (see Gárate 1988).

Methods to estimate sample size from a more logical and less wasteful statistical approach require knowledge about the variable to be estimated, its coefficient of variation, and the desired accuracy of measurement together with the level of significance associated to it.

The first requirement, although both obvious and fundamental, has been ignored many times in the past. The majority of household O–D surveys have been designed on the basis of very vague objectives, such as 'to reproduce the travel patterns in the

Table 3.1   Sample sizes recommended in traditional surveys

Population of area	Sample size (dwelling units)	
	Recommended	Minimum
Under 50 000	1 in 5	1 in 10
50 000–150 000	1 in 8	1 in 20
150 000–300 000	1 in 10	1 in 35
300 000–500 000	1 in 15	1 in 50
500 000–1 000 000	1 in 20	1 in 70
Over 1 000 000	1 in 25	1 in 100

area'. What is the meaning of this? Is it the elements of the O–D matrix which are required, and if this is the case, are they required by purpose, mode and time of day, or is it just the flow trends between large zones which are of interest?

If only area-wide trip generation rates are required, it can be shown that samples of some 1000 individuals guarantee 90% confidence levels for a 5% tolerance (error) on the values. However, the situation changes dramatically if the interest centres on the number of trips on each cell of a typical O–D matrix (i.e. with many zones; obviously the level of detail affects this). For example, if each cell had around 1000 trips it can be shown that a sample of 4.3% guarantees errors of less than 25% with 90% confidence; however, for volumes of 20 to 30 trips between zones (which are very common in practice), the same level of accuracy would require 100% samples (i.e. the whole population).

The above examples show that sample size selection is not an easy task. Firstly, it is necessary to have great clarity about the survey objectives; secondly, a decision must be reached about how much effort should be spent in order to achieve a given level of accuracy in the results (Brög and Ampt 1982).

The second element (coefficient of variation of the variable to be measured) was an unknown in the past, but now it may be estimated using information from the large number of household O-D surveys which have been conducted in recent years. Finally, the accuracy level (percentage error acceptable to the analyst) and its confidence level are context-dependent matters to be decided by the analyst on the basis of personal experience. Any sample may become too large if the level of accuracy required is too strict. It can be said that this aspect is where the 'art' of sample size determination lies.

Once these three factors are known, the sample size (n) may be computed using the following formula (M.E. Smith 1979):

$$n = \frac{CV^2 Z_\alpha^2}{E^2} \tag{3.16}$$

where CV is the coefficient of variation, E is the level of accuracy (expressed as a proportion) and Z_α is the value of the standard normal variate for the confidence level (α) required.

Example 3.6: Assume that we want to measure the number of trips per household in a certain area, and that we have data about the coefficient of variation of this variable for various locations in the USA as follows:

Area	CV
Average for U.S.A. (1969)	0.87
Pennsylvania (1967)	0.86
New Hampshire (1964)	1.07
Baltimore (1962)	1.05

As all the values are near to one, we will choose this figure for convenience. As mentioned above, the decision about accuracy and confidence level is the most difficult; equation (3.16) shows that if we postulate levels which are too strict, sample size increases exponentially. On the other hand, it is convenient to fix strict levels in this case because the number of trips per household is a very crucial variable (i.e. if this number is badly wrong, the accuracy of subsequent models will be severely compromised). In this example we will ask for 0.05 level of accuracy at a 95% level.

For $\alpha = 95\%$ the value of Z_α is 1.645, therefore we get:

$$n = 1.0(1.645)^2/(0.05)^2 = 1084$$

that is, it would suffice to take a sample of approximately 1100 observations to ensure trip rates with a 5% tolerance 95% of the time. The interested reader may consult M.E. Smith (1979) for other examples of this approach.

3.3.2.3 Other Important Types of Surveys

Roadside Interviews   These provide useful information about trips not registered in household surveys (i.e. external–external trips in a cordon survey). They are often a better method for estimating trip matrices than home interviews as larger samples are possible. For this reason, the data collected are also useful in validating and extending the household-based information.

Roadside interviews involve asking a sample of drivers and passengers of vehicles (e.g. cars, public transport, goods vehicles) crossing a roadside station, a limited set of questions; these include at least origin, destination and trip purpose. Other information such as age, sex and income is also desirable but seldom asked due to time limitations; however, well trained interviewers can easily add at least part of these data from simple observation of the vehicle and occupants (with obvious difficulties in the case of public transport).

The conduct of these interviews requires a good deal of organisation and planning to avoid unnecessary delays, ensure safety and deliver quality results. The identification of suitable sites, co-ordination with the police and arrangements for lighting and supervision are important elements in the success of these surveys. We shall concentrate here on issues of sample size and accuracy.

To determine the sample size the following expression can be used:

$$n > \frac{p(1 - p)}{\left(\dfrac{e}{z}\right)^2 + \dfrac{p(1 - p)}{N}} \tag{3.17}$$

where n is the number of passengers to survey, p is the proportion of trips with a given destination, e is an acceptable error (expressed as a proportion), z is the standard Normal variate value for the required confidence level, and N is the population size (i.e. observed passenger flow at a roadside station). It can be seen that for a given N, e and z, the value $p = 0.5$ yields the highest (i.e. most conservative) value for n in (3.17). Taking this value and considering $e = 0.1$ (i.e. a maximum error of 10%) and $z = 1.96$ (corresponding to a confidence level of 95%), the values shown in Table 3.2 are obtained.

Table 3.2   Variation of sample size with hourly flow

N (passengers/hour)	n (passengers/hour)	$100n/N$ (%)
100	49	49.0
200	65	32.5
300	73	24.3
500	81	16.2
700	85	12.1
900	87	9.7
1100	89	8.1

Example 3.7: An examination of historical data during preparatory work for a roadside interview revealed that flows across the survey station varied greatly throughout the day. Given this, it was considered too complex to try to implement the strategy of Table 3.2 in the field. Therefore, the following simplified table was developed:

Estimated hourly flow passengers/hour	Sample size %
900 or more	10.0 (1 in 10)
700 to 899	12.5 (1 in 8)
500 to 699	16.6 (1 in 6)
300 to 499	25.0 (1 in 4)
200 to 299	33.3 (1 in 3)
1 to 199	50.0 (1 in 2)

The fieldwork procedure requires stopping at random the corresponding number of vehicles, interviewing all their passengers and asking origin, destination and trip purpose. In the case of public-transport trips, given the practical difficulties in stopping the vehicles for the time required to interview all passengers, the survey may be conducted with the vehicles in motion. For this to work it is necessary to define road sections rather than stations and the number of interviewers to be used depends on the observed vehicle-occupancy factors at the section. Even this approach may be unworkable if the vehicles are overloaded.

Cordon Surveys   These provide useful information about external–external and external–internal trips. Their objective is to determine the number of trips that enter, leave and/or cross the cordoned area, thus helping to complete the information coming from the household O–D survey. The main one is taken at the external cordon, although surveys may be conducted at internal cordons as well. In order to reduce delay they sometimes involve stopping a sample of the vehicles passing a control station (usually with police help), to which a short mail-return questionnaire is given. In some Dutch studies a sample of licence plates is registered at the control station and the questionnaires are sent to the corresponding addresses stored in the Incomes and Excise computer. An important problem here is that return-mail surveys are known to produce biased results: this is because less than 50% of questionnaires are usually returned and it has been shown that the type of person who returns them is different to those that do not (see Brög and Meyburg 1980). This is why in many countries roadside surveys often ask a rather limited number of questions (i.e. occupation, purpose, origin, destination and modes available) to encourage better response rates.

Screen-line Surveys   Screen lines divide the area into large natural zones (e.g. at both sides of a river or motorway), with few crossing points between them. The procedure is analogous to that of cordon surveys and the data also serve to fill gaps in and validate (see Figure 3.4) the information coming from the household and cordon surveys. Care has to be taken when aiming to correct the household survey data in this way, because it might not be easy to conduct the comparison without introducing bias.

Travel Diary Surveys   These are a special type of household survey which request similar kinds of information as the O–D survey but in greater detail; they are applied separately to each member of the household travelling at the time of the study. Travel diaries are meant to be carried and self-completed by the subjects during the day; for this reason they must satisfy the following design objectives:

- Ease of transport. To accomplish this a small format allowing their storage in pockets or handbags is required.
- Ease of understanding to the user. With this end in mind it is necessary to devise a way of superimposing their pages in order to have the instructions permanently in view (see for example Ortúzar and Hutt 1988).
- Ease of completion. To try and accomplish this, as is usual with self-completion questionnaires, pre-codified options are offered wherever possible in order to minimise the need to register written text.

A travel diary survey is normally conducted with two objectives in mind: first, to aid the general correction process of the O–D survey, as will be discussed in section

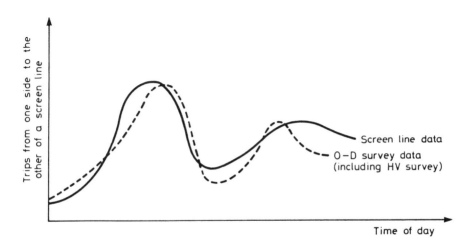

Figure 3.4   Household survey data consistency check

3.3.3; and second to provide a data bank suitable to estimate disaggregate modal choice models. This method of data collection implies the following steps:

- A first visit to each household in the sample (which should be selected taking care to avoid households contacted for the O–D survey), in order to present and explain the diaries, and to gather the same socioeconomic data as in the case of the O–D survey. Interviewees are briefly trained in the use of the instrument and asked to fill it with complete details of their travel data for the following day; more that one diary is given to each traveller if the exercise is to be repeated during the week (see Brög and Ampt 1982).
- A second visit the day following the last surveyed day (i.e. 24 hours later in the case of one-day diaries), in order to collect the completed forms for each traveller. It is often necessary to help some people complete the forms at this stage.

Travel diaries are processed in two stages: at first only the socioeconomic information and trip rates by purpose are registered, as this is the data needed for the O–D survey correction process. A second stage involves considering the data for each trip and incorporating very precise measures of the level-of-service variables for the reported chosen option and for the rest of the alternatives in each individual's choice set; as will be discussed in Chapter 8, the data are needed in this form for disaggregate choice model estimation.

3.3.3 O–D Survey Data Correction, Expansion and Validation

3.3.3.1 Data Correction

The need to correct the O–D survey data in order to achieve results which are not only representative of the whole population, but also reliable and valid, has been discussed at length (see Brög and Erl 1982; Wermuth 1981). It is now accepted that simply expanding the sample, which has been the method most commonly used in practice, is not appropriate; Brög and Ampt (1982) identify a series of correction steps as follows.

Correction by Household Size   Although the unit of analysis is the individual, samples are usually selected from lists of addresses; therefore it is possible to over-sample households of bigger size and under-consider households of small size. To solve this type of problem, sample family size should be compared with census family size and corrected accordingly.

Sociodemographic Correction   This might be necessary if differences in the distribution of the variables sex and age are detected between sample and population (i.e. census). It is important to check that the definitions of family and household are consistent in both cases. The correction must be done after having corrected for household size.

Non-response Correction   This problem is caused by possible variations in travel behaviour between those that do and do not answer the survey (i.e. people that travel more are more difficult to interview because they are out more often). It may be possible to estimate correction factors on the basis of the number of visits required to complete the questionnaire at different types of household. Again, this correction should be done after the previous two, and it is important to note that its application may induce significant changes in the data (Brög *et al.* 1982).

Correction for Non-reported Trips   This problem arises because the traditional type of home interview survey normally tends to underestimate non-mandatory trips. For this reason it is interesting to check the number of trips by purpose obtained in the O–D survey, with those of the travel diaries where more detailed information about each journey and especially about non-mandatory trips, should have been gathered. A correction method proposed by Ortúzar and Hutt (1988) considers the following steps:

- Divide the household into categories (say defined by income, number of cars and family size); the total number of categories is limited by the condition that each one must have at least 30 observations from the travel diary survey (i.e. to ensure that their mean trip rate is distributed normal).
- Calculate the average number of trips by purpose (and its variance) for each category, and for both the O–D survey and travel diary data; let the means be $\bar{X}_a$ and $\bar{X}_b$ and the variances S_a and S_b respectively. Calculate $D = \bar{X}_a - \bar{X}_b$.
- The *minimum detectable difference* (d) between the means of a certain variable X in two samples with sizes N_a and N_b, for an 80% probability of finding that their actual difference (D) is significant at the 95% level, is given by (Skelton 1982):

$$d = 2.8 \left(\frac{S_a}{N_a} + \frac{S_b}{N_b} \right)^{1/2} \tag{3.18}$$

- If $D > d$, the difference is significant; therefore if the average trip rate in that category is smaller in the O–D survey than in the travel diary, it has to be factored to equal the average trip rate for the diaries. If the reverse occurs no correction is performed (i.e. the factor is one).
- If $D \leqslant d$, the difference is not significant and no correction is required.

3.3.3.2 Sample Expansion

Once the data have been corrected it is necessary to expand them in order to represent the total population; to achieve this expansion factors are defined for each study zone as the ratio between the total number of addresses in the zone (A) and the number obtained as the final sample. However, often data on A are outdated leading to problems in the field. The following expression is fairly general in this sense:

$$F_i = \frac{A - A(C + CD/B)/B}{B - C - D}$$

where F_i is the expansion factor for zone i, A is the total number of addresses in the original population list, B is the total number of addresses selected as the original sample, C is the number of sampled addresses that were non-eligible in practice (e.g. demolished, non-residential), and D is the number of sampled addresses where no response was obtained. As can be seen, if A was perfect (i.e. $C = 0$) the factor would simply be $A/(B - D)$ as defined above. On the other hand, if $D = 0$ it can be seen that the formula takes care of subtracting from A the proportion of non-eligible cases, in order to avoid a bias in F_i.

3.3.3.3 Validation of Results

Data obtained from O–D surveys are normally submitted to three validation processes. The first simply considers on site checks of the completeness and coherence of the data; this is usually followed by their coding and digitising in the office. The second is a computational check of valid ranges for most variables and in general of the internal consistency of the data. Once these processes are completed, the data is assumed to be free of obvious errors.

To carry out the last process it is necessary to count traffic at cordons and screen lines during the O–D survey period. Later, the corrected and expanded survey data are contrasted with the information obtained from the counts (both of vehicles and pedestrians) suitably transformed by means of occupancy rates also measured on site. This last stage, which really deserves the title of validation as the survey data are compared with independent and more reliable information (all vehicles are counted, not just a sample), usually presents some practical problems; for example in the case of car trips, information about routes chosen is normally lacking and if assignment is not done carefully, large errors may creep in.

3.3.4 Stated-preference Surveys

The previous discussion has been conducted under the implicit assumption that any choice data corresponded to *revealed-preference* (RP) information; this means data about actual or observed choices made by individuals. It is interesting to note that we are seldom in a position to actually observe choice; normally we just manage to obtain data on what people report they do (or more often, what they have done on the previous day).

In terms of understanding travel behaviour RP data have limitations:

- observations of actual choices may not provide sufficient variability for constructing good models for evaluation and forecasting. For example, attributes like travel time and fare may be so correlated in the sample that it may be very difficult to separate their effects in model estimation and therefore for forecasting purposes.
- the observed behaviour may be dominated by a few factors making it very difficult to detect the relative importance of other variables. This is a particular problem with secondary qualitative variables like public-transport information services, security, decor; but these attributes cost money and we would like to find out how much travellers value them before allocating resources among them.
- the difficulties in collecting responses for policies which are entirely new, for

example a completely new mode (perhaps a people mover) or cost-recovery system (e.g. electronic road pricing).

These limitations would be surmounted if we could undertake real-life controlled experiments within cities or transport systems. The opportunities for doing this in practice are very limited. *Stated-preference* (SP) surveys provide an approximation to this, a sort of quasi-experiment based on hypothetical situations set up by the researcher.

What distinguishes RP from SP surveys is that in the latter case individuals are asked about what they would choose to do (or how would they rank/rate certain options) in one or more hypothetical situations. The degree of artificiality of these situations may vary, according to the needs and rigour of the exercise:

- the *decision context* may be a hypothetical or a real one; in other words, the respondent may be asked to consider an actual journey or one that she might consider undertaking in the future;
- the *alternatives* offered are often hypothetical although one of them may well be an existing one, for example the mode just chosen by the respondent including all its attributes;
- the *response* elicited from the individual may take the form of *choices* or just *preferences* expressed in a number of ways.

A very basic problem with SP data collection is how much faith can we put on individuals actually doing what they stated they would do when the case arises (for example, after introducing a new option). In fact, experience in the 1970s was not very good in this sense, with large differences between predicted and actual choice (e.g. only half the people doing what they said they would) found in many studies (see Ortúzar 1980a).

The situation improved considerably in the 1980s and recently very good agreement with reality has been reported from models estimated using SP data (Louviere 1988). However, this has occurred because SP data-collection methods have improved enormously and are now very demanding, not only in terms of survey design expertise but also in their requirements for trained survey staff and quality-assurance procedures. For a good 'guide' to practice see Pearmain *et al*. (1991).

The main features of an SP survey may be summarised as follows:

(a) it is based on the elicitation of respondents' statements of how they would respond to different hypothetical (travel) alternatives;
(b) each option is represented as a 'package' of different attributes like travel time, price, headway, reliability and so on;
(c) the researcher constructs these hypothetical alternatives so that the individual effect of each attribute can be estimated; this is achieved using *experimental design* techniques that ensure the variations in the attributes in each package are statistically independent from one another;
(d) the researcher has to make sure that interviewees are given hypothetical alternatives they can understand, appear plausible and realistic, and relate to their current level of experience;
(e) the respondents state their preferences towards each option by either *ranking* them in order of attractiveness, *rating* them on a scale indicating strength of

preference, or simply *choosing* the most preferred option from a pair or group of them;

(f)   the responses given by individuals are analysed to provide quantitative measures of the relative importance of each attribute; in many cases choice models can be estimated as part of this analysis as discussed in detail in Chapter 8.

Figure 3.5 provides an example of the presentation of SP alternatives for bus services; each alternative is described in a card in terms of its attributes: headway, in-vehicle time, fare and interchange; the respondent is then asked to rank these cards from the best to the worst alternative and this is simply achieved by putting the cards in that order.

Fare	Interchange	Time on bus	Walk time
70 p	No change	15 mins	10 mins

Fare	Interchange	Time on bus	Walk time
70 p	No change	20 mins	8 mins

Fare	Interchange	Time on bus	Walk time
85 p	No change	15 mins	10 mins

Fare	Interchange	Time on bus	Walk time
85 p	1 change	15 mins	8 mins

Figure 3.5   Example of stated-preference ranking exercise

The power of an SP exercise lies in the freedom to design quasi-experiments to meet the requirements of a wide variety of research needs. This power has to be balanced by the need to ensure the responses provided by the subjects are realistic, that is as close as possible to how they would have responded had these hypothetical options actually existed in practice. This balance must be struck at different stages in the SP exercise:

(a)   identification of the key attributes of each alternative and construction of the 'packages' constituting the options; all essential attributes must be present and the options must be plausible and realistic;

(b) design of the way in which the options will be presented to the respondents and how they will be allowed to express their preferences; the form of presentation of the alternatives must be easy to understand and within the context of the respondents experience and constraints;

(c) development of a sampling strategy to be followed to ensure a rich and representative data set;

(d) appropriate conduct of the survey including supervision and quality-assurance procedures;

(e) use of good model estimation techniques, ideally combining SP and RP data and having in mind the way in which the resulting weights or models will be used to assist decision making.

3.3.4.1 Attributes and Alternatives

A stated-preference experiment has as one of its main elements the construction of a set of hypothetical (although realistic) options, which we will refer to as *technologically feasible alternatives*; these are defined on the basis of the factors assumed to influence most strongly the choice problem under consideration. The design of these technologically feasible alternatives requires four distinct tasks: (a) the identification of the range of choices (i.e. which broad options will be included: rail and car, or different types of services within a mode), (b) the selection of the attributes to be included in each broad option, (c) the selection of the measurement unit for each attribute, and (d) the specification of number and magnitudes of the attribute levels.

(a) The range of options is usually given by the objective of the exercise; however, one should not omit realistic alternatives a user might consider in practice. For example, in studying potential responses of car drivers to new road-pricing initiatives it may not be sensible to consider only alternative modes of travel; changes to departure time or to alternative destinations (to avoid the most expensive road charges) may be very relevant responses. By ignoring them one places the respondent in a more artificial (less realistic) context.

(b) The set and nature of the attributes should also be chosen to ensure realistic responses. The most important attributes must be present and they should be sufficient to describe the technologically feasible alternatives. Care must be applied here as particular combinations of attributes (e.g. a high-quality, high-frequency, low-cost alternative) may not be seen as realistic by respondents thus reducing the value of the whole exercise.

In order to ensure that the right attributes are included and the options described in an easy-to-understand manner it is advantageous to undertake a small number of group discussions with a representative sample of individuals. A trained moderator will make sure all relevant questions regarding perception of alternatives, identification of key attributes and the way in which they are described and perceived by subjects, and the key elements establishing the context of the exercise are all discussed and reported. Discussion groups cost money and in many cases the researcher will be tempted to skip them believing a good understanding of the problem and context already exists. In that case, it will be essential to undertake a carefully monitored pilot survey where any issues of attribute description and alternative presentation can be explored.

(c) The selection of the metric for most attributes is relatively straightforward. However, there are some situations that may require more careful consideration, in particular with respect to qualitative attributes like 'comfort' or 'reliability'. For example travel time reliability can be presented as a distribution of journey times on different days of a working week, or as the probability of being delayed by more of a certain time. For more on this issue see the discussion of stimulus presentation below.

3.3.4.2 Experimental Design

The design of the options and their presentations has, in essence, three steps: (a) the selection of the attribute levels and combinations constituting each alternative (experimental design), (b) the design of the presentation of these alternatives (stimulus presentations), and (c) the specification of the responses to be elicited from the respondents. We discuss the first of these steps next.

As already suggested, most stated-preference exercises use experimental designs to construct the hypothetical alternatives presented to the respondents. An experimental design is usually 'orthogonal'; that is, it ensures that the attribute combinations presented are varied independently from one another. The advantage of this is that the effect of each attribute on the responses is more easily identified.

The number of attributes (a) and the number of levels each one can take (n), determine a factorial design (n^a). Tables exist (see for example Kocur *et al.* 1982) which give the number of hypothetical options needed to test most designs of interest and guarantee orthogonality.

The designs distinguish cases considering *principal effects* only and cases allowing the treatment of *interactions* (i.e. when the effects of two variables are not additive they may enter as the product of the two variables, see Figure 3.6). A problem with

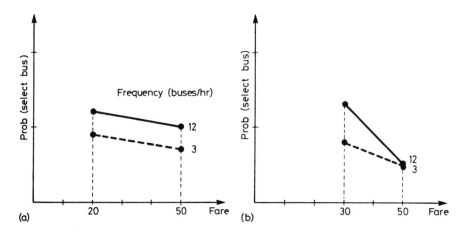

Figure 3.6   Presence and absence of attribute interaction: (a) without intreaction, (b) with interaction

the latter is that they require the construction of many more options; for this reason *fractional factorial designs*, which assume that some or all variable products are negligible, are often used.

Example 3.7: Consider the following experimental design (Pearmain *et al*. 1991) for three attributes for a bus service (fare, frequency and travel time) and each attribute at two levels (Low and High). For a full factorial design we need $2^3 = 8$ options and these are described in both descriptive and numerical terms (the latter representing a standard notation for experimental design) in Table 3.3.

In order to translate this experimental design into a suitable context for a respondent, we must first ascertain the current levels experienced by her. Let us assume she is currently using a service with a travel time of 25 minutes, a headway of 30 minutes and a fare of £0.50; we are interested in improving the service and would like to find out how much more we could charge for it.

The experimental design can now be translated into the alternatives presented in Table 3.4. Note that we now have to decide how to convert 'Low' and 'High' or 'Fast' and 'Slow' into descriptions relevant and plausible to the respondent. Moreover, the size of the changes in the attribute levels must be large enough to elicit detectable changes in stated preferences, but not so large that they would compromise credibility.

This is a full factorial design, in the sense that the analyst would be able to recover

Table 3.3 Experimental design for three attributes at two levels each

	Attributes		
Options	Fare	Travel Time	Frequency
1	Low	Fast	Infrequent
2	Low	Fast	Frequent
3	Low	Slow	Infrequent
4	Low	Slow	Frequent
5	High	Fast	Infrequent
6	High	Fast	Frequent
7	High	Slow	Infrequent
8	High	Slow	Frequent

	Numeric representation: -1 = 'poor', 1 = 'good'		
Options	Attribute 1	Attribute 2	Attribute 3
1	1	1	-1
2	1	1	1
3	1	-1	-1
4	1	-1	1
5	-1	1	-1
6	-1	1	1
7	-1	-1	-1
8	-1	-1	1

Table 3.4 Presentation of public-transport options

Options	Fare (£)	Travel Time (mins)	Frequency (buses/h)
1	0.50	18	2
2	0.50	18	4
3	0.50	25	2
4	0.50	25	4
5	0.80	18	2
6	0.80	18	4
7	0.80	25	2
8	0.80	25	4

all direct and interaction effects. However, if we are also interested in the effect of reducing the level of service we would need to increase the number of levels for travel time and frequency to three thus increasing the number of alternatives to $3^2 \times 2^1 = 18$; so many options will tend to induce fatigue in the respondent and reduce the value of the responses. Fractional factorial designs permit a reduction in the number of options at the cost of being unable to recover one or more interaction effects; see Pearmain *et al.* (1991) for guidelines on how to reduce the number and complexity of the options having regard to the objectives of the exercise and the time and comprehension constraints of the survey instruments.

In fact, a key design issue is complexity. Experience has shown that people give the most reliable responses when asked to consider simultaneous changes in up to three factors only (Huber and Hanson 1986). The complexity of the response task can thus be expected to influence the amount of error in the data; pre-tests, survey monitoring and subsequent debriefing can uncover general problems and checks can be incorporated in the instrument to identify individuals with poor understanding. In this sense interactive survey procedures appear to be particularly appropriate because they allow the detection of such problems and immediately probe further or provide additional instructions.

Example 3.8: Consider a situation with five attributes, two at two levels and the rest at three levels (i.e. a $2^2 \times 3^3$ design). In this case, depending on the number of interactions to be tested, the number of options required would vary as follows:

- 108 to consider all effects (i.e. a full factorial design);
- 81 to consider principal effects and all interactions between pairs of attributes, ignoring effects of a higher order;
- 27 to consider principal effects and interactions between one attribute and all the rest;
- 16 only if no interactions are considered.

Once the factorial design has been decided, the technologically feasible alternatives are constructed (which may, of course, be just hypothetical) and eventually the experiment conducted and the data collected. Fowkes and Wardman (1988) make a series of practical recommendations for the desirable variation of the attribute levels;

their experience indicates that often it may be beneficial to sacrifice some purity in the experimental design (e.g. lose complete orthogonality) if one gains in realism.

Pearmain and Swanson (1990) have been experimenting with adaptive designs for SP surveys using laptop computers; the use of advanced software allows them to modify the experiment in the light of the responses of the subject. Another approach to the same problem has been explored by Holden *et al.* (1992). The general results suggest that there is much to be gained from tailoring the experiments to each respondent but care must be exercised not to lose the desirable properties of the sample and general design. This is, of course, more easily done when using portable computers to conduct the survey. The reader is referred to these two papers for further details.

Whatever the approach to experimental design adopted in an SP survey it is very important to be able to test the survey instrument before applying it on a large scale. This is usually achieved in two stages:

- use simulated data to check that the design allows recovery of all of the parameters of the expected model;
- pre-test the survey instrument using a small stratified sample in order to consider the opinion of the largest possible number of interesting sectors of the population;
- evaluate the results of this pilot both in terms of the quality of the survey instrument and of the intuitive quality of the responses obtained by population strata; correct the instrument before its distribution.

Of course, the pilot survey will also serve to check operational, administrative and quality-assurance arrangements.

3.3.4.3 Stimulus Presentation

In order to guarantee realistic responses by the interviewee it is very important that the attributes of the options are presented in terms similar to those perceived by the traveller. This may call, for example, for the use of high-quality graphic material to convey an impression of what new rolling stock might be like. The researcher must be careful to avoid any implicit bias in the illustrative material used. Graphic illustrations are often preferred to photographs because of the higher control afforded in respect of the details included in them.

Example 3.9: In a study of the role of train frequency over demand for intercity travel (Steer and Willumsen 1983) it was found that although different people perceived the key variable (frequency) in different ways, almost nobody thought about it in terms of trains per hour or per day. Therefore the SP surveys started by ascertaining how was frequency (i.e. the analyst's concept) viewed by the traveller, for instance:

- 'I took the last train that puts me in Newcastle before 11 a.m.; it was the 7:50 from Kings Cross', or
- 'I just turned up at the station and found that the next train to Newcastle was due in 15 minutes'.

The interviewer then converted the different frequency attributes of the experimental design into the same terms, for example: 'To get to Newcastle before 11 a.m. you must now take the 7:30 train' in a low-frequency option, or '. . . the 8:00 train', in a high frequency one. Alternatively, '. . . the next train to Newcastle was in 30 . . .' or '. . . 10 minutes', for each option. Travellers were then asked to choose among alternatives described in terms they were familiar with and which affected their current journey choices.

3.3.4.4 Identification of Preferences

The next issue is how will the respondents be asked to express their preferences for each option offered to them. As suggested before, there are three main ways of collecting information on preferences about alternatives, i.e. asking respondents to rank them in order of preference, to rate them on an arbitrary scale or to choose between them in choice experiments. Other approaches, for example allocating a limited budget to alternative uses, have also been tested but these three are the most frequently employed.

(a) Ranking responses. This approach presents all the options at once to the respondents and they are then asked to rank them in order of preference thus implying a hierarchy of utility values. The main attraction of this approach is that all options are presented together but this also limits the number of alternatives that can be considered without fatiguing the respondent. Furthermore, the researcher needs to be aware that the data provided by this method represent *judgements* by respondents that do not necessarily correspond to the type of choices they face in real life.

(b) Rating techniques have been used for many years by market research practitioners. In this case respondents are asked to express their degree of preference for an option using an arbitrary scale, often between 1 and 10, with specific labels attached to key figures, for example 1 = '*strong dislike*', 5 = '*indifference*', 10 = '*strong preference*'. The responses might then be handled using normal arithmetic operations (calculating averages, ratios, etc). However, it has been shown that responses are not independent from the scales used and the labels attached to them. There is no evidence, therefore, that individual preferences can be usefully elicited and translated into cardinal scales of this type. Nevertheless, the principle can be extended to refine the *choice* experiments discussed below. In this case, the respondent is allowed to express her degree of preference between two alternatives, typically on a 5-point scale: '*definitely choose A*', '*probably choose A*', '*cannot choose*', '*probably choose B*' and '*definitely choose B*'.

(c) Choice experiments require the respondent to select an option either from a pair (binary choice) or a group of them. In its pure form, the respondent only chooses her preferred alternative thus expressing the choice in terms analogous to a revealed-preference survey. In its extended form the respondent is allowed to declare her preferences in a rating scale as outlined above. To increase realism, it may be possible to allow the choice of '*none of the above*' to avoid forcing the selection of a least bad but still unacceptable alternative.

3.3.4.5 Sampling Strategy

As in any other data-collection exercise, issues such as sample composition and size are very important in the design of an appropriate SP experiment. Also, in common with RP studies, a basic requirement is to obtain a sufficiently large and representative sample. On the other hand, SP experiments are statistically efficient, in the sense that each interviewee produces not just one observation but several on the same choice context; therefore, samples are typically smaller than for comparable RP studies (Bradley 1988). In fact, an early rule-of-thumb seems to have stated that around 30 interviews per market segment might be sufficient. However, more recent work suggests that 75–100 interviews per segment would be more appropriate (see Pearmain and Swanson 1990; Bradley and Kroes 1990; and Swanson *et al.* 1992).

Part of the difficulty lies in the nature of the information collected in SP surveys. The fact that each interview may result in 10 or more stated responses to the same number of (hypothetical) choice situations provides information about the variations in responses within each individual. However, for a good representative model we need to incorporate the variations that occur *between* as well as *within* individuals, and only an adequately sized and representative sample can do this.

The problem of sample representativeness may be complicated in SP contexts, precisely because of the additional flexibility offered in principle by the approach (the analyst can control the contexts). For example, if a set of individuals of a very-well-defined type (say frequent users of a given service) are asked about within-mode comparisons of service improvements, it is possible to ensure more easily that the survey context is relevant to all respondents (i.e. that the options posed are technologically feasible). However, a model estimated with this data will be able to provide little evidence about the behaviour of other groups, such as new users, which might be attracted by the same service improvements.

To forecast demand it is thus necessary to survey many types of individuals in order to obtain representative results. If we sample randomly, we may require large samples in order to achieve an adequate number of observations about minority choices. We saw already, that in RP studies choice-based sampling could provide a very cost-efficient method in this case; this is made more interesting by the fact that a simple correction can usually be made in model estimation to avoid statistical bias as we will see in Chapter 8. In SP data, however, choice-based samples may induce additional bias because of the way different types of individuals perceive or interpret the choice context (see the discussion by Bradley 1988).

3.3.4.6 Realism and Complexity

A key element in the success of SP surveys is the degree of realism achieved in the responses. A number of guidelines have been developed by practitioners in the field and these are described, for example, in Pearmain *et al.* (1991). Realism must be preserved in the *context* of the exercise, the *options* that are presented and the *responses* that are allowed. This can be achieved in a number of ways:

- focusing on *specific* rather than general behaviour; for example, respondents should be asked how they would have responded to an alternative on a given occasion, rather than in general; the more abstract the question the less reliable the response;

- using a realistic choice context—one the respondents have had recent personal experience of;
- retaining the constraints on choice required to make the context realistic; this usually means asking respondents to express preferences in respect of a very recent journey without relaxing any of its constraints: e.g. 'if today you would prefer to use the car to visit your dentist in the evening directly from work then retain this restriction in your choices'. Easing these constraints will just produce unrealistically elastic responses;
- using existing (perceived) levels of attributes so that the options are built around existing experience;
- using respondents' perceptions of what is possible to limit the attribute values in the exercise. For example, in considering improved rail services, do not offer options where the station is closer to home than feasible;
- ensuring all relevant attributes are included in the presentation; this is especially important if developing travel choice models and not just measuring the relative importance of different attributes;
- keeping the choice experiments simple, without overloading the respondent. Remember we respond to very complex choices in practice but we do so over a long period of time, acquiring experience about alternatives at our own pace and selecting the best for us. In an SP exercise these choices are compressed on a very short period of time and must, therefore, be suitably simplified;
- allowing respondents to opt for a response outside the set of experimental alternatives. For example, in a mode choice exercise if all options become too unattractive the respondent may decide to change destination, time of travel or not to travel at all; allow her a 'will do something else' alternative. If a computer-based interview is used it could be programmed to branch then to another exercise exploring precisely these other options;
- making sure that all the options are clearly and unambiguously defined. This could be quite difficult when dealing with qualitative attributes like security or comfort: do not express alternatives as 'poor' or 'improved' as this is too vague and prone to different interpretations by respondents. Describe instead what measures or facilities are involved in improving security or ride comfort (closed circuit TV in all stations/attendants present at all times, . . ., air-conditioning in all coaches as in InterCity trains).

3.3.4.7 Use of Computers in SP Surveys

Computers have been used now for several years in the conduct of surveys of many kinds, including stated-preference. Computers do offer very significant advantages over 'paper and pen' methods but they have, given present technology, a few limitations. Let us consider them first.

In the case of SP surveys one is most likely to use portable, preferably notebook size, microcomputers. In the past, their main limitations were battery life and weight but modern machines have practically overcome these problems. The second restriction is screen size and quality. Contemporary portable computers offer a reasonable screen size some 80 columns wide and 25 lines high. High-resolution colour screens permit the display of more information, at a price; however, even clever windows graphics cannot provide the ease of handling afforded by printed cards. Therefore, there is little advantage in using computers for most **ranking** SP exercises.

However, a computer screen is perfectly suited to paired choices in their pure and generalised (with rating) form. It is also possible to display not only the attributes that vary as part of the SP experiment but also other features that remain fixed, such as destination, clock times, or indeed anything else that may be relevant or useful to the respondent.

What makes computer-based interviewing most attractive is, however, the task of tailoring the experiment to the subject. Most SP interviews will include a questionnaire in which information about the respondent and a recent journey (or purchase, etc.) is collected and used to build a subsequent experiment. This questionnaire can be reproduced in software with the added advantages of automatic entry validation and automatic routeing, (see Figure 3.7). With a computerised system the responses to this initial questionnaire can be used to generate the SP experiments and options automatically for each subject, following a specified design. Automatic routeing can be used to select the appropriate experiment for each individual depending on her circumstances. Furthermore, range and logic checks on the responses and pop-up help screens or look-up information windows (e.g. for timetables) can be incorporated to improve the quality of the interview.

The use of computers for SP surveys makes it possible to design much more complex interviews than might be attempted manually, although this complexity may never be apparent to the respondent, or even to the interviewer. Moreover, good software permits randomisation of the order in which the options are offered to each

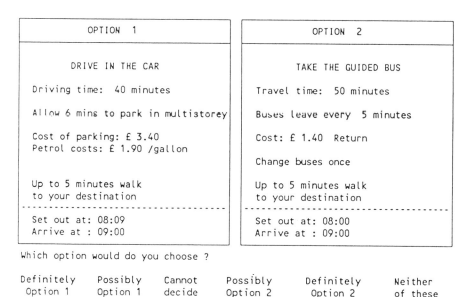

Figure 3.7 Example of computerised questionnaire

individual thus removing a further potential source of bias in the responses. As all responses are stored directly on disk there are no data entry costs nor errors and data are available immediately for processing.

A number of software packages offer excellent facilities for designing and coding very complex interviews with a minimum of understanding of computing itself; among the best known are ALASTAIR (Steer Davies Gleave), MINT (Hague Consulting Group) and ACA (Sawtooth Software). For an example of the type of screens generated by these packages see Figure 3.8. In summary, the practical advantages of computer-based SP interviews are:

- an interesting format that is consistent across interviews and respondents;
- automatic question branching, prompting and response validation;
- automatic data coding and storage;
- the ease with which the SP exercise can be tailored to each individual;
- the reduction in interview time achieved because the interviewer does not have to calculate and prepare written options;
- reduced training and briefing costs;
- the statistical advantages of randomisation.

On the debit side one has the initial costs of investing in hardware, software, insurance and the requirement to provide some back-up services (disks, spare battery packs, modems, technical advice to interviewers and supervisors, etc) on location.

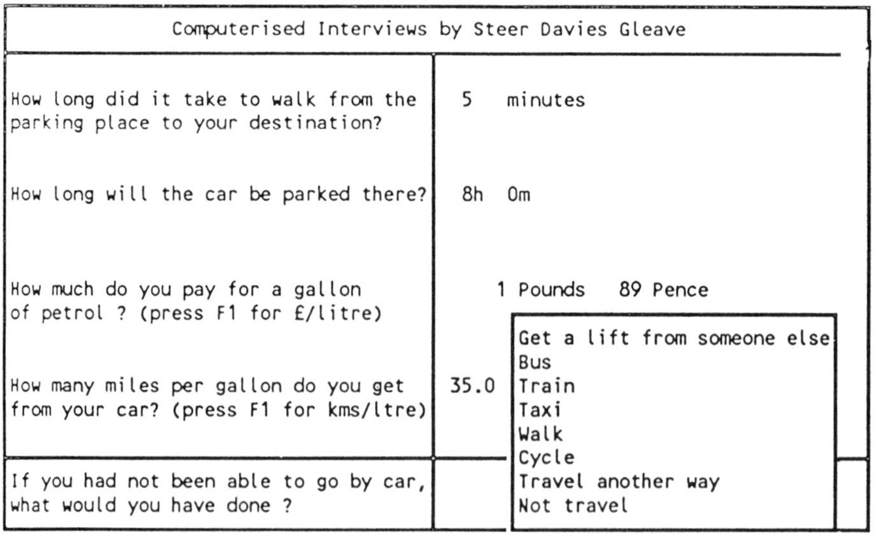

```
┌─────────────────────────────────────────────────────────────────┐
│          Computerised Interviews by Steer Davies Gleave          │
├──────────────────────────────────┬──────────────────────────────┤
│ How long did it take to walk from the │  5    minutes            │
│ parking place to your destination?    │                          │
│                                       │                          │
│ How long will the car be parked there?│  8h   0m                 │
│                                       │                          │
│ How much do you pay for a gallon      │     1 Pounds    89 Pence  │
│ of petrol ? (press F1 for £/litre)    │  ┌─────────────────────┐ │
│                                       │  │Get a lift from someone else│
│                                       │  │Bus                  │ │
│ How many miles per gallon do you get  │ 35.0 │Train             │ │
│ from your car? (press F1 for kms/ltre)│  │Taxi                 │ │
│                                       │  │Walk                 │ │
│                                       │  │Cycle                │ │
│ If you had not been able to go by car,│  │Travel another way   │ │
│ what would you have done ?            │  │Not travel           │ │
└──────────────────────────────────┴──────────────────────────────┘
```

Figure 3.8   Example of choice SP screens from ALISTAIR

3.3.4.8 Quality Issues in SP Surveys

Stated-preference techniques have proved to be a very powerful instrument in research and model development in transport and other fields. Their value depends on the careful application of the guidelines developed so far and discussed in the preceding pages. A key element in this is restricting the artificiality of the exercise to the minimum required. The more the analyst is interested in predicting future behaviour the more important it is to make sure the *decision context* is specific (an actual journey, not a hypothetical one) and the *response space* is behavioural.

One of the dangers of SP techniques is that it is relatively easy to cut corners in order to reduce costs. For example, one can allow the decision context to become less specific and more generic; this makes the sampling easier, the questionnaire simpler and, not surprisingly, the resulting models quite believable as what they would reflect is 'ideal' rather than constrained behaviour. The value of goodness-of-fit indicators in SP surveys is entirely dependent on the quality and realism of the experiment. The problem is that the models resulting from 'cheaper' SP studies will only be found to be flawed much later.

The same is true of the analysis techniques discussed in Chapter 8. Good analysis will often require combining SP and RP data to make sure the resulting models are well anchored (scaled) in the restrictions and noise of real behaviour.

SP surveys can be a very cost-effective way of refining and improving modelling tools but too much emphasis on low cost, at the expense of quality assurance and sound analysis, is likely to lead to disappointments and poor decision support.

3.3.5 Longitudinal Data Collection

Most of the discussion so far has been conducted under implicit assumption that we are dealing with cross-sectional (snap-shot) data. However, as we saw in Chapter 1, travel behaviour researchers are becoming increasingly convinced that empirical cross-sectional models have suffered from lack of recognition of the inter-temporality of most travel choices.

In this part we will attempt to provide a brief sketch of longitudinal or time-series data-collection methods and problems; we will first define various approaches and then we will concentrate on the apparently preferred one: panel data. In Chapter 8 we will touch on the added problems of modelling discrete choices in this case.

We will finally examine some evidence about the likely costs of a panel data-collection exercise in comparison with the more typical cross-sectional approach.

3.3.5.1 Basic Definitions

1. Repeated cross-sectional survey. This is one which makes similar measurements on samples from an equivalent population at different points in time, without ensuring that any respondent is included in more than one round of data collection. This kind of survey provides a series of snapshots of the population at several points in time; however, inferences about the population using longitudinal models may be biased with this type of data and it may be preferable to treat observations as if they were obtained from a single cross-sectional survey (see Duncan *et al*. 1987).

2. Panel survey. This is one in which similar measurements, sometimes called *waves*, are made on the same sample at different points in time. There are several types of panel survey, as exemplified below.
3. Rotating panel survey. This is a panel survey in which some elements are kept in the panel for only a portion of the survey duration.
4. Split panel survey. This is a combination of panel and rotating panel survey.
5. Cohort study. This is a panel survey based on elements from population sub-groups that have shared a similar experience (e.g. birth during a given year).

It is important to distinguish between longitudinal survey and data. The latter consist of periodic measurements of certain variables of interest. Although in principle it is possible to obtain panel data from a cross-sectional survey, measurement considerations usually argue for the use of a panel survey design rather than retrospective questioning to obtain reliable panel data.

3.3.5.2 Representative Sampling

Panel designs are often criticised because they may become unrepresentative of the initial population as their samples necessarily age over time. However, this is only strictly true in cohort study designs considering an unrepresentative sample to start with; for example, if the sample consists of people with a common birth year, individuals joining the population either by birth or immigration are not represented in the design.

In general, a panel design should attempt to maintain a representative sample of the entire population over time. So, it must cope not only with the problems of birth, immigration or individual entry by other means, but also be able to handle the incorporation of whole new families into the population (e.g. children leaving the parental home, couples getting divorced). A mechanism is needed to maintain a representative sample that allows families and individuals to enter the sample with known probabilities, but this is not simple (for details see Duncan *et al.* 1987).

3.3.5.3 Sources of Error in Panel Data

A panel design may add to (or detract from, if it is not done with care) the quality of the data. Although repeated contact and interviewing are generally accepted to lead to better-quality information, panels have typically higher rates of non-response than cross-sectional methods, and run the risk of introducing *contamination* as we discuss below.

Effects on Response Error   Respondents in panel studies have repeated contact with interviewers and questionnaires; this may improve the quality of the data for the following reasons:

- Repeated interviewing over time reduces the amount of time between event and interview, thus tending to improve the quality of the recalled information.

- Repeated contact increases the chances that respondents will understand the purpose of the study; also they may become more motivated to do the work required to produce more accurate answers.
- It has been found that data quality tends to improve in later waves of a panel, probably because of learning, by respondents, interviewers or both.

Non-response Issues   Under the generic non-response label, there are included several important issues which have two basic sources: the loss of a unit of information (attribution) and/or the loss of an item of information. Hensher (1987) discusses in detail how to test and correct for this type of error.

The non-response problems associated with the initial wave of a panel are not different to those of cross-sectional surveys, so very little can be done to adjust for their possible effects. In contrast, plenty of data have been gathered about non-respondents in subsequent waves; this can be used to determine their main characteristics, enabling non-response to be modelled as part of the more general behaviour of interest (see Kitamura and Bovy 1987).

Typical large panel designs spend a great amount of effort attending to the 'care and feeding' of respondents: this involves instructing interviewers to contact respondents many times and writing letters of encouragement specifically tailored to the source of respondents' reluctance.

Response Contamination   Evidence has been reported that initial-wave responses in panel studies may differ from those of subsequent waves; for this reason in some panel surveys the initial interviews are not used for comparative purposes. A crucial question is whether behaviour itself, or just its reporting, is being affected by panel membership. Evidence about this is not conclusive, but it seems to depend on the type of behaviour measured. For example, Traugott and Katosh (1979) found that participants in a panel about voting behaviour increased their voting (i.e. changed behaviour) as time went by; however, it was also found that this was caused partly by greater awareness of the political process and partly by the fact that individuals who were less politically motivated tended to drop out of the panel.

3.3.5.4 Relative Costs of Longitudinal Surveys

Questions about the relative costs of longitudinal studies cannot be answered without reference to the alternatives to them. One obvious comparison is between a single cross-sectional survey, with questions about a previous period, and a two-wave panel. However, if the longitudinal study is designed to keep its basic sample representative each year and if enough resources are devoted to the task, it can also serve as an (annual) source of representative cross-sectional data and thus ought to be compared with a series of such surveys rather than just a single one.

Duncan *et al*. (1987) have made rough calculations on these lines, concluding that in the first case the longitudinal survey would cost between 20 to 25% more than the cross-sectional survey with retrospective questions. However, they also conclude that in the second case the field costs of each successive wave of the cross-sectional study would be between 30 and 70% higher than additional waves of the panel survey, depending on the length of the interview.

Other costs are caused by the need to contact and persuade respondents in the case of panels and by the need to sample again with each fresh cross-section in the other case. Finally, there are other data processing costs associated with panels but these must be weighed against the greater opportunity to check for inconsistencies, analysis of non-response, and so forth.

3.4 NETWORK AND ZONING SYSTEMS

One of the most important early choices facing the transport modeller is that of the level of detail (resolution) to be adopted in a study. This problem has many dimensions: it refers to the schemes to be tested, the type of behavioural variables to be included, the treatment of time, and so on. This section concentrates on design guidelines for two of these choices: zonal and network definition.

We shall see that in these two cases, as in other key elements of transport modelling, the final choices reflect a compromise between two conflicting objectives: accuracy and cost. In principle greater accuracy could be achieved by using a more detailed zoning and network system; in the limit, this would imply recognising each individual household, its location, distance to access points to the network, and so on. With a large enough sample (100% rate over several days) the representation of the current system could be made very accurate indeed. However, the problem of stability over time weakens this vision of accuracy as one would need to forecast, at the same level of detail, changes at the individual household level that would affect transport demand. This is an impossible and unnecessary task. Lesser levels of detail therefore, are not only warranted on the grounds of economy but also on those of accuracy whenever forecasting is involved (recall our discussion in section 3.2).

3.4.1 Zoning Design

A zoning system is used to aggregate the individual households and premises into manageable chunks for modelling purposes. The main two dimensions of a zoning system are the number of zones and their size. The two are, of course, related. The greater the number of zones, the smaller they can be to cover the same study area. It has been common practice in the past to develop a zoning system specifically for each study and decision-making context. This is apparently wasteful if one performs several studies in related areas; moreover, the introduction of different zoning systems makes it difficult to use data from previous studies and to make comparisons of modelling results over time.

The first choice in establishing a zoning system is to distinguish the study area itself from the rest of the world. Some ideas may help in making this choice:

- In choosing the study area one must consider the decision-making context, the schemes to be modelled, and the nature of the trips of interest: mandatory, optional, long or short distance, and so on.
- For strategic studies one would like to define the study area so that the majority of the trips have their origin and destination inside it; however, this may not be

possible for the analysis of transport problems in smaller urban areas where the majority of the trips of interest are through-trips and a bypass is to be considered.

- Similar problems arise with traffic management studies in local areas where again, most of the trips will have their origin, destination or both clearly outside the area of interest. What matters in these cases is whether it is possible to model changes to these trips arising as a result of new schemes.
- The study area should be somewhat bigger than the specific area of interest covering the schemes to be considered. Opportunities for re-routeing, changes in destination and so on, must be allowed for; we would like to model their effects as part of the study area itself.

The region external to the study area is normally divided into a number of *external* zones. In some cases it might be enough to consider each external zone to represent 'the rest of the world' in a particular direction; the boundaries of these different slices of the rest of the world could represent the natural catchment areas of the transport links feeding into the study area. In other cases, it may be advantageous to consider external zones of increasing size with the distance to the study area. This may help in the assessment of the impacts over different types of travellers (e.g. long- and short-distance).

The study area itself is also divided into smaller *internal* zones. Their number will depend on a compromise between a series of criteria discussed below. For example, the analysis of traffic management schemes will generally call for smaller zones, often representing even car parks or major generators/attractors of trips. Strategic studies, on the other hand, will often be carried out on the basis of much larger zones. For example, strategic studies of London have been undertaken using fine zoning systems of about 1000 zones (for about 7.2 million inhabitants) and several levels of aggregation of them down to about 50 zones (at borough level). Examples of zone numbers chosen for various studies are presented in Table 3.5.

Zones are represented in the computer models as if all their attributes and properties were concentrated in a single point called the *zone centroid*. This notional spot is best thought of as floating in space and not physically on any location on a

Table 3.5 Typical zone numbers for studies

Location	Population	Number of zones	Comments
London (1972)	7.2 million	~ 2252	Fine level subzones
		~ 1000	Normal zones at GLTS
		~ 230	GLTS districts
		52	Traffic boroughs
Montréal island (1980)	2.0 million	~ 1260	Fine zones
Ottawa (1978)	0.5 million	~ 120	Normal zones
Santiago (1986)	4.5 million	~ 260	Zones, strategic study
Washington (1973)	2.5 million	~ 1075	Normal zones
		~ 134	District level
West Yorkshire (1977)	1.4 million	~ 1500	Fine zones
		~ 463	Coarse zones

map. Centroids are attached to the network through *centroid connectors* representing the average costs (time, distance) of joining the transport system for trips with origin or destination in that zone. Nearly as important as the cost associated to each centroid connector is the node in the network it connects to. These should be close to natural access/egress points for the zone itself. The role of centroids and centroid connectors in modelling should help in defining zone boundaries.

The following is a list of zoning criteria which has been compiled from experience in several practical studies:

1. Zoning size must be such that the aggregation error caused by the assumption that all activities are concentrated at the centroid is not too large. It might be convenient to start postulating a system with many small zones, as this may be aggregated in various ways later depending on the nature of the projects to be evaluated.
2. The zoning system must be compatible with other administrative divisions, particularly with census zones; this is probably the fundamental criterion and the rest should only be followed if they do not lead to inconsistencies with it.
3. Zones should be as homogeneous as possible in their land use and/or population composition; census zones with clear differences in this respect (i.e. residential sectors with vastly different income levels) should not be aggregated, even if they are very small.
4. Zone boundaries must be compatible with cordons and screen lines and with those of previous zoning systems. However, it has been found in practice that the use of main roads as zone boundaries should be avoided, because this increases considerably the difficulty of assigning trips to zones, when these originate or end at a zonal boundary.
5. The shape of the zones should allow an easy determination of their centroid connectors; this is particularly important for later estimation of intra-zonal characteristics. A zone should represent the natural catchment area of the transport networks and its centroid connector(s) identified so as to represent the main costs to access them.
6. Zones do not have to be of equal size; if anything, they could be of similar dimensions in travel time units, therefore generating smaller zones in congested than in uncongested areas.

It is sometimes advantageous to develop a hierarchical zoning system, as in the London Transportation Studies, where subzones are aggregated into zones which in turn are combined into districts, traffic boroughs and finally sectors. This facilitates the analysis of different types of decisions at the appropriate level of detail. Hierarchical zoning systems benefit from an appropriate zone-numbering scheme where the first digit indicates the broad area, the first two the traffic borough, the first three the district, and so on.

3.4.2 Network Representation

The transportation network is deemed to represent the supply side of the modelling effort, i.e. what the transport system offers to satisfy the movement needs of trip makers in the study area. The description of a transport network in a computer model can be undertaken at different levels of detail and requires the specification of its structure, its properties or attributes and the relationship between those properties and traffic flows. For an early general review of network representation issues, see Lamb and Havers (1970).

3.4.2.1 Network Details

The transport network may be represented at different levels of aggregation in a model. At one extreme one has models with no specific links at all; they are based on continous representations of transport supply (Smeed 1968). These models may provide, for example, a continuous equation of the average traffic capacity per unit of area instead of discrete elements or links. At a slightly higher level of disaggregation one can consider individual roads but include speed-flow properties taken over a much larger area; see for example Wardrop (1968).

Normal practice, however, is to model the network as a *directed graph*, i.e. a system of nodes and links joining them (see Larson and Odoni 1981): most nodes are taken to represent junctions and the links stand for homogeneous stretches of road between junctions; links are characterised by several attributes such as length, speed, number of lanes and so on. Links are normally unidirectional; even if during input a single two-way link is specified for simplicity, it will be converted internally into two one-way links. A subset of the nodes is associated with zone centroids, and a subset of the links to centroid connectors. A typical configuration of this type is presented in Figure 3.9.

A problem with this scheme is that 'at-node' connectivity is offered to each link joining it at no cost. In practice, some turning movements at junctions may be much more difficult to perform than others; indeed, some turning movements may not be allowed at all. In order to represent these features of real road networks better, it is possible to penalise and/or ban some turning movements. This can be done manually by expanding the junction providing separate (sometimes called dummy) links for each turning movement and associating a different cost to each. Alternatively, some commercial computer programs are capable of performing this expansion in a semi-automatic way, that is following simple instructions from the user about difficult or banned movements.

The level of disaggregation can be increased further when detailed traffic simulation models are used. In these cases additional links are used at complex junctions to account for the performance of reserved lanes, give-way lines, and so on.

Sometimes networks are subsets of larger systems; they may be cordoned off from them thus defining access or cordon points where the network of interest is connected to the rest of the world. These points are sometimes called 'gateways' and dummy links may be used to connect them to external zones.

A key decision in setting up a network is how many levels to include in the road hierarchy. If more roads are included, the representation of reality should be better; there is again a problem of economy versus realism which forces the modeller to select some links for exclusion. Moreover, it does not make much sense to include a large number of roads in the network and then make coarse assumptions about turning movements and delays at junctions. It is not sensible either to use a very

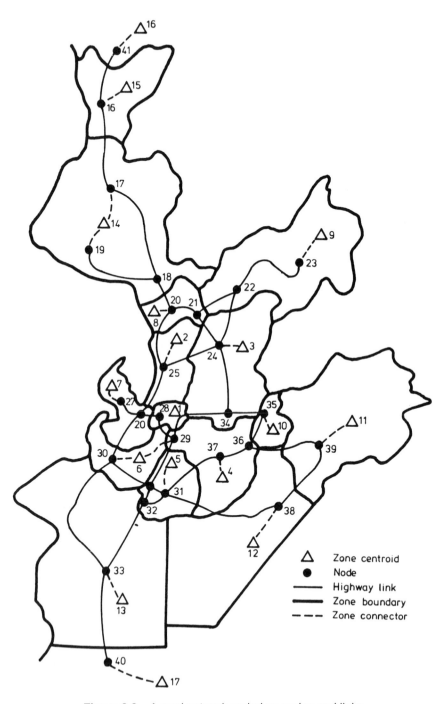

Figure 3.9   A road network coded as nodes and links

detailed network with a coarse zoning system as then *spatial aggregation* errors (i.e. in terms of centroid connections to the network) will reduce the value of the modelling process. This is particularly important in the case of public-transport networks, as we will see in Chapter 11. What matters is to make route choice and flows as realistic as possible within the limitations of the study.

Jansen and Bovy (1982) investigated the influence of network definition and detail over road assignment accuracy. Their conclusion was that the largest errors were obtained at the lower levels in the hierarchy of roads. Therefore, one should include in the network at least one level below the links of interest: for example, in a study of A (trunk) roads one should also include B (secondary) roads.

In the case of public-transport networks an additional level of detail is required. The modeller must specify the network structure corresponding to the services offered. These will be coded as a sequence of nodes visited by the service (bus, rail), normally with each node representing a suitable stop or station. Junctions without bus stops can, therefore, be excluded from the public-transport network. Two types of extra links are often added to public-transport networks. These are walk links, representing the parts of a journey using public transport made on foot, and links to model the additional costs associated with transfering from one service (or mode) to another.

3.4.2.2 Link Properties

The level of detail provided about the attributes of links depends on the general resolution of the network and on the type of model used. At the very minimum the data for each link should include:

- its length;
- its travel speeds—either free-flow speeds or an observed value for a given flow level;
- the capacity of the link, usually in passenger car equivalent units (PCU) per hour.

In addition to this a cost-flow relationship is associated to each link as discussed below. In some cases, more elaborate models are used to relate delay to traffic flow, but these require additional information about links, for example:

- type of road (e.g. expressway, trunk road, local street);
- road width, or number of lanes, or both;
- an indication of the presence or otherwise of bus lanes, or prohibitions of use by certain vehicles (e.g. lorries);
- banned turns, or turns to be undertaken only when suitable gaps in the opposing traffic become available, and so on;
- type of junction and junction details including signal timings;
- storage capacity for queues and their presence at the start of a signal phase.

Some research results have identified other attributes of routes as important to drivers, for example tolls, signposting and fuel consumption; see for example Outram and Thompson (1978) and Wootton *et al*. (1981). Work in the Netherlands has shown that (weighted) time and distance explained only about 70% of the routes actually chosen. The category of the road (motorway, A road, B road), scenic quality, traffic signals and capacity helped to explain additional routes. As our understanding of how these attributes influence route choice improves, we will be able to develop more accurate assignment models. The counterpart of this improvement will be the need to include other features of roads, like their scenic quality, number of junctions of each type, and so on.

3.4.2.3 Network Costs

Most current assignment techniques assume that drivers seek to minimise a linear combination of time and distance, sometimes referred to as generalised cost for route choice. This is known to be a simplifying assumption as there may be differences not only in the perception of time, but also about its relative importance compared with other route features. However, the great majority of network models in use today deal only with travel time and distance.

When modelling travel time as a function of flow one must distinguish two different cases. The first case is when the assumption can be made that delay on a link depends only on the flow on the link itself; this is typical of long links away from junctions and therefore it has been used in most inter-urban assignment models so far. The second case is encountered in urban areas where the delay on a link depends in an important way on flow on other links, for example for non-priority traffic at a give-way or roundabout junction.

The introduction of very general flow-delay formulations is not difficult until one faces the next issue, equilibration of demand and supply. There, the mathematical treatment of the first case (often called the separable cost function case) is simpler than the second; however, there are now techniques for balancing demand and supply in the case of link-delay models depending on flows on several links, i.e. when the effect of each link flow cannot be separated. Chapter 11 will provide a fuller discussion of cost-flow relationships.

EXERCISES

3.1 We require to estimate the population of a certain area for the year 2000 but we only have available reliable census information for 1970 and 1980, as follows:

$$P_{70} = 240 \pm 5 \quad \text{and} \quad P_{80} = 250 \pm 2$$

To estimate the future population we have available the following model:

$$P_n = P_b t^d$$

where P_n is the population in the forecast year n, P_b the population in the base year b, t is the population growth rate and $d = (n - b)/10$, is the number of decades to extrapolate growth.

Assume that the data from both censuses is independent and that the model does not have any specification error; in that case,

(a) Find out with what level of accuracy is it possible to forecast the population in the year 2000;

(b) You are offered the census information for 1990, but you are cautioned that its level of accuracy is worse than that of the previous two censuses:

$$P_{90} = 265 \pm 8$$

Find out whether it is convenient to use this value.

(c) Repeat the analysis assuming that the specification error of the model is proportional to d, and that it can be estimated as $12d\%$.

3.2 Consider the following modal-split model between two zones i and j (but we will omit the zone indices to alleviate notation):

$$P_1(\Delta t/\theta) = \frac{\exp(-\theta t_1)}{\exp(-\theta t_1) + \exp(-\theta t_2)} = \frac{1}{1 + \exp\{-\theta(t_2 - t_1)\}} = \frac{1}{1 + \exp(-\theta \Delta t)}$$

$$P_2(\Delta t/\theta) = 1 - P_1 = \frac{\exp(-\theta \Delta t)}{1 + \exp(-\theta \Delta t)}$$

where t_k is the total travel time in mode k, and θ a parameter to be estimated.

During the development of a study, travel times were calculated as the average of five measurements (observations) for each mode, at a cost of $1 per observation, and the following values were obtained:

$$t_1 = 12 \pm 2 \text{ min} \quad t_2 = 18 \pm 3 \text{ min}$$

(a) If the estimated value for θ is 0.1, compute a confidence interval for P_1.

(b) Assume you would be prepared to pay $3 per each percentage point of reduction in the error of P_1; find out whether in that case it would be convenient for you to take 10 extra observations in each mode whereby the following values for t_k would be obtained:

$$t_1 = 12 \pm 1 \text{ min} \quad t_2 = 17.5 \pm 1.5 \text{ min}$$

3.3 Consider an urban area where 100 000 people travel to work; assume you possess the following information about them:

(i) General information:

Mode	Average number of cars per household	Family income (1000$/year)
Car	2.40	120
Underground	1.60	60
Bus	0.20	10
Total	0.55	25

(ii) Population distribution

Family income (1000$/year)	Cars per household			
	0	1	2+	Total
Low (< 25)	63.6	15.9	0.0	79.5
Medium (25–75)	6.4	3.7	2.4	12.5
High (> 75)	0.0	2.4	5.6	8.0
Total	70.0	22.0	8.0	100.0

You are required to collect a sample of travellers to estimate a series of models (with a maximum of 8 parameters) which guarantee a negligible specification error if you have available at least 50 observations per parameter. You are also assured that if you take a random sample to 20% of the travellers the error will be negligible and there will be no bias.

Your problem is to choose the most convenient sampling method (random, stratified or choice-based), and for this you have available also the following information:

Hourly cost of an interviewer .... $2 per hour
Questionnaire processing cost .... $0.3 per form
Time required to classify an interviewee .... 4 min
Time required to complete an interview .... 10 min

You are also given the following table containing recommended choice-based sample sizes:

Subpopulation size	% to be interviewed
< 10 000	25
10 000–15 000	20
15 000–30 000	15
30 000–60 000	10
> 60 000	5

3.4 Consider the following results of having collected stratified (based on income I) and choice-based samples of a certain population:

Stratified sample

Mode	Low I	Medium I	High I
Car	3.33	18.00	20.00
Bus	33.34	7.20	4.00
Underground	3.33	4.80	6.00
Total	40.00	30.00	30.00

Choice-based sample

Mode	Low I	Medium I	High I	Total
Car	6.67	20.00	13.33	40.00
Bus	17.24	2.07	0.69	20.00
Underground	16.67	13.33	10.00	40.00

(a) If you know that the income-based proportions in the population are 60, 25 and 15% respectively for low, medium and high income, find an equivalent table for a random sample. Is it possible to validate your answer?

(b) Compute the weighting factors that would be necessary to apply to the observations in the choice-based sample in order to estimate a model for the choice between car, bus and underground using standard software (i.e. that developed for random samples).

4 Trip Generation Modelling

This is the first of several chapters dealing with first-generation aggregate demand models. As we saw in Chapter 1, the trip generation stage of the classical transport model aims at predicting the total number of trips generated by (O_i) and attracted to (D_j) each zone of the study area. This has been usually considered as the problem of answering a question such as: *how many trips* originate at each zone? However, the subject has also been viewed sometimes as a *trip frequency* choice problem: how many shopping (or other purpose) trips will be carried out by this person type during an average week? This is usually undertaken using discrete choice models, as discussed in Chapters 7 to 9, and it is then cast in terms like: what is the probability that this person type will undertake zero, one, two or more trips with this purpose per week?

In this chapter we will only consider the first approach (i.e. predicting the totals O_i and D_j from data on household socioeconomic attributes), which has been the most widely used in practice up to the end of the 1980s. Readers interested in the discrete choice approach, should consult Ben Akiva and Lerman (1985) and, of course Chapters 7 to 9 in this book.

We will start by defining some basic concepts and will proceed to examine some of the factors affecting the generation and attraction of trips. Then we will review the main modelling approaches, starting with the simplest growth-factor technique. Before embarking on more sophisticated approaches we will present a reasonable review of linear regression modelling, which complements well the previous statistical themes presented in Chapters 2 and 3.

We will then consider zonal and household-based linear regression trip generation models, giving some emphasis to the problem of non-linearities which often arise in this case. We will also address for the first time the problem of aggregation (e.g. obtaining zonal totals), which has a trivial solution here precisely because of the linear form of the model. Then we will move to cross-classification models, where we will examine not only the classical category analysis specification but also more contemporary approaches including the person category analysis model. The chapter ends with three shorter sections: the first discusses the problem of predicting future values for the explanatory variables of the model, and the other two consider some additional topics of interest such as elasticity, stability and updating of trip generation models.

4.1 INTRODUCTION

4.1.1 Some Basic Definitions

Journey   This is a one-way movement from a point of origin to a point of destination. Now, although the word 'trip' is literally defined as 'an outward and return journey, often for a specific purpose' (McLeod and Hanks 1986), in transport modelling both terms are used interchangeably. We are usually interested in all vehicular trips, but walking trips longer than a certain study-defined threshold (say 300 metres or three blocks) are often considered; finally, trips made by infants of less than five years of age will usually be ignored, as we saw in Chapter 3.

Home-based (HB) Trip   This is one where the home of the trip maker is either the origin or the destination of the journey.

Non-home-based (NHB) Trip   This, conversely, is one where neither end of the trip is the home of the traveller.

Trip Production   This is defined as the home end of an HB trip or as the origin of an NHB trip (see Figure 4.1).

Trip Attraction   This is defined as the non-home end of an HB trip or the destination of an NHB trip (see Figure 4.1).

Trip Generation   This is often defined as the total number of trips generated by households in a zone, be they HB or NHB. This is what most models would produce and the task then remains of allocating NHB trips to other zones as trip productions.

During the 1980s a series of other terms, such as sojourns, tours and trip chains, were added to the transport modelling kit; these correspond better to the idea that the demand for travel is a *derived* demand (i.e. it depends strongly on the demand for other activities) but have been used mainly by discrete choice modellers in practice (see Daly *et al*. 1983).

4.1.2 Classification of Trips

4.1.2.1 By Trip Purpose

It has been found in practice that better trip generation models can be obtained if trips by different purposes are identified and modelled separately. In the case of HB trips, five categories have been usually employed:

- trips to work;
- trips to school or college (education trips);
- shopping trips;
- social and recreational trips;
- other trips.

The first two are usually called compulsory (or mandatory) trips and all the others are called discretionary (or optional) trips. The latter category encompasses all trips

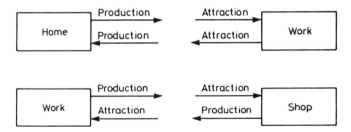

Figure 4.1   Trip productions and attractions

made for less routine purposes, such as health, bureaucracy (need to obtain a passport or a certificate) and trips made as an accompanying person. NHB trips are normally not separated because they only amount to 15–20% of all trips.

4.1.2.2 By Time of Day

Trips are often classified into peak and off-peak period trips; the proportion of journeys by different purposes usually varies greatly with time of day.

Table 4.1 summarises data from the Greater Santiago 1977 Origin Destination Survey (DICTUC, 1978); the morning (AM) peak period (the evening peak period is sometimes assumed to be its mirror image) occurred between 7:00 and 9:00 and the representative off-peak period was taken between 10:00 and 12:00. Some comments are in order with respect to this table. Firstly, note that although the vast majority (87.18%) of trips in the AM peak are compulsory (i.e. either to work or education), this is not the case in the off-peak period. Secondly, a typical trait of a developing country emerges from the data: the large proportion of trips for bureaucratic reasons in both periods. Thirdly, a problem caused by faulty classification, or lack of forward thinking at the data-coding stage, is also clearly revealed: the *return to home* trips (which account for 41.65% of all off-peak trips) are obviously trips with another purpose; the fact that they were returning home is not as important as to why they left home in the first place. In fact, these data needed recoding in order

Table 4.1   Example of trip classification

	AM Peak		Off Peak	
Purpose	No.	%	No.	%
Work	465 683	52.12	39 787	12.68
Education	313 275	35.06	15 567	4.96
Shopping	13 738	1.54	35 611	11.35
Social	7 064	0.79	16 938	5.40
Health	14 354	1.60	8 596	2.74
Bureaucracy	34 735	3.89	57 592	18.35
Accompanying	18 702	2.09	6 716	2.14
Other	1 736	0.19	2 262	0.73
Return to home	24 392	2.72	130 689	41.65

to obtain adequate information for trip generation modelling (see Hall *et al*. 1987). This kind of problem used to occur before the concepts of trip productions and attractions replaced concepts such as origins and destinations, which did not explicitly address the generating capacity of home-based and non-home-based activities.

4.1.2.3 By Person Type

This is another important classification, as individual travel behaviour is heavily dependent on socioeconomic attributes. The following categories are usually employed:

- income level (e.g. nine strata in the Santiago survey);
- car ownership (typically three strata: 0, 1 and 2 or more cars);
- household size and structure (e.g. six strata in most British studies).

It is important to note that the total number of strata can increase very rapidly (e.g. 162 in the above example) and this may have strong implications in terms of data requirements, model calibration and use. For this reason trade-offs, adjustments and aggregations are usually required (see the discussion in Daly and Ortúzar 1990).

4.1.3 Factors Affecting Trip Generation

In trip generation modelling we are typically interested not only in person trips but also in freight trips. For this reason models for four main groups (i.e. personal and freight, trip productions and attractions) are usually required. In what follows we will briefly consider some factors which have been found important in practical studies. We will not discuss freight trip generation modelling, however (although a little had been done by the end of the 1980s), but postpone a discussion on the general topic of freight modelling until Chapter 13.

4.1.3.1 Personal Trip Productions

The following factors have been proposed for consideration in many practical studies:

- income;
- car ownership;
- household structure;
- family size;
- value of land;
- residential density;
- accessibility.

The first four (income, car ownership, household structure and family size) have been considered in several household trip generation studies, while value of land and residential density are typical of zonal studies. The last one, accessibility, has rarely been used although most studies have attempted to include it. The reason is that it offers a way to make trip generation elastic (responsive) to changes in the transport system; we will come back to this issue in section 4.6.

4.1.3.2 Personal Trip Attractions

The most widely used factor has been roofed space available for industrial, commercial and other services. Another factor used has been zonal employment, and certain studies have attempted to incorporate an accessibility measure. However, it is important to note that in this case not much progress has been reported.

4.1.3.3 Freight Trip Productions and Attractions

These normally account for few vehicular trips; in fact, at most they amount to 20% of all journeys in certain areas of industrialised nations, although they can still be significant in terms of their contribution to congestion. Important variables include:

- number of employees;
- number of sales;
- roofed area of firm;
- total area of firm.

To our knowledge, neither accessibility nor type of firm have ever been considered as explanatory variables; the latter is curious because it would appear logical that different products should have different transport requirements.

4.1.4 Growth-factor Modelling

Since the early 1950s several techniques have been proposed to model trip generation. Most methods attempt to predict the number of trips produced (or attracted) by household or zone as a function of (generally linear) relations to be defined from available data. Prior to any comparison of results across areas or time, it is important to be clear about the following aspects mentioned above:

- what trips to be considered (e.g. only vehicle trips and walking trips longer than three blocks);
- what is the minimum age to be included in the analysis (i.e. five years or older).

In what follows we will briefly present a technique which may be applied to predict the future number of journeys by any of the four categories mentioned above. Its basic equation is:

$$T_i = F_i t_i \tag{4.1}$$

where T_i and t_i are respectively future and current trips in zone i, and F_i is a growth factor.

The only problem of the method is the estimation of F_i, the rest is trivial. Normally the factor is related to variables such as population (P), income (I) and car ownership (C), in a function such as:

$$F_i = \frac{f(P_i^d, I_i^d, C_i^d)}{f(P_i^c, I_i^c, C_i^c)} \tag{4.2}$$

where f can even be a direct multiplicative function with no parameters, and the superscripts d and c denote the design and current years respectively.

Example 4.1: Consider a zone with 250 households with car and 250 households without car. Assuming we know the average trip generation rates of each group:

> car-owning households produce: 6.0 trips/day
>
> non-car-owning households produce: 2.5 trips/day

we can easily deduce that the current number of trips per day is:

$$t_i = 250 \times 2.5 + 250 \times 6.0 = 2125 \text{ trips/day}$$

Let us also assume that in the future all households will have a car; therefore, assuming that income and population remain constant (which is a safe hypothesis in the absence of other information), we can estimate a simple multiplicative growth factor as:

$$F_i = C_i^d/C_i^c = 1/0.5 = 2$$

and applying equation (4.1) we can estimate the number of future trips as:

$$T_i = 2 \times 2125 = 4250 \text{ trips/day}$$

However, the method is very crude, as we will demonstrate. If we use our information about average trip rates and make the assumption that these will remain constant (which is actually the main assumption behind one of the most popular forecasting methods, as we will see below), we can estimate the future number of trips as:

$$T_i = 500 \times 6 = 3000$$

which means that the growth factor method would overestimate the total number of trips of approximately 42%. This is very serious because trip generation is the first stage of the modelling process; errors here are carried through the entire process and may invalidate work on subsequent stages.

Growth factor methods are therefore only used in practice to predict the future number of *external* trips to an area; this is because they are not too many in the first place (so errors cannot be too large) and also because there are no simple ways to predict them. In the following sections we will discuss other (superior) methods which can also be used in principle to model personal and freight trip productions and attractions. However, we will just make explicit reference to the case of personal trip productions as this is the area not only where there is more practical experience, but also where the most interesting findings have been reported.

4.2 REGRESSION ANALYSIS

The next subsection provides a brief introduction to linear regression. The reader familiar with this subject can proceed directly to subsection 4.2.2.

4.2.1 The Linear Regression Model

4.2.1.1 Introduction

Consider an experiment consisting in observing the values that a certain variable $\mathbf{Y} = \{Y_i\}$ takes for different values of another variable $\mathbf{X}$. If the experiment is not deterministic we would observe different values of Y_i for the same value of X_i.

Let us call $f_i(Y|X)$ the probability distribution of Y_i for a given value X_i; thus, in general we could have a different function f_i for each value of $\mathbf{X}$ as shown in Figure 4.2. However, such a completely general case is intractable; to make it more manageable certain hypotheses about population regularities are required. Let us assume that:

1. The probability distributions $f_i(Y|X)$ have the same variance σ^2 for all values of $\mathbf{X}$.
2. The means $\mu_i = E(Y_i)$ form a straight line known as the *true regression line* and given by:

$$E(Y_i) = a + bX_i \tag{4.3}$$

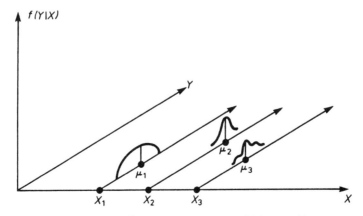

Figure 4.2   General distributions of *Y* given *X*

where the population parameters a and b, defining the line, must be estimated from sample data.

3. The random variables **Y** are statistically independent; this means, for example, that a large value of Y_1 does not tend to make Y_2 large.

The above *weak set of hypotheses* (see for example Wonnacott and Wonnacott 1977) may be written more concisely as:

The random variables Y_i are statistically independent with mean $a + bX_i$ and variance σ^2.

With these Figure 4.2 changes to the distribution shown in Figure 4.3.

It is sometimes convenient to describe the deviation of Y_i from its expected value as the error or disturbance term e_i, so that the model may also be written as:

$$Y_i = a + bX_i + e_i \qquad (4.4)$$

Note that we are not making any assumptions yet about the shape of the distribution of **Y** (and **e**, which is identical except that their means differ) provided it has a finite variance. These will be needed later, however, in order to derive some formal tests for the model. The error term is as usual composed of measurement and specification errors (recall the discussion in Chapter 3).

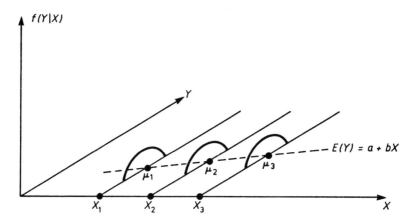

Figure 4.3   Distribution of Y assumed in linear regression

4.2.1.2 Estimation of a and b

Figure 4.4 can be labelled the *fundamental graph* of linear regression. It shows the true (dotted) regression line $E(Y) = a + bX$, which is of course unknown to the analyst, who must estimate it from sample data about **Y** and **X**. It also shows the estimated regression line $\hat{Y} = \hat{a} + \hat{b}X$; as is obvious, this line will not coincide with the previous one unless the analyst is extremely lucky (though he will never know it). In general the best he can hope is that the parameter estimates will be close to the target.

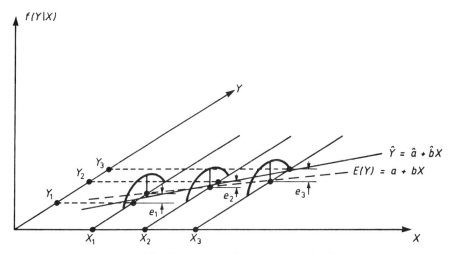

Figure 4.4   True and estimated regression lines

It is important to distinguish between the errors e_i, which are not known and pertain to the true regression line, and the differences ϵ_i, between observed (Y_i) and fitted values $(\hat{Y}_i)$, which are a crucial element of the most attractive line-fitting method: least squares estimation.

If we define a new variable $x_i = X_i - \bar{X}$, where $\bar{X}$, is the mean of **X**, it is easy to show that the previous regression lines keep their slope (b and $\hat{b}$ respectively) but obviously change their intercepts (a and $\hat{a}$ respectively) in the new axes (Y, x). The change is convenient because the new variable **x** has the following property: $\sum_i x_i = 0$.

The least square estimators are given by:

$$\hat{a} = \bar{Y} \tag{4.5}$$

which ensures that the fitted line goes through the *centre of gravity* $(\bar{X}, \bar{Y})$ of the sample of n observations, and

$$\hat{b} = \frac{\sum_i x_i Y_i}{\sum_i x_i^2} \tag{4.6}$$

These estimators have the following interesting properties:

$$E(\hat{a}) = a \qquad \mathrm{Var}(\hat{a}) = \sigma^2/n$$

$$E(\hat{b}) = b \qquad \mathrm{Var}(\hat{b}) = \sigma^2/\sum_i x_i^2$$

In passing, the formula for the variance of $\hat{b}$ has interesting implications in terms of experimental design. For example, if the **X** are too close together, as in Figure

4.5a, their deviations from the mean $\bar{X}$ will be small and consequently the sum of x_i will be small; for this reason the variance of $\hat{b}$ will be large and so $\hat{b}$ will be an unreliable estimator. In the contrary case, depicted in Figure 4.5b, even though the errors ϵ are of the same size as previously, $\hat{b}$ will be a reliable estimator.

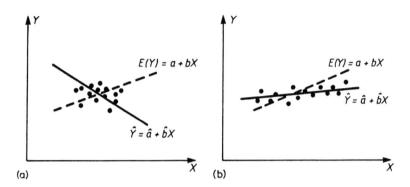

Figure 4.5 Goodness-of-fit and experimental design: (a) unreliable (X_i close), (b) reliable (X_i spread out)

4.2.1.3 Hypothesis Tests for $\hat{b}$

In order to carry out these hypothesis tests we need to know the distribution of $\hat{b}$ and for this we require to add the *strong hypotheses* that the variables $\mathbf{Y}$ are distributed Normal. Now, as $\hat{b}$ is just a linear combination of the Y_i it follows that it is also distributed $N(b, \sigma^2/\sum_i x_i^2)$. This means we can standardise it in the usual way, obtaining

$$z = \frac{\hat{b} - b}{\sigma/\sqrt{(\sum_i x_i^2)}} \qquad (4.7)$$

which is distributed $N(0, 1)$; it is also useful to note that z^2 is generally known as a *quadratic form* and is distributed χ^2 with one degree of freedom. However we do not know σ^2, the variance of $\mathbf{Y}$ with respect to the true regression. A natural estimator is to use the *residual variance* s^2 around the fitted line:

$$s^2 = \frac{\sum_i (Y_i - \hat{Y}_i)^2}{n - 2}$$

We divide by $(n - 2)$ to obtain an unbiased estimator, because two degrees of freedom have been used to calculate $\hat{a}$ and $\hat{b}$ which define $\hat{Y}_i$ (see Wonnacott and Wonnacott 1977).

However, if we substitute s^2 by σ^2 in (4.7) the standardised $\hat{b}$ becomes distributed Student (or t) with $(n-2)$ degrees of freedom:

$$t = \frac{\hat{b} - b}{s/\sqrt{(\sum_i x_i^2)}} \qquad (4.8)$$

The denominator of (4.8) is usually called *standard error* of $\hat{b}$ and is denoted by s_b, hence $t = (\hat{b} - b)/s_b$.

The t-test   A typical null hypothesis is H_0: $b = 0$; in the case (4.8) reduces to:

$$t = \hat{b}/s_b \qquad (4.9)$$

and this value needs to be compared with the critical value of the Student statistics for a given significance level α and the appropriate number of degrees of freedom. One problem is that the alternative hypothesis H_1 may imply unilateral ($b > 0$) or bilateral (b not equal 0) tests; this can only be determined examining the phenomenon under study.

Example 4.2: Assume we are interested in studying the effect of income (I) in the number of trips by non-car-owning households (T), and that we can use the following relation:

$$T = a + bI$$

As in theory we can conclude that any influence must be positive (i.e. higher income always means more trips) in this case we should test H_0 against the unilateral alternative hypothesis H_1: $b > 0$. If H_0 is true, the t-value from (4.9) is compared with the value $t_{\alpha;d}$, where d are the appropriate number of degrees of freedom, and the null hypothesis is rejected if $t > t_{\alpha;d}$ (see Figure 4.6).

On the other hand, if we were considering incorporating a variable whose effect in either direction was not evident (for example, number of female workers, as these

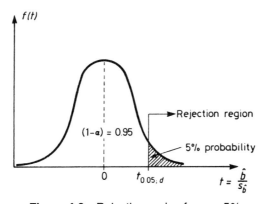

Figure 4.6   Rejection region for $\alpha = 5\%$

may or may not produce more trips than their male counterparts), the null hypothesis should be the bilateral $H_1: b \neq 0$, and H_0 would be rejected if 0 is not included in the appropriate confidence interval for $\hat{b}$.

The F-test for the Complete Model   Figure 4.7a shows the set of values $(\hat{a}, \hat{b})$ for which null hypotheses such as the one discussed above are accepted individually. If we were interested in testing the hypothesis that both estimators are for example equal to 0, we could have a region such as that depicted in Figure 4.7b; i.e. accepting that each parameter is 0 individually does not necessarily mean accepting that both should be 0 together.

Now, to make a two-parameter test it is necessary to know the joint distribution of both estimators. In this case, as their marginals are normal, the joint distribution is also bivariate normal. The F-statistic used to test the trivial null hypothesis H_0: $(a, b) = (0, 0)$, provided as one of the standards in commercial computer packages, is given by:

$$F = \left(n\hat{a}^2 + \sum_i x_i^2 \hat{b}^2 \right) \Big/ 2s^2$$

H_0 is accepted if F is less than or equal to the critical value $F_\alpha(2, n - 2)$. Unfortunately the test is not very powerful (i.e. it is nearly always rejected), but similar ones may be constructed for more interesting null hypotheses such as $(a, b) = (\bar{Y}, 0)$.

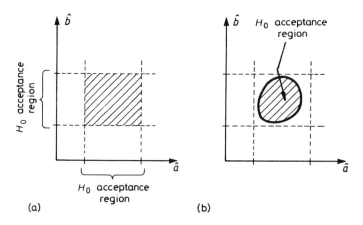

Figure 4.7   Acceptance regions for null hypothesis: (a) both parameters individually, (b) both parameters together

4.2.1.4 The Coefficient of Determination R^2

Figure 4.8 shows the regression line and some of the data points used to estimate it. If no values of **x** were available, the best prediction of Y_i would be $\bar{Y}$. However, the figure shows that for x_i the error of this method would be high: $(Y_i - \bar{Y})$. When x_i is known, on the other hand, the best prediction for Y_i is $\hat{Y}_i$ and this reduces the error

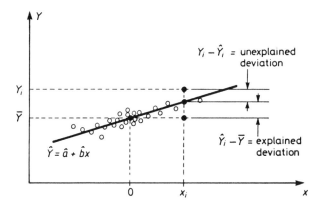

Figure 4.8 Explained and unexplained deviations

to just $(Y_i - \hat{Y}_i)$, i.e. a large part of the original error has been explained. From Figure 4.8 we have:

$$(Y_i - \bar{Y}) \quad = \quad (\bar{Y}_i - \bar{Y}) \quad + \quad (Y_i - \bar{Y}_i), \qquad \forall\, i$$

total deviation   explained deviation   unexplained deviation

If we square the total deviations and sum over all values of i, we get the following:

$$\sum_i (Y_i - \bar{Y})^2 \quad = \quad \sum_i (\hat{Y}_i - \bar{Y})^2 \quad + \quad \sum_i (Y_i - \hat{Y}_i)^2 \tag{4.10}$$

total variation   explained variation   unexplained variation

Now, because $(\hat{Y}_i - \bar{Y}) = \hat{b}x_i$ it is easy to see that the explained variation is a function of the estimated regression coefficient $\hat{b}$. The process of decomposing the total variation into its parts is known as *analysis of variance* of the regression, or ANOVA (note that variance is just variation divided by degrees of freedom).

The coefficient of determination is defined as the ratio of explained to total variation:

$$R^2 = \frac{\sum (\hat{Y}_i - \bar{Y})^2}{\sum (Y_i - \bar{Y})^2} \tag{4.11}$$

It has limiting values of 1 (perfect explanation) and 0 (no explanation at all); intermediate values may be interpreted as the percentage of the total variation explained by the regression. The index is trivially related to the sample correlation R, which measures the degree of association between X and Y (see Wonnacott and Wonnacott 1977).

4.2.1.5 Multiple Regression

This is an extension of the above for the case of more explanatory variables and, obviously, more *regressors* ($\hat{b}$ parameters). The solution equations are similar, although more complex, but some extra problems arise which are usually important, such as the following:

1. Multicollinearity. The occurs when there is a linear relation between the explanatory variables; in this case the equations for the regressors $\hat{\mathbf{b}}$ are not independent and cannot be solved uniquely.
2. How many regressors to include. To make a decision in this case, several factors have to be taken into consideration:
   - Are there strong theoretical reasons to include a given variable, or is it important for policy testing with the model?
   - Is the variable significant (i.e. is H_0 rejected in the t-test?) and is the estimated sign of the coefficient consistent with theory or intuition?

   If in doubt, one way forward is to take out the variable and re-estimate the regression in order to see the effect of its removal on the rest of the coefficients; if this is not too important the variable can be left out for *parsimony* (the model is simpler and the rest of the parameters can be estimated more accurately). Commercial software packages provide an 'automatic' procedure for tackling this issue (the stepwise approach); however, this may induce some problems, as we will comment below. We will come back to this general problem in section 8.3 (Table 8.1) when discussing discrete choice model specification issues.
3. Coefficient of determination. This has the same form as (4.11). However, in this case the inclusion of another regressor always increases R^2; to eliminate this problem the corrected R^2 is defined as:

$$\bar{R}^2 = [R^2 - k/(n - 1)][(n - 1)/(n - k - 1)] \tag{4.12}$$

where n stands for sample size as before and k is the number of regressors $\hat{\mathbf{b}}$.

In trip generation modelling the multiple regression method has been used both with aggregate (zonal) and disaggregate (household and personal) data. The first approach has been practically abandoned in the case of trip productions, but it is still the premier method for modelling trip attractions.

4.2.2 Zonal-based Multiple Regression

In this case an attempt is made to find a linear relationship between the number of trips produced or attracted by zone and average socioeconomic characteristics of the households in each zone. The following are some interesting considerations:

1. Zonal models can only explain the variation in trip making behaviour between zones. For this reason they can only be successful if the inter-zonal variations adequately reflect the real reasons behind trip variability. For this to happen it would be necessary that zones not only had an homogeneous socioeconomic composition, but represented as wide as possible a range of conditions. A major problem is that the main variations in person trip data occur at the intra-zonal level.

2. Role of the intercept. One would expect the estimated regression line to pass through the origin; however, large intercept values (i.e. in comparison to the product of the average value of any variable and its coefficient) have often been obtained. If this happens the equation may be rejected; if on the contrary, the intercept is not significantly different from zero, it might be informative to re-estimate the line, forcing it to pass through the origin.

3. Null zones. It is possible that certain zones do not offer information about certain dependent variables (e.g. there can be no HB trips generated in non-residential zones). Null zones must be excluded from analysis; although their inclusion should not greatly affect the coefficient estimates (because the equations should pass through the origin), an arbitrary increment in the number of zones which do not provide useful data will tend to produce statistics which overestimate the accuracy of the estimated regression.

4. Zonal totals versus zonal means. When formulating the model the analyst appears to have a choice between using aggragate or total variables, such as trips per zone and cars per zone, or rates (zonal means), such as trips per household per zone and cars per household per zone. In the first case the regression model would be:

$$Y_i = \theta_0 + \theta_1 X_{1i} + \theta_2 X_{2i} + \ldots + \theta_k X_{ki} + E_i$$

whereas the model using rates would be:

$$y_i = \theta_0 + \theta_1 x_{1i} + \theta_2 x_{2i} + \ldots + \theta_k x_{ki} + e_i$$

with $y_i = Y_i/H_i$; $x_i = X_i/H_i$; $e_i = E_i/H_i$ and H_i the number of households in zone i.

Both equations are identical, in the sense that they seek to explain the variability of trip making behaviour between zones, and in both cases the parameters have the same meaning. Their unique and fundamental difference relates to the error-term distribution in each case; it is obvious that the constant variance condition of the model cannot hold in both cases, unless H_i was itself constant for all zones i.

Now, as the aggregate variables directly reflect the size of the zone, their use should imply that the magnitude of the error actually depends on zone size; this *heterocedasticity* (variability of the variance) has indeed been found in practice. Using multipliers, such as $1/H_i$, allows heterocedasticity to be reduced because the model is made independent of zone size. In this same vein, it has also been found that the aggragate variables tend to have higher intercorrelation (i.e. multicollinearity) than the mean variables. However, it is important to note that models using aggregate variables often yield higher values of R^2, but this is just a spurious effect because zone size obviously helps to explain the total number of trips (see Douglas and Lewis 1970). What is certainly unsound is the mixture of means and aggregate variables in a single model.

To end this theme it is important to remark that even when rates are used, zonal regression is conditioned by the nature and size of zones (i.e. the spatial aggregation problem). This is clearly exemplified by the fact that inter-zonal variability diminishes with zone size as shown in Table 4.2, constructed with data from Perth (Douglas and Lewis 1970).

Table 4.2 Inter-zonal variation of personal productions for two different zoning systems

Zoning system	Mean value of trips/household/zone	Inter-zonal variance
75 small zones	8.13	5.85
23 large zones	7.96	1.11

4.2.3 Household-based Regression

Intra-zonal variation may be reduced by decreasing zone size, especially if zones are homogeneous. However, smaller zones imply a greater number of them and this has two consequences:

- more expensive models in terms of data collection, calibration and operation;
- greater sampling errors, which are assumed non-existent by the multiple linear regression model.

For these reasons it seems logical to postulate models which are independent of zonal boundaries. At the beginning of the 1970s it was decided that the most appropriate analysis unit in this case was the household (and not the individual); it was argued that a series of important interpersonal interactions inside a household could not be incorporated even implicitly in an individual model (e.g. car availability, that is, who has use of the car). We will challenge this thesis in section 4.3.3.

In a household-based application each home is taken as an input data vector in order to bring into the model all the range of observed variability about the characteristics of the household and its travel behaviour. The calibration process, as in the case of zonal models, proceeds stepwise, testing each variable in turn until the best model (in terms of some summary statistics for a given confidence level) is obtained. Care has to be taken with automatic stepwise computer packages because they may leave out variables which are slightly worse predictors than others left in the model, but which may prove much easier to forecast.

Example 4.3: Consider the variables trips per household (Y), number of workers (X_1) and number of cars (X_2). Table 4.3 presents the results of successive steps of a step-wise model estimation; the last row also shows (in parenthesis) values for the t-ratio (equation 4.9). Assuming large sample size, the appropriate number of degrees of freedom $(n - 2)$ is also a large number so the t-values may be compared with the critical value 1.645 for a 95% significance level on a one-tailed test (we

Table 4.3 Example of stepwise regression

Step	Equation	R^2
1	$Y = 2.36X_1$	0.203
2	$Y = 1.80X_1 + 1.31X_2$	0.325
3	$Y = 0.91 + 1.44X_1 + 1.07X_2$	0.384
	$\quad\quad\;\;(3.7)\quad\;(8.2)\quad\quad(4.2)$	

know the null hypothesis is unilateral in this case as Y should increase with both X_1 and X_2).

The third model is a good equation in spite of its low R^2. The intercept 0.91 is not large (compare it with 1.44 times the number of workers, for example) and the explanatory variables are significantly different from zero (H_0 is rejected in all cases). The model could probably benefit from the inclusion of other variables.

An indication of how good these models are may be obtained from comparing observed and modelled trips for some groupings of the data (see Table 4.4). This is better than comparing totals because in such case different errors may compensate and the bias would not be detected. As can be seen, the majority of cells show a reasonable approximation (i.e. errors of less than 30%). If large bias were spotted it would be necessary to adjust the model parameters; however, this is not easy as there are no clear-cut rules to do it, and it depends heavily on context.

Table 4.4   Comparison of trips per household (observed/estimated).

	Number of workers in household			
No. of cars	0	1	2	3 or more
0	0.9/0.9	2.1/2.4	3.4/3.8	5.3/5.6
1	3.2/2.0	3.5/3.4	3.7/4.9	8.5/6.7
2 or more	—	4.1/4.6	4.7/6.0	8.5/7.8

4.2.4 The Problem of Non-linearities

As we have seen, the linear regression model assumes that each independent variable exerts a linear influence on the dependent variable. It is not easy to detect non-linearity because apparently linear relations may turn out to be non-linear when the presence of other variables is allowed in the model. Multivariate graphs are useful in this sense; the example of Figure 4.9 presents data for households stratified by car ownership and number of workers. It can be seen that travel behaviour is non-linear with respect to family size.

It is important to mention that there is a class of variables, those of a qualitative nature, which usually shows non-linear behaviour (e.g. type of dwelling, occupation of the head of the household, age, sex). In general there are two methods to incorporate non-linear variables into the model:

1. Transform the variables in order to linearise their effect (e.g. take logarithms, raise to a power). However, selecting the most adequate transformation is not an easy or arbitrary exercise, so care is needed; also, if we are thorough, it can take a lot of time and effort;
2. Use dummy variables. In this case the independent variable under consideration is divided into several discrete intervals and each of them is treated separately in the model. In this form it is not necessary to assume that the variable has a linear effect, because each of its portions is considered separately in terms of its effect on travel behaviour. For example, if car ownership was treated in this way,

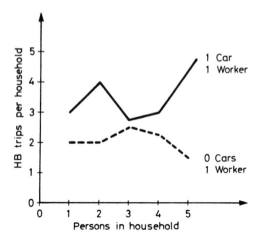

Figure 4.9   An example of non-linearity

appropriate intervals could be 0, 1 and 2 or more cars per household. As each sampled household can only belong to one of the intervals, the corresponding dummy variable takes a value of 1 in that class and 0 in the others. It is easy to see that only $(n - 1)$ dummy variables are needed to represent n intervals.

Example 4.4: Consider the model of Example 4.3 and assume that variable X_2 is replaced by the following dummies:

Z_1, which takes the value 1 for households with one car and 0 in other cases;

Z_2, which takes the value 1 for households with two or more cars and 0 in other cases.

It is easy to see that non-car-owning households correspond to the case where both Z_1 and Z_2 are 0. The model of the third step in Table 4.3 would now be:

$$Y = 0.84 + 1.41X_1 + 0.75Z_1 + 3.14Z_2 \qquad R^2 = 0.387$$
$$\quad (3.6) \quad (8.1) \qquad (3.2) \qquad (3.5)$$

Even without the better R^2 value, this model is preferable to the previous one just because the non-linear effect of X_2 (or Z_1 and Z_2) is clearly evident and cannot be ignored. Note that if the coefficients of the dummy variables were for example, 1 and 2, and if the sample never contained more than two cars per household, the effect would be clearly linear. The model is graphically depicted in Figure 4.10.

Looking at Figure 4.10, the following question arises: would it not be preferable to estimate separate regressions for the data on each group as in that case we would not require each line to have the same slope (the coefficient of X_1)? The answer is in general no unless we had a reasonable amount of data for each class. The fact is that the model with dummies uses all the data, while each separate regression would use only part of it, and this is in general disadvantageous. It is also interesting to mention that the use of dummy variables tends to reduce problems of multicollinearity in the data (see Douglas and Lewis 1971).

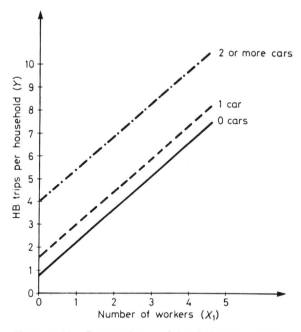

Figure 4.10   Regression model with dummy variables

4.2.5 Obtaining Zonal Totals

In the case of zonal-based regression models, this is not a problem as the model is estimated precisely at this level. In the case of household-based models, though, an aggregation stage is required. Nevertheless, precisely because the model is linear the aggregation problem is trivially solved by replacing the average zonal values of each independent variable in the model equation and then multiplying it by the number of households in each zone. However, it must be noted that the aggregation stage can be a very complex matter in non-linear models, as we will see in Chapter 9.

Thus, for the third model of Table 4.3 we would have:

$$T_i = H_i(0.91 + 1.44\overline{X}_{1i} + 1.07\overline{X}_{2i})$$

where T_i is the total number of HB trips in the zone, h_i is the total number of households in it and $\overline{X}_{ji}$ is the average value of variable X_j for the zone.

On the other hand, when dummy variables are used, it is also necessary to know the number of households in each class for each zone; for instance, in the model of Example 4.4 we require:

$$T_i = H_i(0.84 + 1.41\overline{X}_{1i}) + 0.75H_{1i} + 3.14H_{2i}$$

where H_{ji} is the number of households of class j in zone i.

This last expression allows us to appreciate another advantage of the use of dummy variables over separate regressions. To aggregate the models, in this latter case, it would be necessary to estimate the average number of workers per household (X_1) for each car-ownership group in each zone, and this may be complicated.

4.2.6 Matching Generations and Attractions

It might be obvious to some readers that the models above do not guarantee, by default, that the total number of trips originating (the *origins* O_i) at all zones will be equal to the total number of trips attracted (the *destinations* D_j) to them, that is the following expression does not necessarily hold:

$$\sum_i O_i = \sum_j D_j \qquad (4.13)$$

The problem is that this equation is implicitly required by the next sub-model (i.e. trip distribution) in the structure; it is not possible to have a trip distribution matrix where the total number of trips (T) obtained by summing all rows is different to that obtained when summing all columns (see Chapter 5).

The solution to this difficulty is a pragmatic one which takes advantage of the fact that normally the trip generation models are far 'better' (in every sense of the word) than their trip attraction counterparts. The first normally are fairly sophisticated household-based models with typically good explanatory variables. The trip attraction models, on the other hand, are at best estimated using zonal data. For this reason, normal practice considers that the total number of trips arising from summing all origins O_i is in fact the correct figure for T; therefore, all destinations D_j are multiplied by a factor f given by:

$$f = T / \sum_j D_j \qquad (4.14)$$

which obviously ensure that their sum adds to T.

4.3 CROSS-CLASSIFICATION OR CATEGORY ANALYSIS

4.3.1 The Classical Model

4.3.1.1 Introduction

Up to the late 1960s most transportation planning studies in the USA developed trip generation equations based on linear regression analysis, particularly when modelling personal trip productions. In fact, the regression model was favoured as the central method in the Federal Highway Administration's guide to trip generation analysis (FHWA 1967).

At the end of the 1960s an alternative method for modelling trip generation appeared and quickly became established as the preferred one in the United Kingdom. The method was known as *category analysis* in the UK (Wootton and Pick 1967) and *cross-classification* in the USA; there it went through a similar development process as the linear regression model, with earliest procedures being at the zonal level and subsequent models based on household information.

The method is based on estimating the response (e.g. the number of trip productions per household for a given purpose) as a function of household attributes. Its basic assumption is that trip generation rates are relatively stable over time for certain household stratifications. The method finds these rates empirically and for this it typically needs a large amount of data; in fact, a critical element is the number of households in each class. Although the method was originally designed to use census data in the UK, a serious problem of the approach remains the need to forecast the number of households in each strata in the future.

4.3.1.2 Variable Definition and Model Specification

Let $t^p(h)$ be the average number of trips with purpose p (and at a certain time period) made by members of households of type h. Types are defined by the stratification chosen; for example, a cross-classification based on m household sizes and n car ownership classes will yield mn types h. The standard method for computing these cell rates is to allocate households in the calibration data to the individual cell groupings and total, cell by cell, the observed trips $T^p(h)$ by purpose group. The rate $t^p(h)$ is then the total number of trips in cell h, by purpose, divided by the number of households $H(h)$ in it. In mathematical form it is simply as follows:

$$t^p(h) = T^p(h)/H(h) \tag{4.15}$$

The 'art' of the method lies in choosing the categories such that the standard deviations of the frequency distributions depicted in Figure 4.11 are minimised.

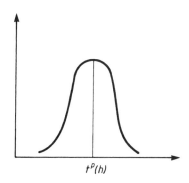

$t^p(h)$

Figure 4.11   Trip-rate distribution for household type

The method has, in principle, the following advantages:

1. Cross-classification groupings are independent of the zone system of the study area.
2. No prior assumptions about the shape of the relationship are required (i.e. they do not even have to be monotonic, let alone linear).
3. Relationships can differ in form from class to class (e.g. the effect of changes in household size for one or two car-owning households may be different).

And in common with traditional cross-classification methods it has also several disadvantages:

1. The model does not permit extrapolation beyond its calibration strata, although the lowest or highest class of a variable may be open-ended (e.g. households with two or more cars and five or more residents).
2. There are no statistical goodness-of-fit measures for the model, so only aggregate closeness to the calibration data can be ascertained.
3. Unduly large samples are required, otherwise cell values will vary in reliability because of differences in the numbers of households being available for calibration at each one. For example, in the Monmouthshire Land Use/Transportation Study (see Douglas and Lewis 1971) the following distribution for 108 categories (six income levels, three car ownership levels and six household structure levels) was found, using a sample of 4000 households:

Table 4.5   Household frequency distribution

	No. of categories				
	21	69	9	7	2
No. of households surveyed	0	1–49	50–99	100–199	200+

Accepted wisdom suggests that at least 50 observations per cell are required to estimate the mean reliably; thus, this criterion would be satisfied in only 18 of the 108 cells for a sample of 4000 households. There may be some scope for using stratified sampling to guarantee more evenly distributed sample sizes in each category. This involves, however, additional survey costs.
4. There is no effective way to choose among variables for classification, or to choose best groupings of a given variable; the minimisation of standard deviations hinted at in Figure 4.11 requires an extensive 'trial and error' procedure which may be considered infeasible in practical studies.

4.3.1.3 Model Application at Aggregate Level

Let us denote by n the person type (i.e. with and without a car), be $a_i(h)$ the number of households of type h in zone i, and by $H^n(h)$ the set of households of type h containing persons of type n. With this we can write the trip productions with purpose p by person type n in zone i, O_i^{np}, as follows:

$$O_i^{np} = \sum_{h \in H^n(h)} a_i(h) t^p(h) \qquad (4.16)$$

To verify how the model works it is possible to compare these modelled values with observed values from the calibration sample. Inevitable errors are due to the use of averages for the $t^p(h)$; one would expect a better stratification (in the sense of minimising the standard deviation in Figure 4.11) to produce smaller errors.

There are various ways of defining household categories. The first application in the UK (Wootton and Pick 1967), which was followed closely by subsequent applications, employed 108 categories as follows: six income levels, three car ownership levels (0, 1 and 2 or more cars per household) and six household structure groupings, as in Table 4.6.

The problem is clearly how to predict the number of households in each category in the future. The method most commonly used (see Wilson 1974) consists in, firstly, defining and fitting to the calibration data, probability distributions for income (I), car ownership (C) and household structure (S); secondly, using these to build a joint probability function of belonging to household type $h = (I, C, S)$. Thus, if the joint distribution function is denoted by $\phi(h) = \phi(I, C, S)$, the number of households in zone i belonging to class h, $a_i(h)$, is simply given by:

$$a_i(h) = H_i \phi(h) \qquad (4.17)$$

where H_i is the total number of households in the zone. This household estimation model may be partially tested by running it with the base-year data used in calibration. The total trips estimated with equation (4.16), but with simulated values for $a_i(h)$, can then be checked against the actual observations.

One further disadvantage of the method can be added at this stage:

5. If is is required to increase the number of stratifying variables, it might be necessary to increase the sample enormously. For example, if another variable was added to the *original* application discussed above and this was divided into three levels, the number of categories would increase from 108 to 324 (and recall the discussion on Table 4.5).

Table 4.6   Example of household structure grouping

Group	No. employed	Other adults
1	0	1
2	0	2 or more
3	1	1 or less
4	1	2 or more
5	2 or more	1 or less
6	2 or more	2 or more

4.3.2 Improvements to the Basic Model

4.3.2.1 Multiple Classification Analysis (MCA)

MCA is an alternative method to define classes and test the resulting cross-classification which provides a statistically powerful procedure for variable selection and classification. This allows us to overcome several of the disadvantages cited above for other types of cross-classification methods. The interested reader is refered to Stopher and McDonald (1983) for full details as only a summary is provided below.

Consider a model with a continuous dependent variable (such as the trip rate) and two discrete independent variables, such as household size and car ownership. A grand mean can be estimated for the dependent variable over the entire sample of households. Also, group means can be estimated for each row and column of the cross-classification matrix; each of these can be expressed in turn as deviations from the grand mean. Observing the signs of the deviations, a cell value can be estimated by adding to the grand mean the row and column deviations corresponding to the cell. In this way, some of the problems arising from too few observations on some cells can be compensated.

Example 4.5: Table 4.7 presents data collected in a study area and classified by three car-ownership and four household-size levels. The table presents the number of households observed in each cell (category) and the mean number of trips calculated over rows, cells and the grand average.

Table 4.7   Number of households per cell and mean trip rates

Household size	0 car	1 car	2+ cars	Total	Mean trip rate
1 person	28	21	0	49	0.47
2 or 3 persons	150	201	93	444	1.28
4 persons	61	90	75	226	1.86
5 persons	37	142	90	269	1.90
Total	276	454	258	988	
Mean trip rate	0.73	1.53	2.44		1.54

As can be seen, the values range from 0 (it is unlikely to find households with one person and more than one car) to 201. Although we are cross-classifying by only two variables in this simple example, there are already four cells with less than the conventional minimum number (50) of observations required to estimate mean trip rate and variance with some reliability.

We would like to use now the mean row and column values to estimate average trip rates for each cell, including that without observations in this sample. We can compute the deviation (from the grand mean) for zero cars as $0.73 - 1.54 = -0.81$; for one car as $1.53 - 1.54 = -0.01$, and for two cars or more $2.44 - 1.54 = 0.90$; similarly, we can calculate the deviations for each of the four household size groups as: $-1.07, -0.26, 0.32$ and 0.36. If the variables are not correlated with these values we can work out the full trip-rate table; for example, the trip rate for one person household and one car is $1.54 - 1.07 - 0.01 = 0.46$ trips. In the case of one person and no car, the rate turns out to be negative and equal to -0.34 ($1.54 - 1.06 - 0.82$); this has no meaning and therefore the actual rate is forced to zero. Table 4.8 depicts the full trip-rate table together with its row and column deviations.

Table 4.8   Trip rates calculated by multiple classification

HH size	Car ownership level			Deviations
	0 car	1 car	2+ cars	
1 person	0.00	0.46	1.37	-1.07
2 or 3 persons	0.46	1.27	2.18	-0.26
4 persons	1.05	1.85	2.76	0.32
5 persons	1.09	1.89	2.80	0.36
Deviations	-0.81	-0.01	0.90	

In contrast to standard cross-classification models, deviations are not only computed for households in, say, the cell one person–one car; rather, car deviations are computed over all household sizes and vice versa. Thus, if interactions are present these deviations should be adjusted to account for interaction effects. This can be done by taking a weighted mean for each of the group means of one independent variable over the groupings of the other independent variables, rather than a simple mean (which would in fact be equivalent to assuming that variation is random over the data in a group). These weighted means will in general tend to decrease the sizes of the adjustments to the grand mean when interactions are present. Nevertheless, the cell means of a multiway classification will still be based on means estimated from all the available data, rather than being based on only those items of data falling in the multiway cell.

The most important statistical goodness-of-fit measures associated with MCA are (Stopher and McDonald 1983):

- an F-statistic to assess the entire cross-classification scheme (recall the discussion in section 4.2);
- a correlation ratio statistic for assessing the contribution of each classification variable (see Stopher 1975); and
- an R^2 measure for the complete cross-classification model.

These measures allow the analyst to compare different classification schemes and to assess their fit to the base-year data.

Apart from the statistical advantages, it is important to note that cell values are no longer based on only the size of the data sample within a given cell; rather, they are based on a grand mean derived from the entire data set, and on two (or more) class means which are derived from all data in each class relevant to the cell in question.

Example 4.6: Table 4.9 provides a set of rates computed in the standard category analysis procedure (i.e. by using individual cell means). These values may be compared with those of Table 4.8.

Two points of interest emerge from the comparison. First, there are rates available even for empty cells in the MCA case. Second, some counterintuitive progressions, apparent in Table 4.9 (e.g. the decrease of rate values for 0 and 1 car-owning households when increasing household size from 4 to 5 or more), are removed in Table 4.8. Note that they could have arisen by problems of small sample size at least in one case.

Table 4.9 Trip rates for same study area calculated using category analysis

HH size	Car ownership level		
	0 car	1 car	2+ cars
1 person	0.12	0.94	
2 or 3 persons	0.60	1.38	2.16
4 persons	1.14	1.74	2.60
5 persons	1.02	1.69	2.60

4.3.2.2 Regression Analysis for Household Strata

A mixture of cross-classification and regression modelling of trip generation may be the most appropriate approach on certain occasions. For example, in an area where the distribution of income is unequal it may be important to measure the differential impact of policies on different income groups; therefore it may be necessary to model travel demand for each income group separately throughout the entire modelling process. Assume now that in the same area car ownership is increasing fast and, as usual, it is not clear how correlated these two variables are; a useful way out may be to postulate regression models based on variables describing the size and make-up of different households, for a stratification according to the two previous variables.

Example 4.7: Table 4.10 presents the 13 income and car-ownership categories (C_i) defined in ESTRAUS (1989) for the Greater Santiago 1977 origin–destination data. As can be seen, the bulk of the data corresponds to households with no cars and low income. Also note that categories 7 and 10 have rather few data points; this is,

Table 4.10 Stratification of the 1977 Santiago sample

Household income (US$/month)	Household car ownership			Total
	0	1	2+	
< 125	6 564 (C_1)	215 (C_2)		6 779
125–250	4 464 (C_3)	627 (C_4)		5 091
250–500	1 532 (C_5)	716 (C_6)	87 (C_7)	2 334
500–750	305 (C_8)	436 (C_9)	118 (C_{10})	859
> 750	169(C_{11})	380 (C_{12})	301 (C_{13})	790
Total	12 974	2 373	506	15 853

unfortunately, a general problem of this approach. Even smaller samples for very low income and high car ownership led to the aggregation of some categories at this range.

The independent variables available for analysis (i.e. after leaving out the stratifying variables) included variables of the *stage in the family cycle* variety, which we will discuss in section 4.4. However, after extensive specification searches it was found that the most significant variables were: number of workers (divided into four classes depending on earnings and type of job), number of students and number of residents.

Linear regression models estimated with these variables for each of the 13 categories were judged satisfactory on the basis of correct signs, small intercepts, reasonable significance levels and R^2 values (e.g. between 0.401 for category 4, and 0.682 for category 7; see Hall *et al.* 1987).

4.3.3 The Person-category Approach

4.3.3.1 Introduction

This is an interesting alternative to the household-based models discussed above, which was originally proposed by Supernak (1979). It has been argued that this approach offers the following advantages (Supernak *et al.* 1983):

1. A person-level trip generation model is compatible with other components of the classical transport demand modelling system, which is based on tripmakers rather than on households.
2. It allows a cross-classification scheme that uses all important variables and yields a manageable number of classes; this in turn allows class representation to be forecast more easily.
3. The sample size required to develop a person-category model can be several times smaller than that required to estimate a household-category model.
4. Demographic changes can be more easily accounted for in a person-category model as, for example, certain key demographic variables (such as age) are virtually impossible to define at household level.
5. Person categories are easier to forecast than household categories as the latter require forecasts about household formation and family size; these tasks are altogether avoided in the case of person categories. In general the bulk of the trips are made by people older than 18 years of age; this population is easier to forecast 15 to 20 years ahead as only migration and survival rates are needed to do so.

The major limitation that a person-category model may have relates precisely to the main reason why household-based models were chosen to replace zonal-based models at the end of the 1960s; this is the difficulty of introducing household interaction effects and household money costs and money budgets into a person-based model. However, Supernak *et al.* (1983) argue that it is not clear how vital these considerations are and how they can be effectively incorporated even in a household-based model; in fact, from our discussion in sections 4.2.3 and 4.3.1 it is clear that this is done in an implicit fashion only.

4.3.3.2 Variable Definition and Model Specification

Let t_j be the trip rate, that is, the number of trips made during a certain time period by (the average) person in category j; t_{jp} is the trip rate by purpose p. T_i is the total number of trips made by the inhabitants of zone i (all categories together). N_i is the number of inhabitants of zone i, and α_{ji} is the percentage of inhabitants of zone i belonging to category j. Therefore the following basic relationship exists:

$$T_i = N_i \sum_j \alpha_{ji} t_j \tag{4.18}$$

As in other methods, trips are divided into home-based (HB) and non-home-based (NHB), and can be further divided by purpose (p) which may apply to both HB and NHB trips.

Model development entails the following stages:

1. Consideration of several variables which are expected to be important for explaining differences in personal mobility. Also, definition of plausible person categories using these variables.
2. Preliminary analysis of trip rates in order to find out which variables have the least explanatory power and can be excluded from the model. This is done by comparing the trip rates of categories which are differentiated by the analysed variable only and testing whether their differences are statistically significant.
3. Detailed analysis of trip characteristics to find variables that define similar categories. Variables which do not provide substantial explanation of the data variance, or variables that duplicate the explanation provided by other better variables (i.e. easier to forecast or more policy responsive) are excluded. The exercise is conducted under the constraint that the number of final categories should not exceed a certain practical maximum (for example, 15 classes).

For this analysis the following measures may be used: the coefficient of correlation (R_{jk}), slope (m_{jk}) and intercept (a_{jk}) of the regression $t_{jp} = a_{jk} + m_{jk} t_{kp}$. The categories j and k may be treated as similar if these measures satisfy the following conditions (Supernak *et al*. 1983):

$$R_{jk} > 0.900$$
$$0.75 < m_{jk} < 1.25 \tag{4.19}$$
$$a_{jk} < 0.10$$

These conditions are quite demanding and may be changed.

4.3.3.3 Model Application at the Aggregate Level

Zonal home-based productions are computed in a straightforward manner using equation (4.18), or a more disaggregated version explicitly including trip purpose if desired. However, the estimation of trip attractions in general and NHB trip productions at the zonal level is more involved and requires the development of *ad hoc* methods heavily dependent on the type of information available at each application (see Supernak 1979 for a Polish example).

4.4 FORECASTING VARIABLES IN TRIP GENERATION ANALYSIS

The choice of variables used to predict (household) trip generation rates has long been an area of concern for transportation planners; these variables typically include household numbers, household size (and/or structure), number of vehicles owned and income. However, interest arose in the early 1980s on research aimed at enriching trip generation models with theories and methods from the behavioural sciences. The major hypothesis behind this work was that the social circumstances in which individuals live should have a considerable bearing on the opportunities and constraints they face in making activity choices; the latter in turn, may lead to differing travel behaviour. For example, it is clear that whether a person lives alone or not should affect the opportunities to coordinate and trade-off activities with others in order to satisfy their travel necessities. Thus, a married couple with young pre-school children will generally find themselves less mobile than a similar couple without children or with older children who require less intensive care. Elderly and retired persons living with younger adults are likely to be more active outside the home than elderly people living alone or with persons roughly their own age.

At the household level the situation is quite similar: households of unrelated individuals, for example, tend to follow a pattern of activities which is less influenced by the presence of other household members (and which normally leads to more frequent trips) than is the case of households of related individuals (obviously with similar size, and other characteristics). This is both because of the reduced coordination among the different members and because their activity patterns typically involve fewer home-centred activities.

One way of introducing these notions into the modelling of trip generation is to develop a set of household types that effectively captures these distinctions and then add this measure to the equations predicting household behaviour. One possible approach considers the age structure of the household and its lifestyle. The approach is consistent with the idea that travel is a *derived demand* and that travel behaviour is part of a larger allocation of time and money to activities in separate locations. For example, the concept of *life style* may be operationalised as the allocation of varying amounts of time to different (activity) purposes both within and outside the home, where travel is just part of this time allocation (see Allaman *et al*. 1982). It appears that the time allocation of individuals varies systematically across various segments of the population, such as age, sex, marital status and even race; this may be because different household structures place different demands on individuals.

One set of hypotheses that can be tested empirically is whether the major break points (or stages) in the life (or family) cycle are consistent with major changes of time allocation. For example, the break points may be:

- the appearance of pre-school children;
- the time when the youngest child reaches school age;
- the time when a youth leaves home and either lives alone, with other young adults, or marries;
- the time when all the children of a couple have left home but the couple has not yet retired;
- the time when all members of a household have reached retirement age.

It is usually illuminating to compare households at one stage of this life cycle with households of the immediately preceeding stage.

The concepts of life style and stage of family cycle are important from two points of view: first, that of identifying stable groupings (based on age or sex) with different activity schedules and consequently demands for travel; second, that of allowing the tracing of systematic changes which may be based on demographic variations (e.g. changes in age structure, marital or employment status). Numerous demographic trends of significance in terms of travel behaviour have been receiving increasing attention since the early 1980s (see Spielberg *et al*. 1981). One of the most significant for predicting travel behaviour is the changing ratio of households to population, particularly in industrialised nations. Although the rate of population growth fell steadily during the 1980s, the rate of household formation increased in some cases. This is due, among other reasons, to increases in the number of single-parent households and the number of persons who are setting up individual households. Therefore, travel forecasting methodologies which implicitly assume stable ratios of households to population (as is often the case) may be severely affected by this structural shift in the demographic composition of society.

Another trend which has been well discussed is the overall ageing of the population, again particularly in industrialised nations. This is important because age tends to be associated with a decline in mobility and a change in life style. It is interesting to note though, that differences in trip generation by age may reflect in part the so-called *cohort effects*. This means that older people may travel less, simply because they always did so, rather than because of their age. However, this effect may be largest for people over 65 and declining trip generation rates for other age groups probably reflect a true decrease in the propensity of travel.

Finally, another trend worth noting is the increase in the proportion of women joining the labour force. Its significance for transportation planning and forecasting stems from two effects. The first is simply the direct employment effect, where time allocation and consequently travel behaviour are profoundly influenced by the requirements of actually being employed. The second one is more subtle and concerns changes in household roles and their impacts on life style, particularly for couples with children.

To end this section it is interesting to mention that the ideas discussed above led to a proposal for incorporating a household structure variable in trip generation modelling, which was tested with real data (Allaman *et al*. 1982). The household structure categories proposed were based on the age, sex, marital status and last name of each household member. These variables allowed the determination of the presence or absence of dependents in the household, the number and type of adults present, and the relationship among household members. However, although models using this variable were pronounced a considerable improvement over traditional practice by Allaman *et al*. (1982), further tests with a different data set performed by McDonald and Stopher (1983) led to its rejection. This was not only on the basis of statistical evidence but also on policy sensitivity (i.e. it is difficult to use household structure as a policy variable) and ease of forecasting grounds (i.e. forecasting at zonal level, particularly to obtain a distribution of households by household structure category, appears to be very problematic). McDonald and Stopher (1983) argue that in these two senses a variable of the housing type variety should be preferred and it is bound to be easier to use by a local government planning agency.

4.5 STABILITY AND UPDATING OF TRIP GENERATION PARAMETERS

4.5.1 Temporal Stability

Transport models, in general, are developed to assist in the formulation and evaluation of transport plans and projects. Although on many occasions use has been made of descriptive statistics for examining travel trends, most developments have used cross-sectional data to express the amount of travel in terms of explanatory factors; these need to be both plausible and easy to forecast for the model to be policy sensitive in the design-year. A key (often implicit) assumption of this approach is that the model parameters will remain constant (or stable) between base and design years.

Several studies have examined this assumption in a trip generation context, finding in general that it cannot be rejected when trips by all modes are considered together (see Kannel and Heathington 1973; Smith and Cleveland 1976), even in the case of the rather crude zonal-based models (although these are not recommended anyway, for reasons similar to those discussed in section 4.2.2; see Downes and Gyenes 1976). However, more recent analyses have reported different results. For example, Hall *et al*. (1987) compared observed trip rates and regression coefficients of models fitted to household data collected for Santiago in 1977 and 1986, and found them significantly different. Copley and Lowe (1981) reported that although trip rates by bus for certain types of household categories seemed reasonably stable over time, car trip rates appeared to be highly correlated with changes in real fuel prices. The latter has the following potential implications:

1. If there is negative correlation between car trip rates and fuel prices, the usual assumption of constant trip rates in a period of rapidly increasing petrol prices could lead to serious over-provision of highway facilities. If, on the other hand, fuel prices were to fall in real terms, the constant trip rates assumption would lead to under-provision (which is precisely what was experienced in the UK and other industrialised countries towards the end of the 1980s).

2. Furthermore, the balance between future investments in public and private transport facilities may be judged incorrectly if based on the assumption of constant trip rates over time.

Clearly then, the correct estimation of the effect of fuel prices on trip rates (and of any other similar *longitudinal* effects) is of fundamental importance for policy analysis. Unfortunately it cannot be tackled with the cross-sectional data sets typically available for transportation studies.

4.5.2 Geographic Stability

Temporal stability is often difficult to examine because data (of similar quality) are required for the same area at two different points in time. Thus on many occasions it may be easier to examine geographic stability (or transferability) as data on two different locations might become available (for example, if two institutions located in different areas decide to conduct a joint research project). Geographic transferability should be seen as an important attribute of any travel demand model for the following reasons:

1. It would suggest the existence of certain repeatable regularities in travel behaviour which can be picked up and reflected by the model;
2. It would indicate a higher probability that temporal stability also exists; this, as we saw, is essential for any forecasting model;
3. It may allow to reduce substantially the need for costly full-scale transportation surveys on different metropolitan areas (see the discussion on Chapter 9).

It is clear that not all travel characteristics can be transferable between different areas or cities; for example, the average work trip duration is obviously context dependent, that is, it should be a function of area size, shape and the distributions of workplaces and residential zones over space. However, transferability of trip rates should not be seen as unrealistic: trips reflect needs for individuals' participation in various activities outside the home and if trip rates are related to homogeneous groups of people, they can be expected to remain stable and geographically transferable.

The transferability of trip generation models (typically trip rates on a household-category analysis framework) has been tested relatively rarely, producing normally unsatisfactory results (see Caldwell and Demetski 1980; Daor 1981); the few successful examples have considered only part of the trips, for example trips made by car (see Ashley 1978). On the other hand, Supernak (1979, 1981) has reported the successful transferability of the personal-category trip generation model, both for Polish and American conditions.

More recently, Rose and Koppelman (1984) have examined the transferability of a discrete choice trip generation model, allowing for adjustment of modal constants using local data. One of their conclusions is that context similarity appeared to be an important determinant of model transferability; also, because their results showed considerable variability, they caution that great care must be taken in order to ensure that the transferred model is usable in the new context.

4.5.3 Bayesian Updating of Trip Generation Parameters

Assume we want to estimate a trip generation model but lack funds to collect appropriate survey data; a possible (but inadequate) solution is to use a model estimated for another (hopefully similar) area directly. However, it would be highly desirable to modify it in order to reflect local conditions more accurately.

This can be done by means of Bayesian techniques for updating the original model parameters using information from a small sample in the application context. Bayesian updating considers a *prior* distribution (i.e. that of the original parameters to be updated), new information (i.e. to be obtained from the small sample) and a *posterior* distribution corresponding to the updated model parameters for the new context. Updating techniques are very important in a continuous planning framework; we will see this theme appearing at various part of this book.

Consider, for example, the problem of updating trip rates by household categories; following Mahmassani and Sinha (1981) we will employ the notation in Table 4.11.

Table 4.11 Bayesian updating notation for trip generation

Variable	Prior information	New information
Mean trip rate	t_1	t_s
No. of observations	n_1	n_s
Trip rate variance	S_1^2	S_s^2

The mean trip rate of a category (or cell), is of course the average of a sample of household trip rates. According to the Central Limit Theorem, if the number of observations in a cell is at least 30, the sample distribution of the cell (mean) trip rates may be considered distributed Normal independently of the distribution of the household trip rates. Therefore, the prior distribution of the cell trip rates for the original model is $N(t_1, S_1^2/n_1)$, because t_1 and S_1^2/n_1 are unbiased estimators of its mean and variance. Similarly, the cells for the small sample (new information) may be considered distributed Normal with parameters t_s and S_s^2/n_s.

Bayes' theorem states that if the prior and sample distributions are Normal with known variances σ^2, then the posterior (updated) distribution of the mean trip rates is also Normal with the following parameters:

$$t_2 = \frac{1/\sigma_1^2}{1/\sigma_1^2 + 1/\sigma_s^2} t_1 + \frac{1/\sigma_s^2}{1/\sigma_1^2 + 1/\sigma_s^2} t_s \qquad (4.20)$$

$$\sigma_2^2 = \frac{1}{1/\sigma_1^2 + 1/\sigma_s^2} \qquad (4.21)$$

which, substituting by the known values S^2 and n, yield:

$$t_2 = \frac{n_1 S_s^2 t_1 + n_s S_1^2 t_s}{n_1 S_s^2 + n_s S_1^2} \qquad (4.22)$$

$$\sigma_2^2 = \frac{S_1^2 S_s^2}{n_1 S_s^2 + n_s S_1^2} \qquad (4.23)$$

It is important to emphasise that this distribution is not that of the individual trip rates of each household in the corresponding cell, but that of the mean of the trip rates of the cell. In fact the distribution of the individual rates is not known; the only information we have is that they share the same (posterior) mean t_2.

Example 4.8: The mean trip rate, its variance and the number of observations for two household categories, obtained in a study undertaken 10 years ago are shown below:

	Household categories	
Variable (prior data)	1	2
Trips per day	8	5
No. of observations	65	300
Trip rate variance	64	15
Mean trip variance	0.98	0.05

It is felt that these values might be slightly out of date for direct use today, but there are not enough funds to embark on a full-scale survey. A small stratified sample is finally taken, which yields the values shown below:

	Household categories	
Variable (new data)	1	2
Trips per day	12	6
No. of observations	30	30
Trip rate variance	144	36
Mean trip variance	4.80	1.20

The reader can check that by applying equations (4.22) and (4.23) it is possible to estimate the following trip rate values and variances:

	Household categories	
Posterior	1	2
Trip rate (trips/day)	8.68	5.04
Variance	0.82	0.05

4.6 INELASTICITY OF TRIP GENERATION

As we mentioned in Chapter 1, the classical specification of the urban transport planning (four-stage) model incorporates an iterative process between trip distribution and assignment which leaves trip generation unaltered. This is true even in the case of modern forms which attempt to solve the complex supply–demand equilibration problem appropriately, as we will discuss in Chapter 11. A major disadvantage of this approach is that changes to the network are assumed to have no effects on trip productions and attractions. For example, this would mean that the extension of an underground line to a location which had no service previously would not generate more trips between that zone and the rest. Although this assumption may hold for compulsory trips, it may not hold in the case of discretionary trips (e.g. consider the case of shopping trips and a new line connecting a low-income zone with the city's central market, which features more competitive prices than the zone's local shops).

To solve this problem, modellers have attempted to incorporate a measure of accessibility (i.e. ease or difficulty of making trips to/from each zone) into trip generation equations; the aim is to replace $O_i^n = f(H_i^n)$ by $O_i^n = f(H_i^n, A_i^n)$, where H_i^n are household characteristics and A_i^n is a measure of accessibility by person type.

Typical accessibility measures take the general form:

$$A_i^n = \sum_j f(E_j^n, C_{ij})$$

where E_j^n is a measure of attraction of zone j and C_{ij} the generalised cost of travel between zones i and j. A typical analytical expression used to this end has been:

$$A_i^n = \sum_j E_j^n \exp(-\beta C_{ij})$$

where β is a calibration parameter from the gravity model, as discussed in Chapter 5.

Unfortunately this procedure has not produced the expected results, at least in the case of aggregate modelling applications, because the estimated parameters of the accessibility variable have either been non-significant or with the wrong sign. This issue will remain highly topical for many years as it is clearly related to two interesting and unresolved problems: model dynamics and modelling with longitudinal rather than with cross-sectional data (recall the discussion in Chapter 1).

EXERCISES

4.1 Consider a zone with the following characteristics:

Household type	No.	Income ($/month)	Inhabitants	Trips/day
0 cars	180	4 000	4	6
1 car	80	18 000	4	8
2 or more cars	40	50 000	6	11

Due to a decrease in import duties and a real income increase of 30% it is expected that in five years time 50% of households without car would acquire one. Estimate how many trips would the zone generate in that case; check whether your method is truly the best available.

4.2 Consider the following trip attraction models estimated using a standard computing package (t-ratios are given in parentheses);

$$Y = 123.2 + 0.89X_1 \qquad\qquad R^2 = 0.900$$
$$ (5.2) \quad (7.3)$$

$$Y = 40.1 + 0.14X_2 + 0.61X_3 + 0.25X_4 \qquad R^2 = 0.925$$
$$ (6.4) \quad (1.9) \qquad (2.4) \qquad (1.8)$$

$$Y = -1.7 + 2.57X_1 - 1.78X_4 \qquad\qquad R^2 = 0.996$$
$$ (-0.6) \quad (9.9) \qquad (-9.3)$$

where Y are work trips attracted to the zone, X_1 is total employment in the zone, X_2 is industrial employment in the zone, X_3 is commercial employment in the zone and X_4 is service employment.

Choose the most appropriate model, explaining clearly why (i.e. considering all its pros and cons).

4.3 Consider the following two AM peak work trip generation models, estimated by household linear regression:

$$y = 0.50 + 2.0x_1 + 1.5x_2 \qquad\qquad R^2 = 0.589$$
$$ (2.5) \quad (6.9) \quad (5.6)$$

$$y = 0.01 + 2.3x_1 + 1.1Z_1 + 4.1Z_2 \qquad R^2 = 0.601$$
$$ (0.9) \quad (4.6) \quad (1.9) \qquad (3.4)$$

where y are household trips to work in the morning peak, x_1 is the number of workers in the household, x_2 is the number of cars in the household, Z_1 is a dummy variable which takes the value of 1 if the household has one car and Z_2 is a dummy which takes the value of 1 if the household has two or more cars.
(a) Choose one of the models explaining clearly the reasoning behind your decision.
(b) Graphically depict both models using appropriate axis.
(c) If a zone has 1000 households (with an average of two workers per household), of which 50% has no cars, 35% has only one car and the rest exactly two cars, estimate the total number of trips generated by the zone, O_i, with both models. Discuss your results.

4.4 The following table presents data collected in the last household O–D survey (made ten years ago) for three particular zones:

Zone	Residents/HH	Workers/HH	Mean Income	Population
I	2.0	1.0	50 000	20 000
II	3.0	2.0	70 000	60 000
III	2.5	2.0	100 000	100 000

Ten years ago two household-based trip generation models were estimated using this data. The first was a linear regression model given by:

$$y = 0.2 + 0.5x_1 + 1.1Z_1 \quad R^2 = 0.78$$

where y are household peak hour trips, x_1 is the number of workers in the household and Z_1 is a dummy variable which takes the value of 1 for high income ($> 70\,000$) households and 0 in other cases.

The second was a category analysis model based on two income strata (low and high income) and two levels of family structure (1 or less and 2 or more workers per household). The estimated trip rates are given in the following table:

Family structure	Income	
	Low	High
1 or less	0.8	1.0
2 or more	1.2	2.3

If the total number of trips generated today during the peak hour by the three zones are given by:

Zone	Peak hour trips
I	8 200
II	24 300
III	92 500

and it is estimated that the zone characteristics (income, number of households and family structure) have remained stable, decide which model is best. Explain your answer.

5 Trip Distribution Modelling

We have seen how trip generation models can be used to estimate the total number of trips emanating from a zone (generations) and those attracted to each zone (attractions). Generations and attractions provide an idea of the level of trip making in a study area but this is often not enough for modelling and decision making. What is needed is a better idea of the pattern of trip making, from where to where do trips take place, the modes of transport chosen and, as we shall see in Chapter 10, the routes taken.

A number of methods have been put forward over the years to distribute trips among destinations; some of the simplest are only suitable for short-term, tactical studies where no major changes in the accessibility provided by the network are envisaged. Others seem to respond better to changes in network cost and are therefore suggested for longer-term strategic studies or for tactical ones involving important changes in relative transport prices. The chapter starts by detailing additional definitions and notation used; these include the idea of generalised costs of travel. The next section introduces methods which respond only to relative growth rates at origins and destinations; these are suitable for short-term trend extrapolation. Section 5.3 discusses a family of synthetic models, the best known being the gravity model. Approaches to model generation, in particular the entropy-maximising formalism, are presented in section 5.4. An important aspect of the use of synthetic models is their calibration, that is the task of fixing their parameters so that the base-year travel pattern is well represented by the model; this is examined in section 5.5. Section 5.6 presents a variation on the gravity model calibration theme which enables more general forms for the model. Other synthetic models have also been proposed and the most important of them, the intervening-opportunities model, is explored in section 5.7. Finally, the chapter concludes with some practical issues in distribution modelling.

5.1 DEFINITIONS AND NOTATION

It is now customary to represent the trip pattern in a study area by means of a trip matrix. This is essentially a two-dimensional array of cells where rows and columns represent each of the z zones in the study area (including external zones), as shown in Table 5.1.

The cells of each row i contain the trips originating in that zone which have as destinations the zones in the corresponding columns. The main diagonal corresponds

Table 5.1   A general form of a two-dimensional trip matrix

Generations	Attractions					$\sum_j T_{ij}$
	1	2	3	$\ldots j$	$\ldots z$	
1	T_{11}	T_{12}	T_{13}	$\ldots T_{1j}$	$\ldots T_{1z}$	O_1
2	T_{21}	T_{22}	T_{23}	$\ldots T_{2j}$	$\ldots T_{2z}$	O_2
3	T_{31}	T_{32}	T_{33}	$\ldots T_{3j}$	$\ldots T_{3z}$	O_3
$\vdots$						
i	T_{i1}	T_{i2}	T_{i3}	$\ldots T_{ij}$	$\ldots T_{iz}$	O_i
$\vdots$						
z	T_{z1}	T_{z2}	T_{z3}	$\ldots T_{zj}$	$\ldots T_{zz}$	O_z
$\sum_i T_{ij}$	D_1	D_2	D_3	$\ldots D_j$	$\ldots D_z$	$\sum_{ij} T_{ij} = T$

to intra-zonal trips. Therefore: T_{ij} is the number of trips between origin i and destination j; the total array is $\{T_{ij}\}$ or $\mathbf{T}$; O_i is the total number of trips originating in zone i, and D_j is the total number of trips atttracted to zone j.

We shall use lower case letters, t_{ij}, o_i and d_j to indicate observations from a sample from an earlier study; capital letters will represent our target, or the values we are trying to model for the corresponding modelling period.

The matrices can be further disaggregated, for example, by person type (n) and/or by mode (k). Therefore:

T_{ij}^{kn} are trips from i to j by mode k and person type n;

O_i^{kn} is the total number of trips originating at zone i by mode k and person type n, and so on.

Summation over sub- or superscripts will be indicated implicitly by omission, e.g.

$$T_{ij}^n = \sum_k T_{ij}^{kn}$$

$$T = \sum_{ij} T_{ij} \quad \text{and} \quad t = \sum_{ij} t_{ij}$$

In some cases it may be of interest to distinguish the proportion of trips using a particular mode and the cost of travelling between two points:

p_{ij}^k is the proportion of trips from i to j by mode k;

c_{ij}^k is the cost of travelling between i and j by mode k.

The sum of the trips in a row should equal the total number of trips emanating from that zone; the sum of the trips in a column should correspond to the number of trips attracted to that zone. These conditions can be written as:

$$\sum_j T_{ij} = O_i \qquad\qquad\qquad (5.1a)$$

$$\sum_i T_{ij} = D_j \qquad\qquad\qquad (5.1b)$$

If reliable information is available to estimate both O_i and D_j then the model must satisfy both conditions; in this case the model is said to be doubly constrained. In some cases there will be information only about one of these constraints, for example to estimate all the O_i's, and therefore the model will be said to be singly constrained. Thus a model can be origin or production constrained if the O_i's, are available, or destination or attraction constrained if the D_j's are at hand.

The cost element may be considered in terms of distance, time or money units. It is often convenient to use a measure combining all the main attributes related to the disutility of a journey and this is normally referred to as the *generalised cost of travel*. This is typically a linear function of the attributes of the journey weighted by coefficients which attempt to represent their relative importance as perceived by the traveller. One possible representation of this for mode k is (omitting superscript k for simplicity):

$$C_{ij} = a_1 t_{ij}^v + a_2 t_{ij}^w + a_3 t_{ij}^t + a_4 t_{nij} + a_5 F_{ij} + a_6 \phi_j + \delta \qquad (5.2)$$

where

t_{ij}^v is the in-vehicle travel time between i and j;
t_{ij}^w is the walking time to and from stops (stations);
t_{ij}^t is the waiting time at stops;
t_{nij} is the interchange time, if any;
F_{ij} is the fair charged to travel between i and j;
ϕ_j is a terminal (typically parking) cost associated with the journey from i to j;
δ is a *modal penalty*, a parameter representing all other attributes not included in the generalised measure so far, e.g. safety, comfort and convenience;
$a_{1\ldots6}$ are weights attached to each element of cost; they have dimensions appropriate for conversion of all attributes to common units, e.g. money or time.

If the generalised cost is measured in money units then a_1 is sometimes interpreted as the *value of time* (or more precisely the *value of in-vehicle time*) as its units are money/time. In that case, a_2 and a_3 would be the values of walking and waiting time respectively, and in many practical studies they have been taken to be two or three times the expected value of a_1. If the generalised cost is measured in money units a_5 is normally fixed to one.

The generalised cost of travel, as expressed here, represents an interesting compromise between subjective and objective disutility of movement. It is meant to represent the disutility of travel as perceived by the trip maker; in that sense the value of time should be a perceived value rather than an objective, resource-based, value. However, the coefficients $a_{1\ldots6}$ used are often provided externally to the modelling process, usually specified by government. This presumes stability and transferability of values for which there is, so far, only limited evidence.

As generalised costs may be measured in money or time units it is relatively easy to convert one into the other. For example, if the generalised cost is measured in time units, a_1 would be 1.0, $a_{2...3}$ would probably be between 2.0 and 3.0, and $a_{5...6}$ would represent something like the 'duration of money'.

There are some theoretical and practical advantages in measuring generalised cost in time units. Consider, for example, the effect of income levels increasing with time; this would increase the *value of time* and therefore increase generalised costs and apparently make the same destination more expensive. If, on the other hand, generalised costs are measured in time units, increased income levels would appear to reduce the cost of reaching the same destination, and this seems intuitively more acceptable.

A distribution model tries to estimate the number of trips in each of the matrix cells on the basis of any information available. Different distribution models have been proposed for different sets of problems and conditions. We shall explore first models which are mainly useful in updating a trip matrix, or in forecasting a future trip matrix, where information is only available in terms of future trip rates or growth factors. We shall then study more general models, in particular the gravity model family. We shall finally explore the possibility of developing modal-split models from similar principles.

5.2 GROWTH-FACTOR METHODS

Let us consider first a situation where we have a basic trip matrix **t**, perhaps obtained from a previous study or estimated from recent survey data. We would like to estimate the matrix corresponding to the design year, say 10 years into the future. We may have information about the growth rate to be expected in this 10-year period for the whole study area; alternatively, we may have information on the likely growth in the number of trips originating and/or attracted to each zone. Depending on this information we may be able to use different growth-factor methods in our estimation of future trip patterns.

5.2.1 Uniform Growth Factor

If the only information available is about a general growth rate τ for the whole of the study area, then we can only assume that it will apply to each cell in the matrix:

$$T_{ij} = \tau t_{ij} \quad \text{for each pair } i \text{ and } j \tag{5.3}$$

Of course $\tau = T/t$, i.e. the ratio of expanded over previous total number of trips.

Example 5.1: Consider the simple four-by-four base-year trip matrix of Table 5.2. If the growth in traffic in the study area is expected to be of 20% in the next three years, it is a simple matter to multiply all cell values by 1.2 to obtain a new matrix as in Table 5.3.

The assumption of uniform growth is generally unrealistic except perhaps for very short time spans of, say, one or two years. In most other cases one would expect differential growth for different parts of the study area.

Table 5.2   Base-year trip matrix

	1	2	3	4	$\sum_j$
1	5	50	100	200	355
2	50	5	100	300	455
3	50	100	5	100	255
4	100	200	250	20	570
$\sum_i$	205	355	455	620	1635

Table 5.3   Future estimated trip matrix with $\tau = 1.2$

	1	2	3	4	$\sum_j$
1	6	60	120	240	426
2	60	6	120	360	546
3	60	120	6	120	306
4	120	240	300	24	684
$\sum_i$	246	426	546	744	1962

5.2.2 Singly Constrained Growth-Factor Methods

Consider the situation where information is available on the expected growth in trips originating in each zone, for example shopping trips. In this case it would be possible to apply this origin-specific growth factor (τ_i) to the corresponding rows in the trip matrix. The same approach can be followed if the information is available for trips attracted to each zone; in this case the destination-specific growth factors (τ_j) would be applied to the corresponding columns. This can be written as:

$$T_{ij} = \tau_i t_{ij} \quad \text{for origin-specific factors} \qquad (5.4)$$

$$T_{ij} = \tau_j t_{ij} \quad \text{for destination-specific factors} \qquad (5.5)$$

Example 5.2: Consider the following revised version of Table 5.2 with growth predicted for origins:

Table 5.4   Origin-constrained growth trip table

	1	2	3	4	$\sum_j$	Target O_i
1	5	50	100	200	355	400
2	50	5	100	300	455	460
3	50	100	5	100	255	400
4	100	200	250	20	570	702
$\sum_i$	205	355	455	620	1635	1962

This problem can be solved immediately by multiplying each row by the ratio of target O_i over the base year total ($\sum_j$), thus giving the results in Table 5.5.

Table 5.5 Expanded origin-constrained growth trip table

	1	2	3	4	$\sum_j$	Target O_i
1	5.6	56.3	112.7	225.4	400	400
2	50.5	5.1	101.1	303.3	460	460
3	78.4	156.9	7.8	156.9	400	400
4	123.2	246.3	307.9	24.6	702	702
$\sum_i$	257.7	464.6	529.5	701.2	1962	1962

5.2.3 Doubly Constrained Growth Factors

An interesting problem is generated when information is available on the future number of trips originating and terminating in each zone. This implies different growth rates for trips in and out of each zone and consequently having two sets of growth factors for each zone, say τ_i and Γ_j. The application of an 'average' growth factor, say $F_{ij} = 0.5(\tau_i + \Gamma_j)$ is only a poor compromise as none of the two targets or trip-end constraints will be satisfied. Historically a number of iterative methods have been proposed to obtain an estimated trip matrix which satisfies both sets of trip-end constraints, or the two sets of growth factors, which is the same thing.

All these methods involve calculating a set of intermediate correction coefficients which are then applied to cell entries in each row or column as appropriate. After applying these corrections to say, each row, the totals for each column are calculated and compared with the target values. If the differences are significant, new correction coefficients are calculated and applied as necessary.

The best known of these methods is due to Furness (1965), who introduced 'balancing factors' A_i and B_j as follows:

$$T_{ij} = t_{ij}\tau_i\Gamma_j A_i B_j \tag{5.6}$$

or incorporating the growth rates into new variables a_i and b_j:

$$T_{ij} = t_{ij}a_i b_j \tag{5.7}$$

with $a_i = \tau_i A_i$ and $b_j = \Gamma_j B_j$.

The factors a_i and b_j (or A_i and B_j) must be calculated so that the constraints (5.1) are satisfied. This is achieved in an iterative process which in outline is as follows:

1. set all $b_j = 1.0$ and solve for a_i; in this context, 'solve for a_i' means find the correction factors a_i that satisfy the trip generation constraints;

2. with the latest a_i solve for b_j, e.g. satisfy the trip attraction constraints;
3. keeping the b_j's fixed, solve for a_i and repeat steps (2) and (3) until the changes are sufficiently small.

This method produces solutions within 3 to 5% of the target values in a few iterations. There is not much point in enforcing the constraints to a level greater then the accuracy of the estimated trip end totals. This method is often called a bi-proportional algorithm because of the nature of the corrections involved. The problem is not restricted to transport; techniques to solve it have also been 'invented', among others, by Kruithof (1937) for telephone traffic and Bacharach (1970) for updating input-output matrices in economics. The best treatment of its mathematical properties seems to be due to Bregman; see Lamond and Stewart (1981).

It will be shown below that this method is a special case of entropy-maximising models of the gravity type, if the effect of distance or separation between zones is excluded. But in any case, the Furness method tries to produce the minimum corrections to the base-year matrix **t** necessary to satisfy the future year trip-end constraints.

The most important condition required for the convergence of this method is that the growth rates produce target values T_i and T_j such that

$$\sum_i \tau_i \sum_j t_{ij} = \sum_j \Gamma_j \sum_i t_{ij} = T \qquad (5.8)$$

Enforcing this condition may require correcting trip-end estimates produced by the trip generation models.

Example 5.3: The following table represents a doubly constrained growth factor problem:

Table 5.6   Doubly constrained matrix expansion problem

	1	2	3	4	$\sum_j$	Target O_i
1	5	50	100	200	355	400
2	50	5	100	300	455	460
3	50	100	5	100	255	400
4	100	200	250	20	570	702
$\sum_i$	205	355	455	620	1635	
Target D_j	260	400	500	·802		1962

The solution to this problem, after three iterations on rows and columns (three sets of corrections for all rows and three for all columns), can be shown to be:

Table 5.7   Solution to the doubly constrained matrix expansion problem

	1	2	3	4	$\sum_j$	Target O_i
1	5.25	44.12	98.24	254.25	401.85	400
2	45.30	3.81	84.78	329.11	462.99	460
3	77.04	129.50	7.21	186.58	400.34	400
4	132.41	222.57	309.77	32.07	696.82	702
$\sum_i$	260.00	400.00	500.00	802.00	1962	
Target D_j	260	400	500	802		1962

Note that this estimated matrix is within 1% of meeting the target trip ends, more than enough accuracy for this problem.

5.2.4 Advantages and Limitations of Growth-Factor Methods

Growth-factor methods are simple to understand and make direct use of observed trip matrices and forecasts of trip-end growth. They preserve the observations as much as is consistent with the information available on growth rates. This advantage is also their limitation as they are probably only reasonable for short-term planning horizons.

Growth-factor methods require the same database as synthetic methods, namely an observed (sampled) trip matrix; this is an expensive data item. The methods are heavily dependent on the accuracy of the base-year trip matrix. As we have seen, this is never very high for individual cell entries and therefore the resulting matrices are no more reliable than the sampled or observed ones. Any error in the base-year may well be amplified by the application of successive correction factors. Moreover, if parts of the base-year matrix are unobserved, they will remain so in the forecasts. Therefore, these methods cannot be used to fill in unobserved cells of partially observed trip matrices.

Another limitation is that the methods do not take into account changes in transport costs due to improvements (or new congestion) in the network. Therefore they are of limited use in the analysis of policy options involving new modes, new links, pricing policies and new zones.

5.3 SYNTHETIC OR GRAVITY MODELS

5.3.1 The Gravity Distribution Model

Distribution models of a different kind have been developed to assist in forecasting future trip patterns when important changes in the network take place. They start from assumptions about group trip making behaviour and the way this is influenced by external factors such as total trip ends and distance travelled. The best known of these models is the gravity model, originally generated from an analogy with Newton's gravitational law. They estimate trips for each cell in the matrix without directly using the observed trip pattern; therefore they are sometimes called synthetic as opposed to growth-factor models.

Probably the first rigorous use of a gravity model was by Casey (1955), who suggested such an approach to synthesise shopping trips and catchment areas between towns in a region. In its simplest formulation the model has the following functional form:

$$T_{ij} = \frac{\alpha P_i P_j}{d_{ij}^2} \tag{5.9}$$

where P_i and P_j are the populations of the towns of origin and destination, d_{ij} is the distance between i and j, and α is a proportionality factor.

This was soon considered to be too simplistic an analogy with the gravitational law and early improvements included the use of total trip ends (O_i and D_j) instead of total populations, and a parameter n for calibration as the power for d_{ij}. This new parameter was not restricted to being an integer and different studies estimated values between 0.6 and 3.5.

The model was further generalised by assuming that the effect of distance or 'separation' could be modelled better by a decreasing function, to be specified, of the distance or travel cost between the zones. This can be written as:

$$T_{ij} = \alpha O_i D_j f(c_{ij}) \tag{5.10}$$

where $f(c_{ij})$ is a generalised function of the travel costs with one or more parameters for calibration. This function often receives the name of 'deterrence function' because it represents the disincentive to travel as distance (time) or cost increases. Popular versions for this function are:

$$f(c_{ij}) = \exp\left(-\beta c_{ij}\right) \qquad \text{exponential function} \tag{5.11}$$

$$f(c_{ij}) = c_{ij}^{-n} \qquad \text{power function} \tag{5.12}$$

$$f(c_{ij}) = c_{ij}^{n} \exp\left(-\beta c_{ij}\right) \qquad \text{combined function} \tag{5.13}$$

The general form of these functions for different values of their parameters is shown in Figure 5.1.

5.3.2 Singly and Doubly Constrained Models

The need to ensure that the restrictions (5.1) are met requires replacing the single proportionality factor α by two sets of balancing factors A_i and B_j as in the Furness model, yielding:

$$T_{ij} = A_i O_i B_j D_j f(c_{ij}) \tag{5.14}$$

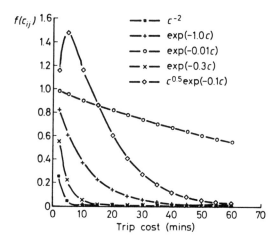

Figure 5.1   Different deterrence functions

In a similar vein one can again subsume O_i and D_j into these factors and rewrite the model as:

$$T_{ij} = a_i b_j f(c_{ij}) \tag{5.15}$$

The expression in (5.14) or (5.15) is the classical version of the doubly constrained gravity model. Singly constrained versions, either origin or destination constrained, can be produced by making one set of balancing factors A_i or B_j equal to one. For an origin-constrained model, $B_j = 1.0$ for all j, and

$$A_i = 1/\sum_j D_j f(c_{ij}) \tag{5.16}$$

In the case of the doubly constrained model the values of the balancing factors are:

$$A_i = 1/\sum_j B_j D_j f(c_{ij}) \tag{5.17}$$

$$B_j = 1/\sum_i A_i O_i f(c_{ij}) \tag{5.18}$$

The balancing factors are, therefore, interdependent; this means that the calculation of one set requires the values of the other set. This suggests an iterative process analogous to Furness's which works well in practice: given set of values for the deterrence function $f(c_{ij})$, start with all $B_j = 1$, solve for A_i and then use these values to reestimate the B_j's; repeat until convergence is achieved.

A more general version of the deterrence function accepts empirical values for it which depend only on the generalised cost of travel. To this end, travel costs are aggregated into a small number (say 10 or 15) of cost ranges or cost bins, indicated by a superscript m. The deterrence function then becomes:

$$f(c_{ij}) = \sum_m F^m \delta_{ij}^m \tag{5.19}$$

where F^m is the mean value for cost bin m, and δ_{ij}^m is equal to 1 if the cost of travelling between i and j falls in the range m, and equal to 0 otherwise.

The formulations (5.11) and (5.12) have one parameter for calibration; formulation (5.13) has two, β and n, and formulation (5.19) has as many parameters as cost bins. These parameters are estimated so that the results from the model reproduce, as closely as possible, the trip length (cost) distribution (TLD) of the observations. A theoretical reason for this requirement is offered below, but meanwhile it is enough to note that the greater the number of parameters, the easier it is to obtain a closer fit with the sampled trip length distribution.

It has been observed, in particular in urban areas, that in the case of motorised trips, the trip length distribution has a shape of the form depicted in Figure 5.2. This shows that there are few short motorised trips, followed by a larger number of medium-length trips; as distance (cost) increases, the number of trips decays again with a few very long trips. The negative exponential and power functions reproduce reasonably well the second part of the curve but not the first. That is one of the reasons behind the combined formulation which is more likely to fit better both parts of the TLD. The greater flexibility of the cost-bin formulation permits an even better fit. However, the approach requires the assumption that the same TLD will be maintained in the future; this is similar to requiring β to be the same for the base and the forecasting years.

It is interesting to note that the bulk of the representational and policy relevance advantages of the gravity model lies in the deterrence function; the rest is very much like the Furness method.

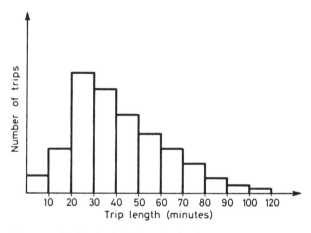

Figure 5.2   Typical trip length distribution in urban areas

5.4 THE ENTROPY-MAXIMISING APPROACH

5.4.1 Entropy and Model Generation

We shall introduce here the entropy-maximisation approach which has been used in the generation of a wide range of models, including the gravity model, shopping models and location models. The approach has a number of followers and detractors but it is generally acknowledged as one of the important contributions to improved modelling in transport. There are several ways of presenting the approach; we have chosen an intuitive rather than strictly mathematical formulation. For a stricter presentation and references to related and alternative approaches, see Wilson (1974).

Consider a system made up of a large number of distinct elements. A full description of such a system requires the complete specification of its *micro* states, as each is distinct and separable. This would involve, for example, identifying each individual traveller, its origin, destination, mode, time of journey, and so on. However, for many practical purposes it may be sufficient to work on the basis of a more aggregate or *meso*-state specification; following our example, a meso state may just specify the **number** of trips between each origin and each destination. In general, there will be numerous and different micro states which produce the same meso state: John Smith and Pedro Pérez living in the same zone, may exchange destinations, generating different micro states but keeping the same meso state.

There is always an even higher level of aggregation, a *macro* state, for example the total number of trips on particular links, or the total trips generated and attracted to each zone. To obtain reliable measures of trip making activity it is often easier to make observations at this higher level of aggregation. In fact, most of our current information about a system is precisely at this level. In a similar way, estimates about the future are usually restricted to macro-state descriptions because of the uncertainties involved in forecasting at more disaggregate levels: for example, it is easier to forecast the population per zone than the number of households in a particular category residing in each zone.

The basis of the method is to accept that, unless we have information to the contrary, all micro states consistent with our information about macro states are equally likely to occur. This is in fact a sensible assumption given our ignorance about meso and micro states. A good way of enforcing consistency with our knowledge about macro states is to express our information as equality constraints in a mathematical programme. As we are interested in the meso-state descriptions of the system, we would like to identify those meso states which are most likely, given our constraints about the macro states.

It is possible to show, see Wilson (1970), that the number of micro states $W\{T_{ij}\}$ associated with the meso state T_{ij} is given by:

$$W\{T_{ij}\} = \frac{T!}{\prod_{ij} T_{ij}!} \qquad (5.20)$$

As it is assumed that all micro states are equally likely, the most probable meso state would be the one that can be generated in a greater number of ways.

Therefore, what is needed is a technique to identify the values of $\{T_{ij}\}$ which maximise W in (5.20). For convenience we seek to maximise a monotonic function of W, namely $\log W$, as both problems have the same maximum. Therefore:

$$\log W = \log \frac{T!}{\prod_{ij} T_{ij}!} = \log T! - \sum_{ij} \log T_{ij}! \tag{5.21}$$

Stirling's (short) approximation for $\log X! = X \log X - X$, can be used to make it easier to optimise:

$$\log W = \log T! - \sum_{ij}(T_{ij} \log T_{ij} - T_{ij}) \tag{5.22}$$

Usually the term $\log T!$ is a constant, therefore it can be omitted from the optimisation problem. The rest of the equation is often referred to as the *entropy function*:

$$\log W' = -\sum_{ij}(T_{ij} \log T_{ij} - T_{ij}) \tag{5.23}$$

Maximising $\log W'$, subject to constraints corresponding to our knowledge about the macro states, enables us to generate models to estimate the most likely meso states, in our case the most likely matrix **T**. The key to this model generation method is, therefore, the identification of suitable micro-, meso- and macro-state descriptions, together with the macro-level constraints that must be met by the solution to the optimisation problem.

In some cases, there may be additional information in the form of prior or old values for the meso states, for example an outdated trip matrix **t**. The problem may be recast with this information and the revised objective function becomes:

$$\log W'' = -\sum_{ij}(T_{ij} \log T_{ij}/t_{ij} - T_{ij} + t_{ij}) \tag{5.24}$$

This is an interesting function in which each element in the summation takes the value zero if $T_{ij} = t_{ij}$ and otherwise is a positive value which increases with the difference between **T** and **t**. Therefore $- \log W''$ is a good measure of the difference between **T** and **t**; it can further be shown that

$$- \log W'' \approx 0.5 \sum_{ij} \frac{(T_{ij} - t_{ij})^2}{t_{ij}} \tag{5.25}$$

where the right-hand side is another good measure of the difference between prior and estimated meso states. Models can be generated minimising $- \log W''$ subject to constraints reflecting our knowledge about macro states. The resulting model is the one with the meso states closest (in the sense of equation (5.24) or approximately (5.25)) to the prior meso states and which satisfies the macro-state constraints.

5.4.2 Generation of the Gravity Model

Consider the definition of micro, meso and macro states from the discussion above. The problem becomes the maximisation of $\log W'$ subject to the following two sets of constraints corresponding to the meso states:

$$O_i - \sum_j T_{ij} = 0 \qquad (5.26a)$$

$$D_j - \sum_i T_{ij} = 0 \qquad (5.26b)$$

These two sets of constraints reflect our knowledge about trip generations and attractions in the zones of the study area. We are only interested in matrix entries that can be interpreted as trips, therefore we need to introduce the additional constraint that:

$$T_{ij} \geq 0 \qquad (5.27)$$

The constrianed maximisation problem can be handled forming the Lagrangian:

$$L = \log W' + \sum_i \alpha_i' \left\{ O_i - \sum_i T_{ij} \right\} + \sum_j \alpha_j'' \left\{ D_j - \sum_i T_{ij} \right\} \qquad (5.28)$$

Taking the first partial derivatives with respect to T_{ij} and equating them to zero we obtain:

$$\frac{\partial L}{\partial T_{ij}} = -\log T_{ij} - \alpha_i' - \alpha_j'' = 0 \qquad (5.29)$$

therefore

$$T_{ij} = \exp(-\alpha_i' - \alpha_j'') = \exp(-\alpha_i')\exp(-\alpha_j'')$$

The values of the Lagrange multipliers are easy to find; making a simple change of variables:

$$A_i' O_i = \exp(-\alpha_i') \quad \text{and} \quad B_j D_j = \exp(-\alpha_j'')$$

we obtain

$$T_{ij} = A_i O_i B_j D_j \qquad (5.30)$$

On the other hand, the use of $-\log W''$ as an objective function generates the model:

$$T_{ij} = A_i O_i B_j O_j t_{ij} \qquad (5.31)$$

which is, of course, the basic Furness model. The version resulting in equation (5.30) corresponds to the case when there is no prior information, e.g. all $t_{ij} = 1$. These two models are close to but not yet the gravity model. What is missing is the deterrence function term. Its introduction requires an additional constraint:

$$\sum_{ij} T_{ij}c_{ij} = C$$

where C is the (unknown) total expenditure in travel in the system (in generalised cost units if they are in use). Restating this constraint as

$$C - \sum_{ij} T_{ij}c_{ij} = 0 \tag{5.32}$$

one can maximise $\log W'$ subject to (5.26), (5.27) and (5.32), and using the same constrained optimisation technique it is possible to obtain the Lagrangian:

$$L = \log W' + \sum_i \alpha_i' \left\{ O_i - \sum_j T_{ij} \right\} + \sum_j \alpha_j'' \left\{ D_j - \sum_i T_{ij} \right\} + \beta \left\{ C - \sum_{ij} T_{ij}c_{ij} \right\} \tag{5.33}$$

Again, taking its first partial derivatives with respect to T_{ij} and equating them to zero gives

$$\frac{\partial L}{\partial T_{ij}} = -\log T_{ij} - \alpha_i' - \alpha_j'' - \beta c_{ij} = 0 \tag{5.34}$$

therefore

$$T_{ij} = \exp\left(-\alpha_i' - \alpha_j'' - \beta c_{ij}\right) = \exp\left(-\alpha_i'\right)\exp\left(-\alpha_j''\right)\exp\left(-\beta c_{ij}\right) \tag{5.35}$$

Making the same change of variables as before one obtains:

$$T_{ij} = A_i O_i B_j D_j \exp\left(-\beta c_{ij}\right) \tag{5.36}$$

which is the classic gravity model. The values for the balancing factors can be derived from the constraints as:

$$A_i = 1 \Big/ \left[\sum_i B_j D_j \exp\left(-\beta c_{ij}\right) \right] \quad \text{and} \quad B_j = 1 \Big/ \left[\sum_i A_i O_i \exp\left(-\beta c_{ij}\right) \right]$$

If one of (5.26a) or (5.26b) are omitted from the constraints a singly constrained gravity model is obtained.

The value of β is related to the satisfaction of condition (5.32). In general C is not known and therefore β is left as a parameter for calibration in order to adjust the model to each specific area. Values of β cannot, therefore, be easily borrowed from one place to another. A useful first estimate for the value of β is one over the average travel cost; in effect, β is precisely measured in inverse of travel cost units.

The use of a different cost constraint, such as (5.37) instead of (5.32),

$$C' - \sum_{ij} T_{ij} \log c_{ij} = 0 \tag{5.37}$$

results in a model of the form

$$T_{ij} = A_i O_i B_j D_j \exp\left(-\beta' \log c_{ij}\right) = A_i O_i B_j D_j c_{ij}^{-\beta'} \tag{5.38}$$

i.e. the gravity model with an inverse power deterrence function!

The reader can verify that the use of constraints (5.32) and (5.37) leads to a gravity model with a combined deterrence function. A further interesting approach is to disaggregate constraint (5.32) into several trip cost groups or bins indicated, as before, by a superscript m:

$$C^m - \sum_{ij} T_{ij} \delta_{ij}^m = 0 \text{ for each } m \tag{5.39}$$

The maximisation of (5.23) subject to (5.26), (5.27) and (5.39) leads to:

$$T_{ij} = A_i O_i B_j D_j \sum_m F^m \delta_{ij}^m = a_i b_j \sum_m F^m \delta_{ij}^m \tag{5.40}$$

which is, of course, the gravity model with a cost-bin deterrence function. This model has some attractive properties, which will be discussed in section 5.6.

5.4.3 Properties of the Gravity Model

As can be seen, entropy maximisation is quite a flexible approach for model generation. A whole family of distribution models can be generated by casting the problem in a mathematical programming framework: the maximisation of an entropy function subject to linear constraints representing our level of knowledge about the system. The use of this formalism has many advantages:

1. It provides a more rigorous way of specifying the mathematical properties of the resulting model. For example, it can be shown that objective function is always convex; it can be shown also that, provided the constraints used (say (5.26) and (5.27)) have a feasible solution space, the optimisation problem has a unique solution even if the set of parameters A_i and B_j is not unique (one is redundant).
2. The use of a mathematical programming framework also facilitates the use of a standard tool-kit of solution methods and the analysis of the efficiency of alternative algorithms.
3. The theoretical framework used to generate the model also assists in providing an improved interpretation of the solutions generated by it. We have seen that the gravity model can be generated from analogies with the physical world or from entropy-maximising considerations; the latter are closely related to information

theory, to error measures and to maximum likelihood in statistics, and the three provide alternative ways of generating the same mathematical form of the gravity model. Although the functional form is the same, each theoretical framework provides a different interpretation to the problem and the solution found. Each may be more appropriate in specific circumstances. We shall come back to this *equifinality issue* in Chapter 8.

4. The fact that the gravity model can be generated in a number of different ways does not make it 'correct'. The appropriateness of the model depends on the acceptability of the assumptions required for its generation and their interpretation. No model is ever appropriate or correct in itself, it can only be more or less suitable to handle a decision question given our understanding of the problem, of the options or schemes to be tested, the information available or collectable at a justifiable cost, and the time and resources securable for analysis; see the discussion on calibration and validation below.

It is interesting to contrast the classical gravity model as in equation (5.36) with Furness's method as derived above in equation (5.31). We can see that one possible interpretation of the deterrence function is to provide a synthetic set of prior entries for each cell in the trip matrix (i.e. use of $\exp(-\beta c_{ij})$ instead of t_{ij}). Both the deterrence function and the prior matrix t_{ij} take the role of providing 'structure' to the resulting trip matrix. This can be seen more clearly if one multiplies and divides the right-hand side of equation (5.31) by T and subsumes this constant in the balancing factors:

$$T_{ij} = Ta_ib_jt_{ij}/T = a'_ib'_jp_{ij} \tag{5.41}$$

where $p_{ij} = t_{ij}/T$, thus giving a better-defined meaning to 'structure' as the proportion of the total trips allocated to each origin-destination pair.

Example 5.4: It is useful to illustrate the gravity model with an example related to the problem of expanding a trip matrix. Consider the cost matrix of Table 5.8 together with the total trip ends as in Table 5.6, and attempt to estimate the parameters a_i and b_j of a gravity model of the type:

$$T_{ij} = a_ib_j \exp(-\beta c_{ij})$$

Table 5.8   A cost matrix and trip-end totals for a gravity model estimation

	Cost matrix (minutes)				
	1	2	3	4	Target O_i
1	3	11	18	22	400
2	12	3	13	19	460
3	15.5	13	5	7	460
4	24	18	8	5	702
D_j	260	400	500	802	1962

Table 5.9   The matrix $\exp(-\beta c_{ij})$ and sums to prepare for a gravity model run

	$\exp(-\beta\ c_{ij})$				
	1	2	3	4	$\sum_j$
1	0.74	0.33	0.17	0.11	1.35
2	0.30	0.74	0.30	0.15	1.49
3	0.21	0.27	0.61	0.50	1.59
4	0.09	0.17	0.45	0.61	1.31
$\sum_j$	1.34	1.51	1.52	1.36	5.74

Base	1	2	3	4	$\sum_j$	Target	Ratio
1	253.12	113.73	56.48	37.86	461.19	400	0.87
2	102.91	253.12	102.91	51.10	510.04	460	0.90
3	72.52	93.12	207.23	169.67	542.54	400	0.74
4	31.00	56.48	153.52	207.23	448.23	702	1.57
$\sum_i$	459.54	516.45	520.15	465.87	1962.00		
Target	260	400	500	802			
Ratio	0.57	0.77	0.96	1.72			

Table 5.10   The resulting gravity model matrix with trip length distribution

	1	2	3	4	$\sum_i$	Target	Ratio	a_i
1	155.73	99.00	64.46	74.17	393.36	400	1.02	1.17
2	57.54	200.22	106.73	90.98	455.56	460	1.01	1.07
3	25.87	47.01	137.16	192.77	402.81	400	0.99	0.68
4	20.86	53.77	191.65	444.08	710.37	702	0.99	1.28
$\sum_j$	260.00	400.00	500.00	802.00	1962.00			
Target	260	400	500	802				
Ratio	1.00	1.00	1.00	1.00				
b_j	179.17	253.50	332.37	570.53				

Ranges (min)							
Cost	1.0–4.0	4.1–8.0	8.1–12.0	12.1–16.0	16.1–20.0	20.1–24	Sum
Trips	355.9	965.7	156.5	179.6	209.2	95.0	1962

given the information that the best value of β is 0.10. The first step would be to build a matrix of the values $\exp(-\beta c_{ij})$, as in Table 5.9.

Then with these values we can calculate the resulting total 'trips' (5.74) and then expand each cell in the matrix by the ratio $1962/5.74 = 341.67$. This produces a matrix of base trips which now has to be adjusted to match trip-end totals. This process is the same as Furness iterations. The values for a_i and b_j are the product of the corresponding correction factors; these factors will then be multiplied by the basic expansion factor 341.67. The resulting gravity model matrix is given in Table 5.10.

The reader may wish to verify that the balancing factors a_i and b_j are only unique to a multiplicative constant. It is also possible to calculate, as usual, the standard balancing factors A_i and B_j dividing each corresponding a_i and b_j by the target values O_i and D_j.

5.5 CALIBRATION OF GRAVITY MODELS

5.5.1 Calibration and Validation

Before using a gravity distribution model it is necessary to calibrate it; this just makes sure that its parameters are such that the model comes as close as possible to reproducing the base-year trip pattern. Calibration is, however, a very different process from validation of a model.

In the case of *calibration* one is conditioned by the functional form and the number of parameters of the chosen model. For example, the classical gravity model has the parameters A_i, B_j, and β (that is $Z + Z + 1$ parameters, Z being the number of zones). The parameters A_i and B_j are calibrated during the estimation of the gravity model, as part of the direct effort to satisfy constraints (5.1). Note that at least one of the A_i or B_j is redundant as there is an additional condition $\sum_i O_i = \sum_j D_j = T$, and therefore one of the (5.1) constraints is linearly dependent on the rest. The parameter β, on the other hand, must be calibrated independently, as we do not have complete information about the total expenditure C in the study area. If we had this information, we could have used it directly without having to estimate β by other means. If the combined deterrence function (5.13) is used, we would have an additional parameter and therefore some additional flexiblity in calibrating the gravity model.

The *validation* task is different. In this case one wants to make sure the model is appropriate for the decisions likely to be tested with it. It may be that the gravity model is not a sufficiently good representation of reality for the purpose of examining a particular set of decisions. It follows from this that the validation task depends on the nature of the policies and projects to be assessed.

A general strategy for validating a model would then be to check whether it can reproduce a known state of the system with sufficient accuracy. As the future is definitively not known, this task is sometimes attempted by trying to estimate some well-documented state in the past, say a matrix from an earlier study. However, it is seldom the case that such a past state is sufficiently well documented. Therefore, less demanding validation tests incorporating data not used during estimation are often employed, for example: to check whether the number of trips across important screenlines or along main roads are well reproduced.

5.5.2 Calibration Techniques

As we have seen, the parameters A_i and B_j are estimated as part of the Furness (bi-proportional) balancing factor operations. The parameter β must be calibrated to make sure that the trip length distribution (TLD) is reproduced as closely as possible. This is a tall order for a single parameter. We shall see later how to improve on this but meantime, what is needed is a practical technique to estimate the best value for β, say β^*.

A naive approach to this task is simply to 'guess' or to 'borrow' a value for β, run the gravity model and then extract the modelled trip length distribution (MTLD). This should be compared with the observed trip length distribution (OTLD). If they are not sufficiently close a new guess for β can be used and the process repeated until a satisfactory fit between MTLD and OTLD is achieved; this would then be taken as the best value β^*. Note that a set of home or roadside interviews will produce OTLDs with much greater accuracy than that of individual cell entries in the trip matrix, because the sampling rate for trip lengths is in effect much higher in this case.

The naive approach is not, however, very practical. Running a doubly constrained gravity model is time consuming and the approach provides no guidance on how to choose a better value for β if the current one is not satisfactory. Conventional curve-fitting techniques are unlikely to work well because the gravity model is not just non-linear but also complex analytically; the A_i's and B_j's are also functions of β through the two sets of equations (5.17) and (5.18).

A number of calibration techniques have been proposed and implemented in different software packages. The most important ones were compared by Williams (1976), who found that a technique due to Hyman (1969) was particularly robust and efficient. We shall describe briefly here Hyman's method.

At any stage in the calibration process a trip matrix $\mathbf{T}(\beta)$, function of the current estimate of β, is available. This matrix also defines a total number of trips $\sum_{ij} T_{ij}(\beta) = T(\beta)$. The method is based on the following requirement for β:

$$c(\beta) = \sum_{ij}[T_{ij}(\beta)c_{ij}]/T(\beta) = c^* = \sum_{ij}(N_{ij}C_{ij})/\sum_{ij}N_{ij} \qquad (5.42)$$

where c^* is the mean cost from the OTLD and N_{ij} is the observed (and expanded) number of trips for each origin destination pair. The method can be described as follows:

1. Start the first iteration making $m = 0$ and an initial estimate of $\beta_0 = 1/c^*$.
2. Make $m = m + 1$; with the current estimate of β_{m-1} calculate a trip matrix using the standard gravity model. Obtain the mean modelled trip cost c_m and compare

it with c^*; if they are sufficiently close stop and accept β_{m-1} as the best value for this parameter; otherwise go to step 3.

3. If $m = 1$, estimate a better value for β_m as:

$$\beta_1 = c_1\beta_0/c^*$$

or if $m > 1$, obtain a better estimate of β as:

$$\beta_{m+1} = \frac{(c^* - c_{m-1})\beta_{m-1} - (c^* - c_m)\beta_m}{c_m - c_{m-1}}$$

4. Repeat steps 2 and 3 as required, i.e. until convergence of the method.

The recalculations in step 3 are made to approximate closer to the equality in (5.42). A few improvements can be introduced to this method, in particular from the computational point of view. Hyman's approach has been shown to be robust and to offer, in general, advantages over alternative algorithms.

5.6 THE TRI-PROPORTIONAL APPROACH

5.6.1 Bi-proportional Fitting

We have seen in section 5.4.2 how Furness's method can be derived from a mathematical programming framework. This non-linear mathematical program can be solved by a number of algorithms, including Newton's method. However, it is possible to show that the method originally proposed by Furness is indeed a practical and efficient algorithm, in particular for large matrices. The method is often referred to as the bi-proportional algorithm as it involves successive corrections by rows and then columns to satisfy the constraints; the algorithm stops when the corrections are small enough, i.e. when the constraints are met within reasonable tolerances.

The conditions necessary for the existence of a unique solution are that constraints (5.26a) and (5.26b) define a feasible solution space in non-negative T_{ij}'s. This requires $\sum_i O_i = \sum_j D_j$ but this is not a sufficient condition. The model has a multiplicative form and therefore it preserves the zeros present in the prior matrix $\{t_{ij}\}$. The existence of many zero entries in the prior matrix may prevent the satisfaction of one or more constraints.

Example 5.5: Consider the case where a previously empty zone k is expected to see development in the future, thus generating and attracting trips. The cell entries for t_{ik} and t_{kj} would have been zero whilst the future O_k and D_k are non-zero. Therefore in this case there are no possible multiplicative correction factors capable of generating a matrix satisfying the constraints for zone k. It may be possible, however, to replace these empty cell values by 'guesses', i.e. suitable values borrowed from similar zones. Nevertheless, the presence of zeros in the prior matrix may cause subtler but no less difficult problems. If we try to solve the problem in Example 5.1 but with the prior matrix in Table 5.11, it will be found that this

Table 5.11 A revised version of the doubly constrained growth factor problem in Table 5.6

	1	2	3	4	$\sum_j$	Target O_i
1	5	50	100	200	355	400
2	0	50	0	0	50	460
3	50	100	5	100	255	400
4	100	200	250	20	570	702
$\sum_i$	155	400	355	320	1230	
Target D_j	260	400	500	802		1962

problem has no feasible solution in non-negative T_{ij}; there are only 11 unknowns and 7 independent constraints but the position of the zeros is such that there is no feasible solution and the bi-proportional algorithm oscillates without converging.

Readers familiar with linear algebra will be able to describe this problem in terms of the rank of the original and an augmented matrix containing the last column in Table 5.7. Furthermore, the reader may verify that after 10 iterations with this problem the corrected matrix stands as:

Table 5.12 The matrix from problem in Table 5.11 after 10 Furness iterations

	1	2	3	4	$\sum_j$	Target O_i
1	3.4	0.7	61.0	355.3	420	400
2	0	388.2	0	0	388	460
3	65.5	2.8	5.9	345.7	420	400
4	191.2	8.3	433.1	101.0	734	702
$\sum_i$	260	400	500	802	1962	
Target D_j	260	400	500	802		1962

Several comments can be made at this stage:

1. The matrix after 10 iterations looks quite different from the prior one, thus casting some doubt about the realism, either of the old matrix, its zeros or the new trip-end totals.
2. The main problem seems to be in the second row, where there is a big difference (about 20%) between target and modelled total. There is no way this row can add up to 460 as the only non-zero cell entry has a maximum of 400 trips. The constraints do not generate a feasible solution space.

3. The problem seems ill-conditioned, e.g. a small change in a cell entry can make the problem a feasible one and produce a fairly different trip matrix. For example, the zero in cell $t_{2,4}$ could have arisen because of the sample used; replacing this zero by a 1 produces the matrix in Table 5.13 after the same 10 iterations. This is a much improved match with a fairly different matrix. In fact, it matches the targets with better than 1% accuracy. There is now a feasible solution space.

Table 5.13 The matrix from problem in Table 5.11 plus a single trip in cell 2, 4 after 10 Furness iterations

	1	2	3	4	$\sum_j$	Target O_i
1	4.1	4.5	76.2	315.4	400	400
2	0	339.2	0	119.1	458	460
3	77.3	17.0	7.2	298.5	400	400
4	178.6	39.3	416.6	68.9	703	702
$\sum_i$	260	400	500	802	1962	
Target D_j	260	400	500	802		1962

Real matrices are often sparse and the occurrence of this type of difficulty cannot be discarded as an academic problem. Failure to converge in a few iterations may well indicate that the presence and location of zeros in the prior matrix prevents the existence of a feasible solution with the new trip ends.

5.6.2 A Tri-proportional Problem

We have already presented the gravity model with a very flexible deterrence function which takes discrete values that are not constrained by a functional form, for each cost bin. This was written in equation (5.40) as:

$$T_{ij} = a_i b_j \sum_m F^m \delta_{ij}^m$$

The main advantages of this model are its flexibility and the ease of calibration. In effect, we can define any number of cost bins and the deterrence function can take any positive value for them; we could even represent situations where, for example, there are few short trips, many intermediate trips, few long trips and again a larger number of long-distance commuting trips.

The calibration of this model requires finding suitable values for the deterrence factor F^m for each cost bin so that the number of trips undertaken for that distance is as close as possible to the observed number. This task is, in fact, very similar to the problem of grossing up a matrix to match trip-end totals. In this case we can start with a unity value for the deterrence factors and then correct these and the parameters a_i and b_j until the trip ends and the TLD constraints are met. It seems natural to extend the bi-proportional algorithm to handle this third dimension (cost bins) and utilise a tri-proportional method to calibrate the model.

The principles behind the technique were proposed by Evans and Kirby (1974). Murchland (1977) has shown that the application of successive corrections on a two-, three- or multi-dimensional space conforms to just one of a group of possible algorithms to solve this type of problems; furthermore, the method is simple to program and does not make excessive demands on computer memory.

Example 5.6: The tri-proportional algorithm can be illustrated with the problem stated in Table 5.8 and with the trip length distribution (cost-bin) targets of Table 5.14.

Table 5.14 TLD target values for a tri-proportional gravity model calibration

	Ranges					
	1.0–4.0	4.1–8.0	8.1–12.0	12.1–16.0	16.1–20.0	20.1–24+
TLD	365	962	160	150	230	95

The model can then be solved using balancing operations to match trip targets by origin, destination and cost bin. After five complete iterations, this results in the matrix and modelled trips by cost bin T_k shown in Table 5.15.

Table 5.15 The matrix from problem in Table 5.14 after five iterations, including values for balancing factors a_i, b_j and F^k

	1	2	3	4	$\sum_j$	a_i
1	161.6	102.5	60.8	72.5	397.4	1.27
2	56.5	199.4	101.2	101.0	458.0	1.13
3	18.9	48.7	116.7	217.1	401.4	0.60
4	23.0	49.5	221.3	411.5	705.3	1.14
$\sum_i$	260	400	500	802	1962	
b_j	0.57	0.70	0.87	1.63		

	Ranges					
	1.0–4.0	4.1–8.0	8.1–12.0	12.1–16.0	16.1–20.0	20.1–24+
TLD	365	962	160	150	230	95
T_k	360.9	966.5	159.0	149.8	230.3	95.5
F^k	224.55	220.13	87.54	102.05	54.66	34.90

Of course, in this case the balancing factors are again not unique, at least up to two arbitrary multiplicative constants. Another way of expressing this is to say the balancing factors have two *degrees of indetermination*, the two multiplicative constants. It is easy to see that if we multiply each a_i by a factor Γ and each b_j by another factor τ and then divide each F^k by $\Gamma\tau$, the modelled matrix will remain unchanged.

5.6.3 Partial Matrix Techniques

The tri-proportional calibration method has been used with a full trip length distribution, i.e. one that has an entry from observations in each cell. It would certainly be advantageous if one could calibrate a suitable gravity model without requiring a complete or full trip matrix. This is particularly important as we know that the cost of collecting data to obtain a complete trip matrix is rather high; furthermore, the accuracy of some of the cell entries is not very high and in calibration we actually use aggregations of the data, namely the TLD and the total trip ends O_i and D_j. Having explored the preferred methods for calibration, it should be clear that the possibility of calibrating gravity models with incomplete or partial matrix does actually exist. For example, we can calibrate a gravity model with exponential cost function just with the total trip ends and a good estimate of the average trip cost, c^*.

The calibration of a gravity model with general deterrence function using the tri-proportional method is even more attractive in this case, as we could use just roadside interviews on cordons and screenlines to obtain good TLDs and trip ends for some but not all the zones in the study area. There would be no need to use trip generation models except for forecasting purposes.

Example 5.7: The basic idea above can be described with the aid of a 3×3 matrix. Consider first a bi-proportional case where the full matrix-updating problem is to adjust a base-year matrix as follows:

				adjusted to		produces		
a	b	c	P			A	B	C
d	e	f	Q			D	E	F
g	h	i	R			G	H	I

S	T	U

In the case of a partial matrix, for example a survey such that entries a and h cannot be observed, we would adjust only to trip ends excluding the corresponding total:

adjusted to   produces

	b	c
d	e	f
g		i

$P{-}S$
Q
$R{-}H$

	B	C
D	E	F
G		I

$S{-}A$	$T{-}H$	U

To fill in the missing cells we could use a gravity model; in the case of this example, one without deterrence function:

$$T_{ij} = a_i b_j$$

The estimated values of a_i and b_j (using data from the observed cells) would then be used to fill in these cells.

An extension to the tri-proportional case is almost trivial. Kirby (1979) has shown that there are two basic conditions required for a valid application of this approach:

1. The gravity model must fit both the available data we have and the data that are not available, i.e. the model must be a good model for the two regions of the matrix: the observed and the unobserved.
2. The two regions of the matrix should not be separable, i.e. it should not be possible to split the matrix into two or more independent matrices, typically:

	Internal	External
Internal	××××××××××××	××××××××
	××××××××××××	××××××××
	××××××××××××	××××××××
	××××××××××××	××××××××
	××××××××××××	××××××××
	××××××××××××	××××××××
	××××××××××××	××××××××
External	××××××××××××	××××××××
	××××××××××××	××××××××
	××××××××××××	××××××××

The problem is that each separate area has the two (or three in the tri-proportional case) degrees of indetermination and therefore the balancing factors cannot produce unique products, and hence trip estimates. This problem is also referred to as the *non-identifiability* of unique products for unobserved cell entries. As the figure above shows, this is likely to occur when roadside interviews take place only on a cordon to a study area. The provision of interviews on a screenline will probably eliminate the problem as it would generate observations for the 'internal-internal' matrix.

5.7 OTHER SYNTHETIC MODELS

The gravity model is by far the most commonly used aggregate trip distribution model. It has a number of theoretical advantages and there is nor lack of suitable software to calibrate and use it. It can be extended further to incorporate more than one person type and it can even be used to model certain types of freight movements. However, the gravity model does not exhaust all the theoretical possibilities. We would like to mention two other approaches which, although they are much less used, offer real alternatives to the gravity model. The first one is the intervening-opportunities model examined here, and the second one the family of direct demand models discussed in Chapter 6.

The basic idea behind the intervening-opportunities model is that trip making is not explicitly related to distance but to the relative accessibility of opportunities for satisfying the objective of the trip. The original proponent of this approach was Stouffer (1940), who also applied his ideas to migration and the location of services and residences. But it was Schneider (1959) who developed the theory in the way it is presented today.

Consider first a zone of origin i and rank all possible destinations in order of increasing distance from i. Then look at one origin–destination pair (i, j), where j is the mth destination in order of distance from i. There are $m - 1$ alternative destinations actually closer (more accessible) to i. A trip maker would certainly consider those destinations as possible locations to satisfy the need giving rise to the journey: these are the *intervening opportunities* influencing a destination choice. Let α be the probability of a trip maker being satisfied with a single opportunity; the probability of her being attracted by a zone with D opportunities is then αD.

Consider now the probability q_i^m of not being satisfied by any of the opportunities offered by the mth destinations away from i. This is equal to the probability of not being satisfied by the first, nor the second, and so on up to the mth:

$$q_i^m = q_i^{m-1}(1 - \alpha D_i^m) \tag{5.43}$$

therefore, omitting the subscript i for simplicity we get

$$\frac{q^m - q^{m-1}}{q^m} = -\alpha D^m \tag{5.44}$$

Now, if we make x_m the cumulative attractions of the intervening opportunities at the mth destination:

$$x_m = \sum_m D^m$$

we can rewrite (5.44) as

$$\frac{q^m - q^{m-1}}{q^m} = -\alpha[x_{m-1} - x_m] \tag{5.45}$$

The limit of this expression for infinitesimally small increments is, of course,

$$\frac{dq^m(x)}{q^m(x)} = -\alpha dx \tag{5.46}$$

Integrating (5.46) we obtain:

$$\log q^m(x) = -\alpha x + \text{constant}$$

or

$$q^m(x) = A_i \exp(-\alpha x) \tag{5.47}$$

where A_i is a parameter for calibration. This relationship expresses the chance of a trip purpose not being satisfied by any of the m destinations from i as a negative exponential function of the accumulated or intervening opportunities at that distance from the origin. The trips T_{ij}^m from i to a destination j (which happens to be the mth away from i) is then proportional to the probability of not being satisfied by any of the $m - 1$ closer opportunities minus the probability of not being satisfied by any of the opportunities up to the mth destination:

$$T_{ij}^m = O_i[q_i(x_{m-1}) - q_i(x_m)]$$
$$T_{ij}^m = O_i A_i[\exp(-\alpha x_{m-1}) - \exp(-\alpha x_m)] \tag{5.48}$$

It is easy to show that the constant A_i must be equal to

$$A_i = 1/[1 - \exp(-\alpha x_m)] \tag{5.49}$$

to ensure that the trip end constraints are satisfied. The complete model then becomes:

$$T_{ij}^m = O_i \frac{[\exp(-\alpha x_{m-1}) \exp(-\alpha x_m)]}{[1 - \exp(-\alpha x_m)]} \tag{5.50}$$

Wilson (1970) has shown that this expression can also be derived from entropy-maximisation considerations.

The intervening-opportunities model is interesting because it starts from different first principles in its derivation: it uses distance as an ordinal variable instead of a continuous cardinal one as in the gravity model. It explicitly considers the opportunities available to satisfy a trip purpose at increased distance from the origin. However, the model is not often used in practice, probably for the following reasons:

- the theoretical basis is less well known and possibly more difficult to understand by practitioners;
- the idea of matrices with destinations ranked by distance from the origin (the nth cell for origin i is not destination n but the nth destination away from i) is more difficult to handle in practice;
- the theoretical and practical advantages of this function over the gravity model are not overwhelming;
- the lack of suitable software.

In Chapter 12 we will discuss a more general version of this model that combines gravity and intervening-opportunities features. This is due to Wills (1986) and lets the data decide which combination of the two models fits reality better. However, the computational complexity of this new model is considerable.

5.8 PRACTICAL CONSIDERATIONS

We have discussed a number of frequently used models to associate origins and destinations and estimate the number of trips between them. However, in doing so we have omitted a number of practical considerations that must necessarily affect the accuracy attainable from the use of such models. These stem from the inherent limitations of our modelling framework and our inability to include detailed descriptions of reality in the models. We shall discuss these features under general headings below.

5.8.1 Sparse Matrices

Observed trip matrices are almost always sparse, i.e. they have a large number of empty cells, and it is easy to see why. A study area with 500 zones (250 000 cells) may have some 2.5 million expected total trips during a peak hour. This yields an average of 10 trips per cell; however, some O–D pairs are more likely to contain trips than others, in particular from residential to high employment areas, thus leaving numerous cells with a very low number of expected trips. Consider now the method used to observe this trip matrix, perhaps roadside interviews. If the sampling rate is 20% (1 in 5) then the chances of making no observations on a particular O–D pair are very high.

This sampled trip matrix will then be expanded, probably using information about the exact sampling ratios in each interview station. The problem generated when expanding empty cells has already been alluded to in section 5.3.4. It may be possible to fill in gaps in the matrix through the use of a partial matrix approach; alternatively, it may be desirable to 'seed' empty cells with a low number and use an alternative matrix expansion method such as that discussed in Chapter 12. It is important to realise, however, that 'observed' trip matrices normally contain a large number of errors and that these will be amplified by the expansion process.

5.8.2 Treatment of External Zones

It may be quite reasonable to postulate the suitability of a synthetic trip distribution model in a study area, in particular for internal-to-internal trips. However, a significant proportion of the trips may have at least one end outside the area. The suitability of a model which depends on trip distance or cost, a variable essentially undefined for external trips, is thus debatable.

Common practice in such cases is to take these trips outside the synthetic modelling process: roadside interviews are undertaken on cordon points at the entrance/exit to the study area. The resulting matrix of external-external (E–E) and external-internal (E–I) trips is then updated and forecast using growth factor methods, in particular those of Furness. However, a number of trip ends from the trip generation/attraction models correspond to the E–I trips and these must be subtracted from the trip-end totals for inclusion as constraints to the synthetic models.

5.8.3 Intra-zonal Trips

A similar problem occurs with intra-zonal trips. Given the limitations of any zoning system, the cost values given to centroid connectors are a very crude but necessary approximation to those experienced in reality. The idea of an intra-zonal trip cost is then poorly represented by these centroid connector costs. Some commercial software allow the user to add/subtract terminal costs to facilitate better modelling of these trips; the idea is that by manipulating these intra-zonal costs one would make the gravity model fit better. However, this is not very good; it is actually preferable to remove intra-zonal trips from the synthetic modelling process and to forecast them using even simpler approaches. This typically assumes that intra-zonal trips are a fixed proportion of the trip ends calculated by the trip generation models.

Moreover, intra-zonal trips are not normally loaded onto the network as they move from a centroid to itself. This makes it less essential to model them in detail. However, in reality, some of these trips use the modelled network. Nevertheless, this problem is probably significative only for rather coarse zoning systems.

5.8.4 Journey Purposes

Different models are normally used for different trip purposes and/or person types. Typically, the journey to work will be modelled using a doubly constrained gravity model while almost all other purposes will be modelled using singly constrained models. This is because it is often difficult to estimate trip attractions accurately for shopping, recreational and social trips and therefore proxies for trip attractiveness are used: retail floor space, recreational areas, population.

Some trip purposes may be more sensitive to cost and therefore deserve the use of different values for the deterrence function.

5.8.5 Generations–Attractions, Origins–Destination

The synthetic models have been developed under the assumption that each trip has a generation and an attraction end. The models essentially link generations to attractions. For home-based trips the generation end is always the home. However, the origin of these trips is only the home for journeys to the place of work (or education, shops, etc.) but on the journey back the destination of the trip is now the home.

Before the resulting trip matrix is assigned onto the network, it must be converted into an origin–destination matrix. In the 24-hour case, the two are practically the same as it is assumed that each generation–attraction trip is made once in each direction during the day. This is of course, an approximation but probably a reasonable one.

However, when a shorter-period O–D matrix is required, some trips will be made in the generation-to-attraction direction while others only in the opposite one. Two different approaches can be used to overcome this problem. The first is to produce a matrix for just a single purpose, typically 'to work', and then assume that these trips follow just one direction of travel, thus producing, for example, the morning journey to work from generation to attraction. Survey data must be used to correct for shift work, flexible working hours and trips for other purposes being made during the morning peak; however, the pattern of the morning peak is still dominated by this journey-to-work purpose. A second approach is to use survey data directly to determine the proportions of the matrices for each purpose which are deemed appropriate for the part of the day under consideration. For example, a typical morning peak matrix may consist of 70% generation-to-attraction movements and only 15% of attraction-to-generation movements.

5.8.6 *K* Factors

The gravity model can provide a reasonable representation of trip patterns provided they can be explained mainly by the size of the generation and attraction power of zones and the deterrence to travel generated by distance (generalised cost). In some circumstances, there may be pairs of zones which have a special association in terms of trip making; for example, a major manufacturer may be located in one zone and most of its employees in another, perhaps as a result of a housing estate developed by the company. In this case, it is likely that more trips will take place between these two points than predicted by any model failing to consider this association, for example the gravity model. This has led to the introduction of an additional set of parameters K_{ij} to the gravity model as follows:

$$T_{ij} = K_{ij} A_i O_i B_{ij} \exp\left(-\beta c_{ij}\right) \tag{5.51}$$

Some practical studies have used these *K* factors in an attempt to improve the calibration of the model. This, of course, they do; with the full set of *K* factors we now have even more flexibility than necessary to reproduce the observed trip matrix; in fact, just the *K* factors are enough to achieve this; the other parameters are surplus to requirement: K_{ij} factors identical to the observed T_{ij} will do the trick; but then we no longer have a model nor any forecasting ability left.

The best advice that can be given in respect of K factors is: do not use them. If a study area has a small number of zone pairs (say, less than 5% of the total) with a special trip making association which is likely to remain in the future, then the use of a few K factors might be justified, sparingly and cautiously. But the use of a model with a full set of K factors cannot be justified. However, see a related issue in section 12.3.2.

5.8.7 Errors in Modelling

It would appear that many of these practical issues reduce the accuracy of the modelling process. This is, in effect, true and it constitutes a reflection of the imperfections in the state of the art in transport modelling. These practical issues are not restricted to distribution modelling; they are present, in one form or another, in other parts of the modelling process.

Because many of the cells in a trip matrix will have small values, say between 0 and 5 in the sample and perhaps 20 to 30 in the expanded or synthesised matrix, their corresponding errors will be relatively large. A small number of studies have tackled the task of calibrating synthetic models and then comparing the resulting trip matrices with observed ones. An investigation by Sikdar and Hutchinson (1981) used data from 28 study areas in Canada to calibrate and test doubly constrained gravity models. The researchers found that the performance of these models was poor, equivalent to a randomly introduced error in the observations of about 75 to 100%; these results reinforce the call for caution in using the results of such models. This should not be entirely surprising; to model a trip matrix with the use of a few parameters (twice the number of zones for an exponential deterrence function) is a very tall order. This is certainly one of the reasons why few studies nowadays make use of the gravity model in its conventional form. In many cases, however, it is desirable to consider how changes in transport costs would influence trip patterns, in particular for more optional purposes like shopping and recreation. In these cases, the idea of using pivot point or incremental versions of the gravity model becomes more attractive, see section 12.3.2.

The treatment of errors in modelling received more attention in the mid-1980s. There seem to be two methods deserving consideration in this field: statistical and simulation approaches. Statistical methods are very powerful but they are not always easy to develop or to implement. They follow the lines suggested in Chapter 3 when discussing the role of data errors in the overall accuracy of the modelling process. Errors in the data are then traced through to errors in the outputs of the models. The UK Department of Transport provides advice in the *Traffic Appraisal Manual* (Department of Transport 1985) on the sensitivity of distribution models to errors in the input data. To some extent the simplest problem is to follow data errors, in particular those due to sampling, through the process of building matrices. A more demanding problem is to follow these errors when a synthetic distribution model is used. One of the advances of the early 1980s was the development of approximate analytical techniques to estimate the output errors due to sampling variabililty. For example, the work of Gunn *et al.* (1980) established approximate expressions for the confidence interval for cell estimates for the tri-proportional formulation of the gravity model. The 95% confidence interval for the number of trips in a cell (i, j) is given by the range $\{C_{ij}/T_{ij} \text{ to } T_{ij}C_{ij}\}$, where C_{ij} is a *confidence factor* given by:

$$C_{ij} = \exp\left(2\left[1/\sum_{ij} n_{ijk} + 1/\sum_{jk} n_{ijk} + 1/\sum_{ki} n_{ijk}\right]^{0.5}\right) \tag{5.52}$$

where n_{ijk} are the number of trips sampled in the observed cells and therefore the summations are over observed cells only. This expression covers only errors due to sampling: data collection and processing errors are likely to increase the range. Moreover, there are other sources of error in the model estimates which are more difficult to quantify; these are mis-specification errors, due to the fact that the model is only a simplified and imperfect representation of reality. Mis-specification errors will, again, increase the range for any confidence interval estimates.

Simulation techniques may play a useful role in cases where analytical expressions for confidence intervals of model output do not exist and are difficult to develop. One can calibrate a model assuming that the data available contain no errors; one would then introduced controlled, but realistic, variability in the data and recalibrate the model. This process could be repeated several times to obtain a range of parameters, each calibrated with a slightly different set of 'survey' data. This process is, of course, quite expensive in time and computer resources and it is therefore attempted mostly for research purposes. However, this type of research can provide useful insights into the stability of model parameters to data errors.

A simpler use of Monte Carlo simulation is in testing the sensitivity of model output to input data in a forecasting mode. One knows that future planning data are bound to contain errors; the use of simulation in this case involves the introduction of reasonable 'noise' into these data and then running the model with each of these future data sets. The results provide an idea of the sensitivity of model output to errors in these planning variables. As no recalibration is involved (the model is assumed to be calibrated with no errors in the base year) the demand on time and resources, although large, is less than in the previous case.

EXERCISES

5.1 A small study area has been divided into four zones and a limited survey has resulted in the following trip matrix:

	1	2	3	4
1	—	60	275	571
2	50	—	410	443
3	123	61	—	47
4	205	265	75	—

Estimates for future total trip ends for each zone are as given below:

Zones	Estimated future origins	Estimated future destinations
1	1200	670
2	1050	730
3	380	950
4	770	995

Use an appropriate growth-factor method to estimate future inter-zonal movements. *Hint*: check conditions for convergence of the chosen method first.

5.2 A study area has been divided into three large zones, A and B on one side of a river and C on the other side. It is thought that travel demand between these zones will depend on whether or not the O–D pair is at the same side of the river. A small sample home interview survey has been undertaken with the following results:

	Destination		
Origin	A	B	C
A	12	10	8
B		5	3
C	4	7	

Blank entries indicate unobserved cells.

Assume a model of the type $T_{ij} = R_i S_j F_k$ where the parameter F_k can be used to represent the fact that the O–D pair is on the same side of the river or not. Calibrate such a model using a tri-proportional algorithm and fill the empty cells in the matrix above.

5.3 A transport study is being undertaken incorporating four cities A, B, C and D. The travel costs between these cities in generalised time units are given below; note that intra-urban movements are excluded from this study:

	Destination			
Origin	A	B	C	D
A	—	1.23	1.85	2.67
B	1.23	—	2.48	1.21
C	1.85	2.48	—	1.44
D	2.67	1.21	1.44	—

Roadside interviews have been undertaken at several sites and the number of drivers interviewed are shown below together with their respective origins and destinations. Blank entries indicate unobserved cells.

	Destinations			
Origin	A	B	C	D
A	–	6		2
B		–	1	4
C	8		–	8
D	6	18	6	–

Assume now that the use of a gravity model of the type $T_{ij} = R_i S_j F_k$ is to be used for this study area with only two cost bins. The first cost bin will cover trips costing between 0 and 1.9 and the second trips costing more than 1.9. Calibrate such a model using a tri-proportional method on this partial matrix. Provide estimates of the parameters R_i, S_j and F_k and of the missing entries in the matrix, excluding intra-urban trips. Are these estimates unique?

6 Modal Split and Direct Demand Models

6.1 INTRODUCTION

In this chapter we shall discuss firstly mode choice as an aggregate problem. It is interesting to see how far we can get using similar approaches to those pursued in deriving and using trip distribution models. We will also examine methods to estimate generation, distribution and modal split simultaneously, the so-called *direct demand* models. These originated in the 1960s, were completely forgotten during the 1970s and early 1980s, but experienced a renaissance towards the end of the 1980s.

The choice of transport mode is probably one of the most important classic models in transport planning. This is because of the key role played by public transport in policy making. Almost without exception public transport modes make use of road space more efficiently than the private car. Furthermore, underground and other rail-based modes do not require additional road space (although they may require a reserve of some kind) and therefore do not contribute to congestion. Moreover, if some drivers could be persuaded to use public transport instead of cars the rest of the car users would benefit from improved levels of service. It is unlikely that all car owners wishing to use their cars could be accommodated in urban areas without sacrificing large parts of the fabric to roads and parking space.

The issue of mode choice, therefore, is probably the single most important element in transport planning and policy making. It affects the general efficiency with which we can travel in urban areas, the amount of urban space devoted to transport functions, and whether a range of choices is available to travellers. The issue is equally important in inter-urban transport as again rail modes can provide a more efficient mode of transport (in terms of resources consumed, including space), but there is also a trend to increase travel by road.

It is important then to develop and use models which are sensitive to those attributes of travel that influence individual choices of mode. We will see how far this necessity can be achieved using aggregate approaches, where alternative policies need to be expressed as modifications to useful if rather inflexible functions like the generalised cost of travel.

6.2 FACTORS INFLUENCING THE CHOICE OF MODE

The factors influencing mode choice may be classified into three groups:

1. Characteristics of the trip maker. The following features are generally believed to be important:
   - car availability and/or ownership;
   - possession of a driving licence;
   - household structure (young couple, couple with children, retired, singles, etc.),
   - income;
   - decisions made elsewhere, for example the need to use a car at work, take children to school, etc;
   - residential density.
2. Characteristics of the journey. Mode choice is strongly influenced by:
   - The trip purpose; for example, the journey to work is normally easier to undertake by public transport than other journeys because of its regularity and the adjustment possible in the long run;
   - Time of the day when the journey is undertaken. Late trips are more difficult to accommodate by public transport.
3. Characteristics of the transport facility. These can be divided into two categories. Firstly, quantitative factors such as:
   - relative travel time: in-vehicle, waiting and walking times by each mode;
   - relative monetary costs (fares, fuel and direct costs);
   - availability and cost of parking.
   Secondly, qualitative factors which are less easy to measure, such as:
   - comfort and convenience;
   - reliability and regularity;
   - protection, security.

A good mode choice model should include the most important of these factors. It is easy to visualise how the concept of generalised cost can be used to represent several of the quantitative factors included under 3.

Mode choice models can be *aggregate* if they are based on zonal (and inter-zonal) information. We can also have *disaggregate* models if they are based on household and/or individual data (see Chapter 7).

6.3 TRIP-END MODAL-SPLIT MODELS

The application of mode choice models over the whole of the population results in trips split by mode, hence modal-split modelling. In the past, in particular in the USA, personal characteristics were thought to be the most important determinants of mode choice and therefore attempts were made to apply modal-split models immediately after trip generation. In this way the different characteristics of the individuals could be preserved and used to estimate modal split: for example, the different groups after a category analysis model. As at that level there was no indication to where those trips might go, the characteristics of the journey and modes were omitted from these models.

This was consistent with a general planning view that as income grew, most people would acquire cars and would want to use them. The objective of transport planning was to forecast this growth in demand for car trips so that investment could be planned to satisfy it. The modal-split models of this time related the choice of mode only to features like income, residential density and car ownership. In some cases the availability of reasonable public transport was included in the form of an accessibility index.

In the short run these models could be very accurate, in particular if public transport was available in a similar way throughout the study area and there was little congestion. However, this type of model is, to a large extent, defeatist in the sense of being insensitive to policy decisions; it appears that there is nothing the decision maker can do to influence the choice of mode. Improving public transport, restricting parking, charging for the use of roads, none of these would have any effect on modal split according to these trip-end models.

6.4 TRIP INTERCHANGE MODAL-SPLIT MODELS

Modal-split modelling in Europe was dominated, almost from the beginning, by post-distribution models; that is, models applied after the gravity or other distribution model. This has the advantage of facilitating the inclusion of the characteristics of the journey and that of the alternative modes available to undertake them. However, they make it more difficult to include the characteristics of the trip maker as they may have already been aggregated in the trip matrix (or matrices).

The first models included only one or two characteristics of the journey, typically (in-vehicle) travel time. It was observed that an S-shaped curve seemed to represent this kind of behaviour better, as in Figure 6.1, showing the proportion of trips by mode 1 (T^{l}_{ij}/T_{ij}) against the cost or time difference.

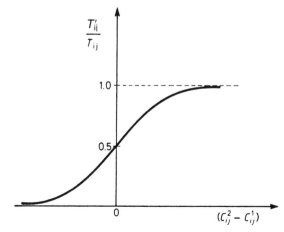

Figure 6.1   Modal-split curve

These were empirical curves, obtained directly from the data and following a similar approach to the curves used to estimate what proportion of travellers would be diverted to use a (longer but faster) bypass route: hence their name of diversion curves. For example, the London Transportation Study (Phase III) used diversion curves for trips to the central area and non-central trips (the former more likely to be made by public transport) and for different trip purposes.

One important limitation of these models is that they can only be used for trip matrices of travellers with a choice available to them. This often means the matrix of car-available persons, although modal split can also be applied to the choice between different public-transport modes.

The models have little theoretical basis and therefore their forecasting ability must be in doubt. They also ignore a number of policy-sensitive variables like fares, parking charges, and so on. Further, as the models are aggregate they are unlikely to model correctly the constraints and characteristics of the modes available to individual households.

6.5 SYNTHETIC MODELS

6.5.1 Distribution and Modal-split Models

The entropy-maximising approach can be used to generate models of distribution and mode choice simultaneously. In order to do this we need to cast the entropy-maximising problem in terms of, for example, two modes as follows:

$$\text{Maximise log } W\{T_{ij}^k\} = -\sum_{ijk}(T_{ij}^k \log T_{ij}^k - T_{ij}^k) \tag{6.1}$$

subject to

$$\sum_{jk} T_{ij}^k - O_i = 0 \tag{6.2}$$

$$\sum_{ik} T_{ij}^k - D_j = 0 \tag{6.3}$$

$$\sum_{ijk} T_{ij}^k c_{ij}^k - C = 0 \tag{6.4}$$

It is easy to see that this problem leads to the solution:

$$T_{ij}^k = A_i O_i B_j D_j \exp(-\beta c_{ij}^k) \tag{6.5}$$

$$P_{ij}^1 = \frac{T_{ij}^1}{T_{ij}} = \frac{\exp(-\beta C_{ij}^1)}{\exp(-\beta C_{ij}^1) + \exp(-\beta C_{ij}^2)} \tag{6.6}$$

where P_{ij}^1 is the proportion of trips travelling from i to j via mode 1. The functional form in (6.6) is known as *logit* and it is discussed in greater detail in the next chapter. However, it is useful to reflect here on some of its properties:

- it generates an S-shaped curve, similar to some of the empirical diversion curves of Figure 6.1;
- if $C_1 = C_2$, then $P_1 = P_2 = 0.5$;
- if $C_2 \gg C_1$, then P_1 tends to 1.0;
- the model can easily be extended to multiple modes

$$P^1_{ij} = \frac{\exp(-\beta C^1_{ij})}{\sum_k \exp(-\beta C^k_{ij})} \tag{6.7}$$

It is obvious that in this formulation β plays a double role. It acts as the parameter controlling dispersion in mode choice and also in the choice between destinations at different distances from the origin. This is probably asking too much of a single parameter, even if underpinned by a known theoretical basis. Therefore a more practical joint distribution/modal-split model has been used in many studies. This has the form:

$$T^{kn}_{ij} = A^n_i O^n_i B_j D_j \exp(-\beta_n K^n_{ij}) \frac{\exp(-\lambda_n C^k_{ij})}{\sum_k \exp(-\lambda_n C^k_{ij})} \tag{6.8}$$

where K^n_{ij} is the *composite cost* of travelling between i and j as perceived by person type n. In principle this composite cost may be specified in different ways; for example, it could be taken to be the minimum of the two costs or, perhaps better, the weighted average of these:

$$K = \sum_k P^k C^k \qquad (i, j \text{ and } n \text{ omitted for simplicity})$$

However, it is interesting to note that most such formulations, which were used in many studies up to the end of the 1970s, are in fact inappropriate.

Example 6.1: Consider the weighted average form above and examine what happens when a new, more expensive mode $(C_2 > C_1)$ is added to an existing unimodal system. In the initial state we would have:

$$K = \sum_k P^k C^k = C^1$$

and in the final state, i.e. after the introduction of mode 2:

$$K^* = P^1 C^1 + P^2 C^2$$

However, by definition $P^1 + P^2 = 1$ and therefore:

$$K^* = (1 - P^2)C^1 + P^2 C^2 = C^1 + P^2(C^2 - C^1)$$
$$K^* = K + P^2(C^2 - C^1)$$

Now, as both P^2 and $(C^2 - C^1)$ are greater than zero, we conclude that $K^* > K$, which is nonsensical as the introduction of a new option, even if it is more expensive, should not increase the composite costs; at worst they should remain the same.

Williams (1977) has shown that the only correct specification, consistent with the prevailing theory of rational choice behaviour (see section 7.2), is:

$$K_{ij}^n = \frac{-1}{\lambda_n} \log \sum_k \exp(-\lambda_n C_{ij}^k) \tag{6.9}$$

where the following restriction must be satisfied:

$$\beta_n \leq \lambda_n \tag{6.10}$$

We will come back to this restriction in Chapter 7. The composite cost measure (6.9) has the following properties:

- $K \leq \text{Min}_k \{C^k\}$

- $\underset{\lambda \to \infty}{\text{Lim}} K = \text{Min}_k \{C_k\}$, that is 'all-or-nothing' mode choice

- $\dfrac{dK}{dC^k} = P^k$

The model ((6.8)–(6.10)) represents the 'state of practice' for aggregate modelling of modal split and distribution, in particular in urban areas. Although a large number of models of this kind has been applied in practice, they are now being replaced by disaggregate models which respond better to the key elements in mode choice and make a more efficient use of data collection efforts; these are discussed in Chapters 7 to 9.

6.5.2 Multimodal-split Models

Figure 6.2 depicts possible model structures for choices involving more than two modes. The N-way structure which became very popular in disaggregate modelling work, as we will see in Chapter 7, is the simplest; however, because it assumes that all alternatives have equal 'weight', it can lead to problems when some of the options are more similar than others (i.e. they are correlated), as demonstrated by the famous blue bus-red bus example (Mayberry 1973).

Example 6.2: Consider a city where 50% of travellers choose car (C) and 50% choose bus (B). In terms of model (6.7), which is an N-way structure, this means that $C_C = C_B$. Let us now assume that the manager of the bus company, in a stroke of marketing genius, decides to paint half the buses red (RB) and half of them blue (BB), but manages to maintain the same level of service as before. This means that $C_{RB} = C_{BB}$, and as the car mode has not changed this value is still equal to C_C. It is interesting to note that model (6.7) now predicts:

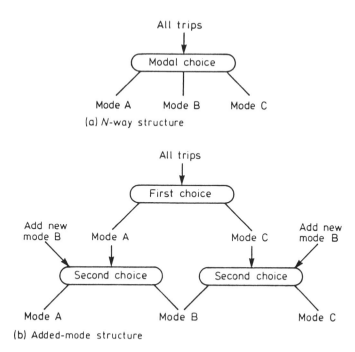

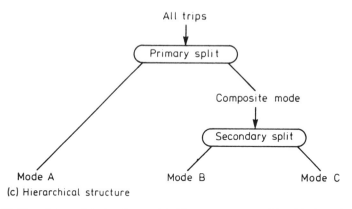

Figure 6.2 Multimodal model structures: (a) *N*-way structure, (b) added-mode structure, (c) hierarchical structure

$$P_{\mathrm{C}} = \frac{\exp\left(-\beta C_{\mathrm{C}}\right)}{\exp\left(-\beta C_{\mathrm{C}}\right) + \exp\left(-\beta C_{\mathrm{RB}}\right) + \exp\left(-\beta C_{\mathrm{BB}}\right)} = 0.33$$

when one would expect P_{C} to remain 0.5, and the buses to share the other half of the market equally between red and blue buses. The example is, of course, exaggerated but serves well to show the problems of the *N*-way structure in the presence of correlated options (in this case completely correlated). We will come back to this in Chapter 7.

The 'added-mode' structure, depicted in Figure 6.2b, was used by many 'pragmatic' practitioners in the later 1960s and early 1970s; however, it has been shown to give different results depending on which mode is taken as the added one (Langdon 1976). Also, work using Monte Carlo simulation has shown that the added mode form with better performance in the base year is not necessarily the one to perform best in the future under certain policy changes (Ortúzar 1980a).

The third possibility, depicted in Figure 6.2c, is the hierarchical or nested structure. Here the options which have common elements (i.e. are more similar than others or correlated) are taken together in a primary split (i.e. public transport). After they have been 'separated' from the uncorrelated option, they are subdivided in a secondary split. In fact, this was the standard practice in the 1960s and early 1970s, but with the short-coming that the composite costs for the 'public-transport' mode were normally taken as the minimum of costs of the bus and rail modes for each zone pair and that the secondary split was achieved through a minimum-cost 'all-or-nothing' assignment. This 'pragmatic' procedure essentially implies an infinite value for the dispersion parameter of the submodal-split function, whereas it has normally been found that it has a value of the same order as the dispersion parameter in the primary split, but satisfying (6.10).

Example 6.3: A hierarchical structure model for the red bus-blue bus problem of Example 6.2 would have the following expression:

$$P_C = \frac{1}{1 + \exp\{-\lambda_1(C_B - C_C)\}}; \quad P_B = 1 - P_C$$

$$P_{R/B} = \frac{1}{1 + \exp\{-\lambda_2(C_{BB} - C_{RB})\}}$$

$$P_{B/B} = 1 - P_{R/B}$$

with

$$C_B = \frac{-1}{\lambda_2} \log\left[\exp(-\lambda_2 C_{RB}) + \exp(-\lambda_2 C_{BB})\right]$$

where P_C is the probability of choosing car, as before, $(1 - P_C)P_{R/B}$ the probability of selecting red bus and $(1 - P_C)P_{B/B}$ the probability of selecting blue bus; λ are the primary and secondary split parameters. It is easy to see that, if $C_B = C_C$, this model correctly assigns a probability of 0.5 to the car option and 0.25 to each of the bus modes.

6.5.3 Calibration of Binary Logit Models

Consider a model of choice between car and public transport with generalised costs of travel, C_{ij}^k, given by an expression such as (5.2). As discussed in Chapter 5, the weights **a** attached to each element of cost are considered given and calibration only involves finding the 'best-fit' values for the dispersion parameter λ and modal penalty δ (assumed associated to the second mode).

Let us assume that we have C_{ij}^1 and C_{ij}^2 as the 'known' part of the generalised cost for each mode and O–D pair. If we also have information about the proportions choosing each mode for each (i, j) pair, P_{ijk}^*, we can estimate the values of λ and δ using linear regression as follows. The modelled proportions **P** for each (i, j) pair, dropping the (i, j) indices for convenience, are:

$$P_1 = \frac{1}{1 + \exp\{-\lambda(C_2 + \delta - C_1)\}}$$

$$P_2 = 1 - P_1 = \frac{\exp\{-\lambda(C_2 + \delta - C_1)\}}{1 + \exp\{-\lambda(C_2 + \delta - C_1)\}}$$

(6.11)

Therefore, taking the ratio of both proportions yields:

$$P_1/(1 - P_1) = 1/\exp\{-\lambda(C_2 + \delta - C_1)\} = \exp\{\lambda(C_2 + \delta - C_1)\}$$

and taking logarithms of both sides and rearranging, we get:

$$\log[P_1/(1 - P_1)] = \lambda(C_2 - C_1) + \lambda\delta$$

(6.12)

where we have observed data for **P** and **C**, and therefore the only unknowns are λ and δ. These values could be calibrated by linear regressions with the left-hand side of (6.12) acting as the dependent variable and $(C_2 - C_1)$ as the independent one; then λ is the slope of the line and $\lambda\delta$ is the intercept. Note that if we assume the weights **a** in the generalised cost function to be unknown, we can still calibrate the model using (6.12) and multiple linear regression. In this case the calibrated weights would include the dispersion coefficient λ. Other, and often better calibration methods are discussed in the next section.

Example 6.4: Data about aggregate mode choice between five zone pairs is presented in the first four columns of Table 6.1; the last two columns of the table give the values needed for the left-hand side of equation (6.12).

This information can be plotted following (6.12) as in Figure 6.3, where it can be deduced that $\lambda \approx 0.72$ and $\delta \approx 3.15$.

Table 6.1 Aggregate binary split data

Zone pair	$P_1(\%)$	$P_2(\%)$	C_1	C_2	$\log[P_1/(1 - P_1)]$
1	51.0	49.0	21.0	18.0	0.04
2	57.0	43.0	15.8	13.1	0.29
3	80.0	20.0	15.9	14.7	1.39
4	71.0	29.0	18.2	16.4	0.90
5	63.0	37.0	11.0	8.5	0.53

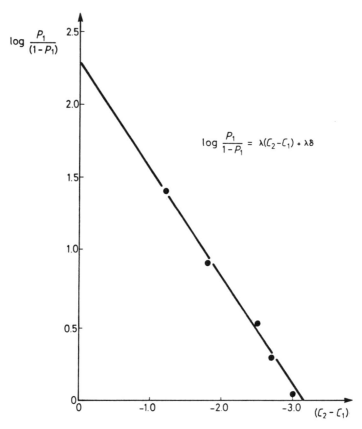

Figure 6.3   Best-fit line for the data in Table 6.1

6.5.4 Calibration of Hierarchical Modal-split Models

This is usually performed in a heuristic or recursive fashion, starting with the submodal split and proceeding upwards to the primary split. A general discussion on the merits of this approach in comparison with the theoretically better simultaneous estimation is postponed until Chapter 7. Within this general approach there are several possible calibration procedures. It has been shown (see for example Domencich and McFadden 1975) that maximum likelihood estimates are preferable to least squares estimates, both on theoretical and practical grounds. This is particularly true when working with large data sets. However, when dealing with aggregate data sources it is usually convenient to group the information into suitable classes for analysis (i.e. cost-difference bins). More importantly, the normally available 'factored-up' data are, by definition, raw sampled data which have been manipulated and multiplied by some empirically derived factors. This can cause discrepancies when several data sources with different factors are employed, but the important point at this stage is that the real data set is very small. Hartley and Ortúzar (1980) compared various procedures, and found that maximum likelihood produced not only the most accurate calibration results but also the more efficient ones in terms of computer time.

Let us consider a trinomial problem involving choice between, for example, car, bus and rail. Let us also assume that the last two modes are suspected of being correlated due to their 'public-transport' nature. The heuristic calibration proceeds as follows. First λ_2 is found for the submodal split (bus vs. rail) as explained in Example 6.4 and its value is used to calculate the public-transport composite costs needed for the primary split using an expression such as that for Example 6.3.

For zone pairs where there is a choice of mode (e.g. trips by both modes are possible), trips are classified into cost-difference bins of a certain minimum size. Those trips with no choice of mode are excluded from the calibration. Between cost bins with trips allocated to them, there can be cost bins without any trips; therefore, bins are aggregated into bigger bins until each bin contains some trips. Finally, a weighted representative cost is calculated for each bin.

Then, if N is the total number of bins, n_k the observed number of trips in cost-difference interval k, r_k the observed number of trips by the first mode in the interval, and

$$P_k = 1/[1 + \exp(-Y_k)]$$

the probability of choosing the first mode in interval k, with $Y_k = ax_k + b$, x_k the representative cost of bin k, and a and b parameters to be estimated (i.e. $\lambda = a$ and the modal penalty $\delta = b/a$), the logarithm of the likelihood function (see Chapter 8 for more details) can be written as:

$$L = \text{Constant} + \sum_k [(n_k - r_k) \log (1 - P_k) + r_k \log P_k] \tag{6.13}$$

The maximisation procedure makes use of the first and second derivatives of (6.13) with respect to the parameters, which in this simple case have straightforward analytical expressions:

$$\frac{\partial L}{\partial a} = \sum_k (r_k - n_k P_k) x_k$$

$$\frac{\partial L}{\partial b} = \sum_k (r_k - n_k P_k)$$

$$\frac{\partial^2 L}{\partial a^2} = -\sum_k n_k P_k (1 - P_k) x_k^2$$

$$\frac{\partial^2 L}{\partial b^2} = -\sum_k n_k P_k (1 - P_k)$$

$$\frac{\partial^2 L}{\partial a \partial b} = -\sum_k n_k P_k (1 - P_k) x_k$$

Knowing the values of the derivatives, any search algorithm will find the maximum without difficulty. Maximisation routines require starting values for the parameters, together with an indication of how far they are from the optimum. The efficiency of calibration typically depends strongly upon the accuracy of these estimates. One procedure for generating close first estimates is to find the equi-probability cost (see Bates *et al*. 1978), where the probability of choosing either mode is 0.5.

Before closing this chapter, one must consider an alternative approach offering to consolidate in a single model the features of two or three of the classic sub-models.

6.6 DIRECT DEMAND MODELS

6.6.1 Introduction

The conventional sequential methodology requires the estimation of relatively well-defined sub-models. An alternative approach is to develop directly a model subsuming trip generation, distribution and mode choice. This is, of course very attractive in itself as it avoids some of the pitfalls of the sequential approach. For example, it has been claimed that gravity models suffer from the problem of having to cope with the errors in trip-end totals and those generated by poorly estimated intra-zonal trips. A direct demand model, as it is calibrated simultaneously for the three sub-models, would not suffer from this drawback.

Direct demand models can be of two types: purely direct, which use a single estimated equation to relate travel demand directly to mode, journey and person attributes; and a quasi-direct approach which employs a form of separability between mode split and total (O–D) travel demand. Direct demand models are closely related to general econometric models of demand and have long been inspired by research in that area.

6.6.2 Direct Demand and Abstract Models

The earliest forms of direct demand models were of the multiplicative kind. The SARC (Kraft 1968) model, for example, estimates demand as a multiplicative function of activity and socioeconomic variables for each zone pair and level-of-service attributes of the modes serving them:

$$T_{ijk} = \phi_k (P_i P_j)^{\theta_{k1}} (I_i I_j)^{\theta_{k2}} \prod_m [(t_{ij}^m)^{\alpha_{km}^1} (c_{ij}^m)^{\alpha_{km}^2}] \tag{6.14}$$

where P is population, I income, t and c travel time and cost of travel between i and j by mode k, and ϕ, θ and α parameters of the model. This complex expression may be rewritten in simpler form, defining the following composite variables (Manheim 1979):

$$L_{ijm} = (t_{ij}^m)^{\alpha_{km}^1}(c_{ij}^m)^{\alpha_{km}^2}$$

$$Y_{ik} = P_i^{\theta_{k1}}I_i^{\theta_{k2}}$$

$$Z_{jk} = P_j^{\theta_{k1}}I_j^{\theta_{k2}}$$

With these changes of variables (6.14) becomes:

$$T_{ijk} = \phi_k Y_{ik} Z_{jk} \prod_m L_{ijm} \tag{6.15}$$

and this transformation eases the interpretation of the model parameters. For example, ϕ_k is just a scale parameter which depends on the purpose of the trips examined. θ_{k1} and θ_{k2} are elasticities of demand with respect to population and income respectively; we would expect them to be of positive sign. α_{km}^1 and α_{km}^2 are demand elasticities with respect to time and cost of travelling; the direct elasticities (i.e when $k = m$) should be negative and the cross-elasticities of positive sign.

The model is very attractive in principle as it handles generation, distribution and modal split simultaneously, including attributes of competing modes and a wide range of level of service and activity variables. Its main problem is the large number of parameters needed to cash in on these advantages. Alternative forms, containing linear and exponential terms in addition to multiplicative ones, have been suggested by Domencich *et al*. (1968).

Another direct demand function originally proposed for the North Eash Corridor Study of the USA is known as the McLynn model (see Manheim 1979) and has the form:

$$T_{ijk} = \phi_k(P_iP_j)^{\theta_1}(I_iI_j)^{\theta_2}\frac{(t_{ij}^k)^{\alpha_k^1}(C_{ij}^k)^{\alpha_k^2}}{\sum_m[(t_{ij}^m)^{\alpha_m^1}(C_{ij}^m)^{\alpha_m^2}]}\left\{\sum_m[(t_{ij}^m)^{\alpha_m^1}(C_{ij}^m)^{\alpha_m^2}]\right\}^\psi \tag{6.16}$$

In order to understand this function, let us consider the following simplifications: only two modes (e.g. 1 and 2) and $\psi = 0$. We will also drop the indices i and j, and define the following composite variables:

$$Y_{ij} = Y = (P_iP_j)^{\theta_1}(I_iI_j)^{\theta_2}$$

$$L_k = \phi_k(t^k)^{\alpha_k^1}(C^k)^{\alpha_k^2}$$

With the above, we can write expressions for modes 1 and 2 as follows:

$$T_1 = YL_1/(L_1 + L_2) \qquad \text{and} \qquad T_2 = YL_2/(L_1 + L_2)$$

Now, the total number of trips is $T = T_1 + T_2 = Y$; then, the modal shares are given by:

$$P_1 = T_1/T = L_1/(L_1 + L_2) = 1/(1 + L_2/L_1)$$
$$P_2 = 1 - P_1 = 1/(1 + L_1/L_2)$$

and these can be plotted as shown in Figure 6.4. Figure 6.4a shows the variation in the proportion using each mode with the ratio of their levels of service; this means, for example, that if we wish to increase the fare of mode 2 without decreasing its current share, we must do something to keep constant the level of service L_2. Figure 6.4b shows that in order to do this we must also vary t_2 (obviously without touching the other mode).

Coming back to the general model (6.16), it is easy to see that the parameters have in general the same meaning but are a lot less than in the SARC model. There is only one new parameter, ψ, which is instructive to analyse. This parameter was introduced in order to allow the term it is affecting to represent the *total potential* of trips by the various modes; in this way, the previous term in the equation would act as the modal split of these trips between the various modes as we just saw.

An examination of some particular cases suggests that the model may not have a solid theoretical basis. For example, if $\psi = 0$, the last term disappears as noted above. If $\psi = 1$, the last term and the term in the denominator of (6.16) cancel, leaving the following function:

$$T_{ijk} = \phi_k(P_iP_j)^{\theta_1}(I_iI_j)^{\theta_2}(t_{ij}^k)^{\alpha_k^1}(C_{ij}^k)^{\alpha_k^2}$$

which effectively says that what happens with one mode does not affect the others, and this is very unrealistic. In fact, it has been shown that only if $0 < \psi < 1$ does the model perform satisfactorily; in this sense it is interesting to note that the 'best-fit'

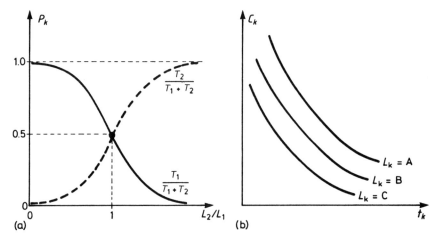

Figure 6.4 Modal split functions in the McLynn model: (a) modal split curves, (b) isocost curves

calibration value obtained in the original study was 0.77 (McLynn and Woronka 1969).

Example 6.5: Consider the following demand function:

$$T_{12} = 10\,000 t_{12}^{\alpha} c_{12}^{\beta} q_{12}^{\mu}$$

where time t is measured in hours, the fare c in dollars and the service frequency q in trips/day. The estimated parameter values are $\alpha = -2$, $\beta = -1$ and $\mu = 0.8$ (note that all the signs are correct according to intuition). The operator wants to increase the fares by 20%; what changes should he make to the level of service in order to keep the same volume of trips if all other things remain equal?

Let us define $L_{12} = L = t^{-2} c^{-1} q^{0.8}$; we know that if L remains constant the total volume T_{12} will not vary (*ceteris paribus*). We also know that the elasticities $E(L, x)$ of the level of service (and hence demand) with respect to each attribute x (time, cost and frequency) are respectively -2, -1 and 0.8.

Now, if only c varies, we have that $L = k/c$, where k is a constant; therefore, a 20% increase in c means a new level of service $L' = k/1.2c$ or $L'/L = 0.833$. That is, a decrease of 16.67% in L. In order to offset this the operator must introduce changes to the travel time, frequency of service or both. Now, from the definition of elasticity (see Chapter 2) we have that:

$$\Delta L^{(c)} \approx E(L, c) L \Delta c / c \approx - L \Delta c / c$$

$$\Delta L^{(t)} \approx E(L, t) L \Delta t / t \approx -2 L \Delta t / t$$

$$\Delta L^{(q)} \approx E(L, q) L \Delta q / q \approx 0.8 L \Delta q / q$$

Therefore if we want $\Delta L^{(c)}$ to be equal to $-\Delta L^{(q)}$, we require:

$$- L \Delta c / c \approx -0.8 L \Delta q / q$$

that is:

$$\Delta q / q \approx 1.25 \Delta c / c \approx 1.25 \times 0.20 = 0.25 \text{ or } 25\%.$$

If we are prepared to vary both frequency and travel time, we would require:

$$\Delta L^{(c)} = -(\Delta L^{(q)} + \Delta L^{(t)})$$

that is

$$2 \Delta t / t = 0.8 \Delta q / q - 0.20$$

a straight line of feasible solutions which is shown in Figure 6.5.

One of the most influential studies of this kind (Quandt and Baumol 1966) proposed an *abstract mode* demand model, the coefficients of which were relieved of their modal subscripts. Its functional form can be generalised as follows:

$$T_{ijm} = \phi_0 \prod_k (A_{ik} A_{jk})^{\phi_k} \prod_h C_{ijh}^{\alpha_h} \prod_h (C_{ijhm}/C_{ijhb})^{\beta_h} \tag{6.17}$$

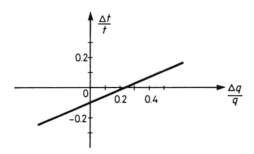

Figure 6.5   Feasible solutions for Example 6.5

where ϕ α and β are parameters for calibration; the index h is used to indicate a cost attribute, for example travel time or fare or headway. A_{ik} are different attributes associated to each zone i, for instance population or income; C_{ijhm} is then the value of the cost attribute h for mode m between i and j and C_{ijhb} is the value of the 'best' attribute h between these two points, say the cheapest fare between them. This dependency on the best mode for modal split and generalised impedance is a weakness of the approach. This may be partly removed, however, by the use of the geometric mean over modes instead of the best mode; see for example Crow *et al.* (1973).

One of the claimed advantages of function (6.17) is that it may model completely new modes without respecification. Yet it can be debated whether the models are sound enough and whether the resulting elasticities, since they correspond to an average of the travel market, are sufficiently representative of any individual mode.

Many different variants of direct demand models have been attempted on a heuristic basis. Its use has been mainly in the inter-urban context with very few applications in urban areas. Usually the logarithms of the number of trips and explanatory variables are taken to make the direct demand model log-linear and therefore estimable using generalised linear model software like GLIM.

Direct demand models are certainly an attractive proposition, in particular in areas where the zones are large, for example inter-urban studies. Timberlake (1988) has discussed the use of direct demand models in developing countries and found them better than conventional approaches. For example, in the Karthoum-Wad Medani Corridor in Sudan, the direct demand model gave a better fit than a gravity model because of the unique traffic characteristics exhibited by Karthoum and Port Sudan in comparison with the rest of the country. The direct demand model was able to accommodate these differences better than the gravity model.

EXERCISES

6.1 A mode choice survey has been undertaken on a corridor connecting three residential areas A, B, C and D with three employment areas U, V and W. The corridor is served by a good rail link and a reasonable road network. The three employment zones are in a heavily congested area and therefore journeys by rail there are often faster than by car. The information collected during the survey is summarised below:

O–D pair	By car				By rail			Proportion by car
	X_1	X_2	X_3	X_4	X_1	X_2	X_3	
A–U	23	3	120	40	19	10	72	0.82
B–U	20	3	96	40	17	8	64	0.80
C–U	18	3	80	40	14	10	28	0.88
D–U	15	3	68	40	14	12	20	0.95
A–V	26	4	152	60	23	10	104	0.72
B–V	19	4	96	60	18	9	72	0.90
C–V	14	4	60	60	11	9	36	0.76
D–V	12	4	56	60	12	11	28	0.93
A–W	30	5	160	80	25	10	120	0.51
B–W	20	5	100	80	16	8	92	0.56
C–W	15	5	64	80	12	9	36	0.58
D–W	10	5	52	80	8	9	24	0.64

where the costs per trip per passenger are as follows:
X_1 = in-vehicle travel time in minutes (line haul plus feeder mode, if any)
X_2 = excess time (walking plus waiting) in minutes
X_3 = out-of-pocket travel costs (petrol or fares), in pence
X_4 = parking costs associated with a one way trip, in pence.

(a) Calibrate a logit modal-split model assuming that the value of travel time is 8 pence per minute and that the value of excess time is twice as much.
(b) Estimate the impact on modal split on each O–D pair of an increase in petrol prices which doubles the perceived cost of running a car (X_3).
(c) Estimate the shift in modal split which could be obtained if no fares were charged on the rail system.

6.2 An inter-urban mode choice study is being undertaken for people with a choice between car and rail. The figures below were obtained as a result of a survey on five origin-destination pairs A to E:

O–D	Elements of cost by each mode				Proportion choosing car
	Car		Rail		
	X_1	X_2	X_1	X_2	
A	3.05	9.90	2.50	9.70	0.80
B	4.05	13.10	2.02	14.00	0.51
C	3.25	9.30	2.25	8.60	0.57
D	3.50	11.20	2.75	10.30	0.71
E	2.45	6.10	2.04	4.70	0.63

where X_1 is the travel time (in hours) and X_2 the out-of-pocket cost (in pounds sterling). Assume that the 'value of time' coefficient is 2.00 per hour and calculate the generalised cost of travelling by each mode.

(a) Calibrate a binary logit modal-split model with these data including the mode specific penalty.
(b) An improved rail service is to be introduced which will reduce travel times by 0.20 of an hour in every journey; by how much could the rail mode increase its fares without losing customers at each O–D pair?
(c) How would you model the introduction of an express coach service between these cities?

6.3 Consider the following trip distribution/modal-split model:

$$V_{ij}^n = A_i O_i B_j D_j \exp(-\beta M_{ij}^n)$$

where

$$M_{ij}^n = -(1/\tau^n) \log \Sigma_k \exp(-\tau^n C_{ij}^k)$$

and $n = 1$ stands for persons with access to car, $n = 2$, persons without access to car, $k = 1$ stands for car and $k = 2$ for public transport.

 If the total number of trips between zones i and j is $V_{ij} = 1000$, compute how many will use car and how many public transport according to the model. The estimated parameter values were found to be: $\tau^1 = 0.10$, $\tau^2 = 0.05$ and $\beta = 0.04$; also, for trips between i and j the modal costs were calculated as: $C_{ij}^1 = 30$ and $C_{ij}^2 = 40$.

6.4 Consider the following modal-split model:

$$P_k = \exp(-\tau C_{ij}^k)/\Sigma_m \exp(-\tau C_{ij}^m)$$

with generalised costs given by the following expression:

$$C_{ij}^k = \Sigma_p \theta_{kp} x_{kp}$$

where θ are parameters weighing the model explanatory variables (time, cost, etc).

(a) Write an expression for the elasticity of P_k with respect to x_{kp}.
(b) Consider now a binary choice situation where the generalised costs have the following concrete expressions:

$$C_{car} = 0.2tt_{car} + 0.1c_{car} + 0.3et_{car}$$

$$C_{bus} = 0.2tt_{bus} + 0.1c_{bus} + 0.3et_{bus} + 0.3$$

where tt is in-vehicle travel time (min), c is travel cost ($) and et is access time (walking and waiting, min). Assume we know the following average data for the modes:

		Variable	
Mode	tt	c	et
Car	20	50	0
Bus	30	20	5

Calculate the proportion of people choosing car if $\tau = 0.4$.

6.5 The railway between the towns of A and B spans 800 km through mountainous terrain. The total one-way travel time, t_r, is 20 hrs and currently the fare, c_r, is 600\$/ton. As the service is used at low capacity t_r is a constant, independent of the traffic volume V_r.

There is a lorry service competing with the railway in an approximately parallel route; its average speed is 50 km/hr and it charges a fare of 950\$/ton. There is a project to build a highway in order to replace the present road; it is expected though that most of its traffic will continue to be heavy trucks.

The level-of-service function of the new highway has been estimated as:

$$t_t = 7 + 0.08V_t \text{ (hours)}$$

where V_t is the total flow of trucks per hour.

On the other hand the railway has estimated its demand function as follows:

$$(V_r/V_t) = 0.83(t_r/t_t)^{-0.8}(c_r/c_t)^{-1.6}$$

and it is expected that the total volume transported between the two towns, $V_r + V_t$, will remain constant and equal to 200 truck loads/hr in the medium term.

(a) Estimate the current modal split (i.e. volumes transported by rail and lorry).
(b) Estimate modal split if the highway is built.
(c) What would be the modal split if:

— the railway decreases its fare to 450\$/ton?
— the lorries were charged a toll of 4\$/ton in order to finance the highway?
— both changes are simultaneous?

7 Discrete Choice Models

In this chapter we provide a rather comprehensive introduction to *discrete choice* (i.e. when individuals have to select an option from a finite set of alternatives) modelling methods. We start with some general considerations and move on to explain the general theoretical framework, random utility theory, in which these models are cast. This serves us to introduce some basic terminology and to present the individual-modeller 'duality' which is so useful to understanding what the theory postulates. Next we introduce the two most popular discrete choice models: multinomial and nested logit, which taken as a family provide the practitioner with a very powerful modelling tool. The implicit assumption throughout is that we possess revealed-preferences cross-sectional data. Finally, we briefly discuss other choice models, such as multinomial probit, which serve to consider the benefits and special problems involved with modelling with panel data, and also other paradigms which offer an alternative perspective to the classical utility-maximising approach.

The problems of model specification and estimation, both with revealed- and stated-preference data, are considered in sufficient detail for practical analysis in Chapter 8; we provide information about certain issues, such as validation samples, which are seldom found in texts on this subject. The problem of aggregation, from various perspectives, and the important question of model updating and transference (particularly for those interested in a continuous planning approach to transport), are tackled in Chapter 9.

7.1 GENERAL CONSIDERATIONS

Aggregate demand (first-generation) transport models, such as those we have discussed in the previous chapters, are either based on observed relations for groups of travellers, or on average relations at a zonal level. On the other hand, disaggregate demand (second-generation) models are based on observed choices made by individual travellers. It is expected that the use of this framework will enable more realistic models to be developed.

In spite of the pioneering work of researchers such as Warner (1962) or Oi and Shuldiner (1962), which made apparent serious deficiencies in the conventional methodologies, first-generation models continued to be used, almost unscathed, in the majority of transport projects until the early 1980s. In fact, only then second-generation models started to be considered as a serious modelling option (see Williams 1981). In general, discrete choice models postulate that:

> *the probability of individuals choosing a given option is a function of their socioeconomic characteristics and the relative attractiveness of the option.*

To represent the attractiveness of the alternatives the concept of *utility* (which is a convenient theoretical construct, tautologically defined as what the individual seeks to maximise) is used. Alternatives, *per se*, do not produce utility: this is derived from their characteristics (Lancaster 1966) and those of the individual; for example, the *observable utility* is usually defined as a linear combination of variables, such as:

$$V_{car} = 0.25 - 1.2\text{IVT} - 2.5\text{ACC} - 0.3\text{C/I} + 1.1\text{NCAR} \qquad (7.1)$$

where each variable represents an attribute of the option or of the traveller. The relative influence of each attribute, in terms of contribution to the overall satisfaction produced by the alternative, is given by its coefficient; for example, a unit change on *access time* (ACC) in (7.1) has approximately twice the impact of a unit change on *in-vehicle travel time* (IVT) and more than seven times the impact of a unit change on the variable *cost/income* (C/I). The variables can also represent characteristics of the individual; for example, we would expect that an individual belonging to a household with a large *number of cars* (NCAR), would be more likely to choose the car option than another belonging to a family with just one vehicle. The *alternative-specific constant* 0.25 in equation (7.1) is normally interpreted as representing the net influence of all unobserved, or not explicitly included, characteristics of the individual or the option in its utility function. For example, it could include elements such as comfort and convenience which are not easy to measure or observe.

In order to predict if an alternative will be chosen, according to the model, the value of its utility must be contrasted with those of alternative options and transformed into a probability value between 0 and 1. For this a variety of mathematical transformations exist which are typically characterised for having an S-shaped plot, such as:

Logit
$$P_1 = \frac{\exp(V_1)}{\exp(V_1) + \exp(V_2)}$$

Probit $P_1 = \displaystyle\int_{-\infty}^{\infty} \int_{-\infty}^{V_2 - V_1 + x_1} \frac{\exp\left\{-\dfrac{1}{2(1-\rho^2)}\left[\left(\dfrac{x_1}{\sigma_1}\right)^2 - \dfrac{2\rho x_1 x_2}{\sigma_1 \sigma_2} + \left(\dfrac{x_2}{\sigma_2}\right)^2\right]\right\}}{2\pi\sigma_1\sigma_2\sqrt{(1-\rho^2)}} dx_2\, dx_1$

where the covariance matrix of the Normal distribution associated to this latter model has the form:

$$\Sigma = \begin{pmatrix} \sigma_1^2 & \rho\sigma_1\sigma_2 \\ \rho\sigma_1\sigma_2 & \sigma_2^2 \end{pmatrix}$$

Discrete choice models cannot be calibrated in general using standard curve-fitting techniques, such as least squares, because their dependent variable P_i is an unobserved probability (between 0 and 1) and the observations are the individual choices (which are either 0 or 1); the only exceptions to this are models for homogeneous groups of individuals, or when the behaviour of every individual is recorded on several occasions, because observed frequencies of choice are also variables between 0 and 1.

Some useful properties of these models have been conveniently summarised by Spear (1977):

1. Disaggregate demand models (DM) are based on theories of individual behaviour and do not constitute physical analogies of any kind. Therefore, as an attempt is made to explain individual behaviour, an important potential advantage over conventional models is that it is more likely that DM models are stable (or transferable) in time and space.
2. DM models are estimated using individual data and this has the following implications:

   - DM models may be more efficient than conventional models in terms of information usage; fewer data points are required as each individual choice is used as an observation. In aggregate modelling one observation is the average of (sometimes) hundreds of individual observations.
   - As individual data are used, all the inherent variability in the information can be utilised.
   - DM models may be applied, in principle, at any aggregation level; however, although this appears obvious, the aggregation processes are not trivial, as we will discuss below.
   - DM models are less likely to suffer from biases due to correlation between aggregate units. A serious problem when aggregating information is that individual behaviour may be hidden by unidentified characteristics associated to the zones; this is known as *ecological correlation*. The example in Figure 7.1 shows that if a trip generation model was estimated using zonal data, we would obtain that the number of trips decreases with income; however, the opposite would be shown to hold if the data were considered at a household level. This phenomenon, which is of course exaggerated in the figure, might occur for example if the land-use characteristics of zone B are conducive to more trips on foot.
3. Disaggregate models are probabilistic; furthermore, as they yield the probability of choosing each alternative and do not indicate which one is selected, use must be made of basic probability concepts such as:

   - The expected number of people using a certain travel option equals the sum over each individual of the probabilities of choosing that alternative:

$$N_i = \sum_n P_{in}$$

   - An *independent* set of decisions may be modelled separately considering each one as a conditional choice; then the resulting probabilities can be multiplied

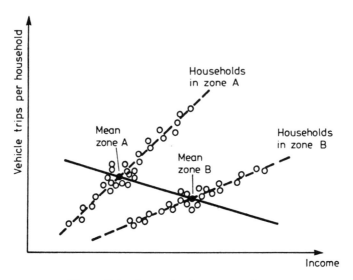

Figure 7.1   Example of ecological fallacy

to yield joint probabilities for the set, such as in:

$$P(f, d, m, r) = P(f)P(d/f)P(m/d, f)P(r/m, d, f)$$

with f = frequency; d = destination; m = mode; r = route.

4. The explanatory variables included in the model can have explicitly estimated coefficients. In principle, the utility function allows any number and specification of the explanatory variables, as opposed to the case of the generalised cost function in conventional models which is generally limited and has several fixed parameters. This has implications such as the following:

- DM models allow for a more flexible representation of the policy variables considered relevant for the study.
- The coefficients of the explanatory variables have a direct marginal utility interpretation (i.e. they reflect the relative importance of each attribute).

In the sections that follow and in the next two chapters we will examine in some detail several interesting aspects of discrete choice models, such as their theoretical base, structure, specification, functional form, estimation and aggregation. Notwithstanding, interested readers are advised that there are at least two good books dealing exclusively with the subject (Ben Akiva and Lerman 1985; Hensher and Johnson 1981).

7.2 THEORETICAL FRAMEWORK

The most common theoretical base, framework or paradigm for generating discrete-choice models is the random utility theory (Domencich and McFadden 1975;

Williams 1977), which basically postulates that:

1. Individuals belong to a given homogeneous population Q, act rationally and possess perfect information, i.e. they always select that option which maximises their net personal utility (the species has even been identified as '*Homo economicus*') subject to legal, social, physical and/or budgetary (both in time and money terms) constraints.

2. There is a certain set $\mathbf{A} = \{A_1, \ldots, A_j, \ldots, A_N\}$ of available alternatives and a set $\mathbf{X}$ of vectors of measured attributes of the individuals and their alternatives. A given individual q is endowed with a set of attributes $\mathbf{x} \in \mathbf{X}$ and in general will face a choice set $\mathbf{A}(q) \in \mathbf{A}$.

In what follows we will assume that the individual's choice set is predetermined; this implies that the effect of the constraints has already been taken care of and does not affect the process of selection among the available alternatives. Choice-set determination will be considered, together with other important but as yet un-resolved issues, in Chapter 8.

3. Each option $A_j \in \mathbf{A}$ has associated a net utility U_{jq} for individual q. The modeller, who is an observer of the system, does not possess complete information about all the elements considered by the individual making a choice; therefore, the modeller assumes that U_{jq} can be represented by two components:

- a measurable, systematic or representative part V_{jq} which is a function of the measured attributes $\mathbf{x}$; and
- a random part ϵ_{jq} which reflects the idiosyncrasies and particular tastes of each individual, together with any measurement or observational errors made by the modeller.

Thus, the modeller postulates that:

$$U_{jq} = V_{jq} + \epsilon_{jq} \tag{7.2}$$

which allows two apparent 'irrationalities' to be explained: that two individuals with the same attributes and facing the same choice set may select different options, and that some individuals may not always select the best alternative (from the point of view of the attributes considered by the modeller).

For the decomposition (7.2) to be correct we need a certain homogeneity in the population under study. In principle we require that *all individuals share the same set of alternatives and face the same constraints* (see Williams and Ortúzar 1982a), and to achieve this we may need to segment the market.

Although we have termed $\mathbf{V}$ *representative* it carries the subscript q because it is a function of the attributes $\mathbf{x}$ and this may vary from individual to individual. Also, without loss of generality it can be assumed that the residuals ϵ are random variables with mean 0 and a certain probability distribution to be specified.

$$V_{jq} = \sum_k \theta_{kj} x_{jkq} \tag{7.3}$$

where the parameters θ are assumed to be constant for all individuals (fixed-coefficients model) but may vary across alternatives. Other possible forms, together with a discussion on how each variable should enter in the utility function, will be presented in Chapter 8.

It is important to emphasise the existence of two points of view in the formulation of the above problem: firstly, that of the individual who calmly weighs all the elements of interest (with no randomness) and selects the most convenient option; secondly, that of the modeller who by observing only some of the above elements needs the residuals ϵ to explain what otherwise would amount to non-rational behaviour.

4. The individual q selects the maximum-utility alternative, that is, the individual chooses A_j if and only if:

$$U_{jq} \geqslant U_{iq}, \quad \forall A_i \in \mathbf{A}(q) \tag{7.4}$$

that is

$$V_{jq} - V_{iq} \geqslant \epsilon_{iq} - \epsilon_{jq} \tag{7.5}$$

As the analyst ignores the value of $(\epsilon_{iq} - \epsilon_{jq})$ it is not possible to determine with certitude if (7.5) holds. Thus the probability of choosing A_j is given by:

$$P_{jq} = \text{Prob} \{ \epsilon_{iq} \leqslant \epsilon_{jq} + (V_{jq} - V_{iq}), \quad \forall A_i \in \mathbf{A}(q) \} \tag{7.6}$$

and as the distribution of the residuals ϵ is not known, it is not possible at this stage to derive an analytical expression for the model. What we do know, however, is that the residuals are random variables with a certain distribution which we can denote by $f(\epsilon) = f(\epsilon_1, \ldots, \epsilon_N)$. In passing let us note that the distribution of $\mathbf{U}$, $f(\mathbf{U})$, is the same but with different mean (i.e. $\mathbf{V}$ rather than 0).

Therefore we can write (7.6) more concisely as:

$$P_{jq} = \int_{R_N} f(\epsilon) \, d\epsilon \tag{7.7}$$

where

$$R_N = \begin{cases} \epsilon_{iq} \leqslant \epsilon_{jq} + (V_{jq} - V_{iq}), \quad \forall A_i \in \mathbf{A}(q) \\ V_{jq} + \epsilon_{jq} \geqslant 0 \end{cases}$$

and different model forms may be generated depending on the distribution of the residuals ϵ.

An important class of random utility models is that generated by utility functions with independent and identically distributed (IID) residuals. In this case $f(\epsilon)$ can be decomposed into:

$$f(\epsilon_1, \ldots, \epsilon_N) = \prod_n g(\epsilon_n)$$

where $g(\epsilon_n)$ is the utility distribution associated with option A_n, and the general expression (7.7) reduces to:

$$P_j = \int_{-\infty}^{\infty} g(\epsilon_j)\, d(\epsilon_j) \prod_{i \neq j} \int_{-\infty}^{V_j - V_i + \epsilon_j} g(\epsilon_i)\, d\epsilon_i \qquad (7.8a)$$

where we have extended the range of both integrals to $-\infty$ (a slight inconsistency) in order to solve them.

A two-dimensional geometric interpretation of this model, together with extensions to the more general case of correlation and unequal variances, are presented and discussed by Ortúzar and Williams (1982). Equation (7.8a) can also be expressed as:

$$P_j = \int_{-\infty}^{\infty} g(\epsilon_j)\, d\epsilon_j \prod_{i \neq j} G(\epsilon_j + V_j - V_i) \qquad (7.8b)$$

with

$$G(x) = \int_{-\infty}^{X} g(x)\, dx$$

and it is interesting to mention that a large amount of effort has been spent in just trying to find out appropriate forms for g which allow (7.8b) to be solved.

Note that the IID residuals requisite means that the alternatives should be, in fact, independent. Mixed-mode options, for example car-rail combinations, will usually violate this condition.

7.3 THE MULTINOMIAL LOGIT MODEL (MNL)

This is the simplest and most popular practical discrete choice model; it can be generated assuming that the random residuals in (7.7) are distributed IID Gumbel (Domencich and McFadden 1975), such that:

$$P_{iq} = \frac{\exp(\beta V_{iq})}{\sum_{A_j \in A(q)} \exp(\beta V_{jq})} \qquad (7.9)$$

where the utility functions usually have the linear in the parameters form (7.3) and the parameter β (which is taken as one in practice as it cannot be estimated separately from the θ') is related to the common standard deviation of the Gumbel variate by:

$$\beta^2 = \pi^2/6\sigma^2 \qquad (7.10)$$

In Chapter 9 we will use (7.10) to discuss the problem of bias in forecasts when use is made of data at different levels of aggregation.

7.3.1 Specification Searches

To decide which variables $x_k \in x$ enter the utility function and whether they are of generic type or specific to a particular alternative, a stepwise process (similar to that of multiple linear regression) is normally employed but starting with a theoretically appealing specification (Ortúzar 1982). Then variations are tested at each step to check whether the variable under scrutiny adds explanatory power to the model; we will examine methods for doing this in Chapter 8.

If for all the individuals q that have available a given alternative A_j we define one of the values of $\mathbf{x}$ equal to one, the coefficient θ corresponding to that variable is interpreted as an alternative specific constant. Although we may specify a constant for every option, it is not possible to estimate their N parameters individually due to the way the model works (as shown in Example 7.1). For this reason one is taken as a reference (fixing its value to 0 without loss of generality) and the remaining $(N - 1)$ values, obtained in the estimation process, are interpreted as relative to the reference. The rest of the variables $\mathbf{x}$ may be of one of two kinds:

- generic, if they appear in the utility function of every alternative and their coefficients can be assumed identical i.e. θ_{jk} may be replaced by θ_k;
- specific, if the assumption of equal coefficients θ_k is not sustainable, a typical example ocurring when the kth variable only appears in V_j.

It must be noted that the most general case considers specific variables only; the generic ones impose on it an equality of coefficients, and this is a condition that may be statistically tested as we will discuss in Chapter 8.

Example 7.1: Consider the following binary logit model:

$$P_1 = \exp(V_1)/[\exp(V_1) + \exp(V_2)] = 1/[1 + \exp(V_2 - V_1)]$$

where the observable utilities are postulated as linear functions of two generic variables x_1 and x_2, and two constants (with coefficients θ_3 and θ_4) as follows:

$$V_1 = \theta_1 x_{11} + \theta_2 x_{12} + \theta_3$$
$$V_2 = \theta_1 x_{21} + \theta_2 x_{22} + \theta_4$$

As can be seen from the model expression, the relevant factor is the difference between both utilities:

$$V_2 - V_1 = \theta_1(x_{21} - x_{11}) + \theta_2(x_{21} - x_{11}) + (\theta_4 - \theta_3)$$

and this allows us to deduce the following conclusions:

- It is not possible to estimate both θ_3 and θ_4, only their difference; for this reason there is no loss of generality if one is taken as 0 and the other estimated relative to it (this of course applies to any number of alternatives).
- If either x_{1j} or x_{2j} have the same value for both options (as in the case of variables representing individual attributes, such as income, age, sex or number of cars in the household), a generic coefficient cannot be estimated as it would always multiply a zero value. This also applies to level-of-service variables which happen to share a common value for two or more options (for example, public-transport fares in a regulated market). In either case they can only appear in some (but not all) options, or need to enter as fully specific variables (i.e. with different coefficients for each alternative).

The problem posed by individual attributes is further compounded by the fact that it is not always easy or clear to decide in which alternative utility(ies) the variable should appear. Consider the case of a variable such as SEX (i.e. 0 for males, 1 for females) in a mode choice study; if we believe, for example, that males always have the upper hand in terms of using the car for commuting purposes, we would not enter the variable in the utilities of both car driver and car passenger, say. However, we may have no insights on whether to enter it or not in the utilities of other modes such as, for example, bus or metro. The problem is that entering the variable in different ways usually yields different estimation results and choosing the optimum may become a hard combinatorial problem, even for a small number of options and attributes. If we lack insight and there are no theoretical grounds for preferring one form over another, the only way out may be trial and error.

7.3.2 Some Properties of the MNL

The model satisfies the axiom of *independence of irrelevant alternatives* (IIA) which can be stated as:

Where any two alternatives have a non-zero probability of being chosen, the ratio of one probability over the other is unaffected by the presence or absence of any additional alternative in the choice set (Luce and Suppes 1965).

As can be seen, in the MNL case the ratio

$$\frac{P_j}{P_i} = \exp\{\beta(V_j - V_i)\}$$

is indeed a constant independent of the rest of the options. Initially this was considered an advantage of the model, as it allows us to treat quite neatly the *new alternative* problem (i.e. being able to forecast the share of an alternative not present at the calibration stage, if its attributes are known); however, nowadays this property is perceived as a disadvantage which makes the model fail in the presence of correlated alternatives (recall the red bus blue bus problem of Chapter 6). We will come back to this in section 7.4.

If there are too many alternatives, such as in the case of destination choice, it can be shown (McFadden 1978) that unbiased parameters are obtained if the model is estimated with a random sample of the available choice set for each individual (for example, seven destination options per individual). Models without this property may require, even if their estimation process is not complex, a large amount of computing time for more than say 50 options. Unfortunately such a figure is not uncommon in a destination-choice context, if one thinks in zoning systems of normal size, even if the combinatorial problem of forming destination/mode choice options is bypassed.

If the model is estimated with information from a sub-area, or with data from a biased sample, it can be shown (Cosslett 1981) that if all individuals have all alternatives available and if the model has a complete set of mode-specific constants, an unbiased model may be obtained just be correcting the constants according to the following expression:

$$K'_i = K_i - \log(q_i/Q_i) \tag{7.11}$$

where q_i is the market share of alternative A_i in the sample and Q_i its market share in the population. All constants must be corrected, including the reference one that is made equal to 0 during estimation.

It is possible to derive fairly simple equations for the direct and cross-elasticities of the model. For example, the direct point elasticity, that is the precentage change in the probability of choosing A_i with respect to a marginal change in a given attribute X_{ikq} is simply given by:

$$E_{P_{iq},X_{ikq}} = \theta_{ik} X_{ikq}(1 - P_{iq}) \tag{7.12}$$

while the cross-point elasticity is also simply given by:

$$E_{P_{iq},X_{jkq}} = -\theta_{jk} X_{jkq} P_{jq} \tag{7.13}$$

that is, the percentage change in the probability of choosing A_i with respect to a marginal change in the value of the kth attribute of alternative A_j, for individual q. Note that as this value is independent from alternative A_i, the cross-elasticities of any option A_i with respect to the attributes X_{jkq} of alternative A_j are equal. This strange result is also due to the IIA property, or more precisely, to the need for IID utility functiions in the model generation.

7.4 THE HIERARCHICAL LOGIT MODEL (HL)

7.4.1 Correlation and Model Structure

In the last section we have discussed the MNL model which, due to its extremely simple covariance matrix, for example in the trinomial case,

$$\Sigma = \sigma^2 \begin{pmatrix} 1 & 0 & 0 \\ 0 & 1 & 0 \\ 0 & 0 & 1 \end{pmatrix}$$

may give rise to problems in either of the following cases:

- when alternatives are not independent (i.e. there are groups of alternatives more similar than others, such as public-transport modes vs. the private car);
- when there are taste variations among individuals (i.e. if the perception of costs varies with income but we have not measured this variable) in which case we require random coefficient models rather than mean-value models as the MNL.

In these two senses the probit model, which can be derived from a multivariate Normal distribution (rather than IID Gumbel), is completely general because it is endowed with an arbitrary covariance matrix. However, as we will see in section 7.5, it is not easy to solve it except for cases with up to three alternatives (see Daganzo 1979).

On the other hand, there are certain situations where even if it were possible to implement it, the full generality afforded by the probit model would be an unnecessary luxury because specific forms for the utility functions suggest themselves. A good example are cases of bi-dimensional choices, such as the combination of destination (D) and mode (M) choice, where alternatives are correlated but taste variations need not be a problem. In these cases the options at each dimension can be denoted as $(D_1, \ldots, D_D)$ and $(M_1, \ldots, M_M)$ with their combination yielding the choice set **A**, whose general element $D_d M_m$ may be a specific destination–mode option to carry out a certain activity.

In this type of context it is interesting to consider functions of the following type (Williams and Ortúzar 1982a):

$$U(d, m) = U_d + U_{dm} \tag{7.14}$$

where for example U_d could correspond to that portion of the utility specifically associated to the destination and U_{dm} to the disutility associated to the cost of travelling. If we write (7.14) following our previous notation we get:

$$U(d, m) = V(d, m) + \epsilon(d, m)$$

where

$$V(d, m) = V_d + V_{dm}$$

and

$$\epsilon(d, m) = \epsilon_d + \epsilon_{dm}$$

It can be shown that if the residuals ϵ are separately IID, under certain conditions the hierarchical or nested logit (HL) model (Williams 1977, Daly and Zachary 1978) is formed:

$$P(d, m) = \frac{\exp\{\beta(V_d + V_d^*)\} \exp(\lambda V_{dm})}{\sum_{d'} \exp\{\beta(V_{d'} + V_{d'}^*)\} \sum_{m'} \exp(\lambda V_{d'm'})} \tag{7.15}$$

with

$$V_d^* = (1/\lambda) \log \sum_{m'} \exp(\lambda V_{dm'})$$

Furthermore, it can easily be shown that if $\beta = \lambda$ (which occurs when $\epsilon_d = 0$) the HL collapses, as special case, to the uniparametric MNL. To understand why this is so, let us first write in full the utility expressions for the first destination in a simple binary mode case:

$$U(1, 1) = V_1 + V_{11} + \epsilon_1 + \epsilon_{11}$$

$$U(1, 2) = V_1 + V_{12} + \epsilon_1 + \epsilon_{12}$$

As can easily be seen, the source of correlation is the residual ϵ_1 which can be found in both $U(1, 1)$ and $U(1, 2)$; therefore when ϵ_d becomes 0, there is no correlation left and the model is indistinguishable from the MNL.

Finally, it can also be shown that for the model to be internally consistent we require that the following condition holds (Williams 1977):

$$\beta \leqslant \lambda \qquad\qquad (7.16)$$

Models that fail to satisfy this requirement have been shown to produce elasticities of the wrong size and/or sign (Williams and Senior 1977).

7.4.2 The HL in Practice

As a modelling tool the HL may be usefully presented in the following fashion (Ortúzar 1980b; Sobel 1980):

1. Its structure is characterised by grouping all subsets of correlated (or more similar) options in hierarchies or nests. Each nest, in turn, is represented by a *composite alternative* which competes with the others available to the individual (the example in Figure 7.2 considers just two nests).

2. The HL may be estimated sequentially using the efficient software developed for the MNL. In Figure 7.2 we would first estimate an MNL for the $A^I(q)$ alternatives of the lower nest, taking care of omitting all those variables (z) which take the same value for this subset of options. These must be introduced later in the superior nest as they affect the choice between the composite alternative N_I and the rest of the options belonging to $A^S(q)$.

3. The introduction of lower nests in their immediate superiors is done by means of the utilities of the composite alternatives which, in general, have two components:

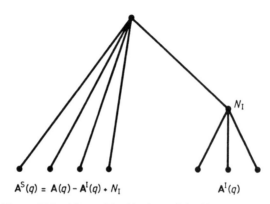

$$A^S(q) = A(q) - A^I(q) + N_I \qquad\qquad A^I(q)$$

Figure 7.2   Hierarchical logit model with two nests

one which consists of the expected maximum utility (EMU) of the lower nest options, and another which considers the vector **z** of attributes which are common to all members of the nest. The EMU portion has the following expression:

$$\text{EMU} = \log \sum_j \exp(W_j) \tag{7.17}$$

where W_j is the utility of alternative A_j in the nest, with the exception of the variables **z** which are common to all options in $\mathbf{A^I}(q)$; note that this is exactly the same form as V_d^* in (7.15) if $\lambda = 1$.

Therefore the composite utility of the nest is:

$$V_I = \phi\text{EMU} + \boldsymbol{\alpha}\mathbf{z} \tag{7.18}$$

where ϕ and $\boldsymbol{\alpha}$ are parameters to be estimated.

4. Having done the above, it is necessary to estimate another MNL for the higher nest; it contains all composite alternatives representing lower hierarchies plus the options which are non-nested at that level.

5. Finally, the probability that individual q selects option $A_j \in A^I(q)$ may be computed as the product of the marginal probability of choosing the composite alternative N_I (in the higher nest) and the conditional probability of choosing option A_j in the lower nest, given that q selected the composite alternative.

Example 7.2: Consider a trinomial modal choice situation involving car (C), bus (B) and metro (M), where the modeller believes that the two latter options are correlated. In this case we would have a lower public-transport (PT) nest which would be modelled by a simple binary logit model of the form:

$$P(\text{M/PT}) = \frac{\exp(W_\text{M})}{\exp(W_\text{M}) + \exp(W_B)}$$

and

$$P(\text{B/PT}) = 1 - P(\text{M/PT})$$

where the utilities **W** contain only those elements which are not common to both modes (i.e. the cost of travel would not enter if both modes charged the same fares).

To separate car from the composite public-transport option at the higher nest, we require another binary logit model:

$$P(\text{C}) = \frac{\exp(V_\text{C})}{\exp(V_\text{C}) + \exp(V_\text{PT})}$$

and

$$P(\text{PT}) = 1 - P(\text{C})$$

where V_c incorporates all the attributes of the car option, i.e. it has exactly the same form as in a MNL.

The public-transport utility is given by:

$$V_{PT} = \phi EMU + \sum_{k'} \alpha_{k'} z_{k'}$$

where

$$EMU = \log[\exp(W_B) + \exp(W_M)]$$

and the summation over k' considers all the common elements $\mathbf{z}$ that were taken out to estimate the binary logit model at the lower nest. Finally, the modelled choice probabilities of each option are given by:

$$P_C = P(C)$$
$$P_B = P(B/PT)P(PT) = P(B/PT)(1 - PC)$$
$$P_M = P(M/PT)P(PT) = (1 - P(B/PT))(1 - PC)$$

6. The internal diagnosis condition (7.16) is expressed in this new notation as:

$$0 < \phi \leqslant 1 \tag{7.19}$$

and let us briefly see why it needs to hold. If $\phi < 0$, an increase in the utility of an alternative in the nest, which should increase the value of EMU, would actually diminish the probability of selecting the nest; if $\phi = 0$, such an increase would not affect the nest's probability of being selected, as EMU would not affect the choice between car and public transport.

On the other hand, if $\phi > 1$ an increase in the utility of an alternative in the nest would tend to increase not only its selection probability but also those of the rest of the options in the nest. Finally, if $\phi = 1$ which is the equivalent to $\beta = \lambda$, the HL model becomes mathematically equivalent to the MNL. In such cases (i.e. when $\phi \approx 1$) it is more efficient to recalibrate the model as an MNL, as the latter has fewer parameters.

HL models are not limited to just two hierarchical levels; in cases with more nesting levels, such as in Figure 7.3, we need at each branch of the structure:

$$0 < \phi_1 \leqslant \phi_2 \leqslant \ldots \leqslant \phi_s \leqslant 1 \tag{7.20}$$

where ϕ_1 correspond to the most inclusive parameter and ϕ_s to that of the highest level.

7. Limitations of the HL:

- In common with the MNL it is not a random coefficients model, so it cannot cope with taste variations among individuals.
- It can only handle as many interdependencies among options as nests have been specified in the structure; furthermore, alternatives in one nest cannot be correlated with alternatives in another nest (this cross-correlation effect, which might be important to test in a mixed-mode modal choice context, for example, can only be handled by the probit model).

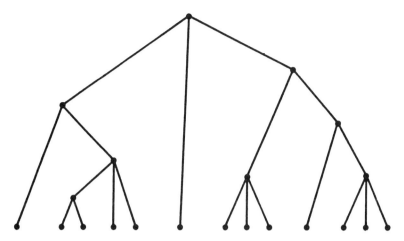

Figure 7.3 Hierarchical logit model with several nests

- The search for the best HL structure may imply the tentative examination of many nesting patterns, as the number of possible structures increases geometrically with the number of options (Sobel 1980). Although *a priori* notions may help in this sense, the modelling exercise might take very much longer than with the simple MNL.
- The sequential estimation method which we have outlined above is simple and feasible, given the availability of software for the MNL, and produces estimates which are consistent (i.e. as the amount of data increases they converge to the true parameter values); however, it has potentially strong problems. For example, if there are not sufficient data to estimate lower nest models, the estimates may be inefficient both in the sense that information is omitted at the lower levels and that the errors thus obtained are passed on to the superior levels. In fact, it has been found empirically that either certain appealing structures cannot be tested, or worse, that the method may lead to the rejection of structures which are demonstrably better (Hensher 1986; Ortúzar *et al.* 1987).

Full information, or simultaneous estimation, is in general much more complex and time consuming (in CPU terms), and up to the mid-1980s it was necessary to write purpose-specific software to achieve it (Small and Brownstone 1982; Hensher 1986). However, there is now a powerful computer package that allows for simultaneous estimation in a very flexible and practical way (Daly 1987; 1992). In fact, given the computational efficiency of the algorithm there seems little reason to use sequential estimation unless forced to do so.

7.5 OTHER CHOICE MODELS

7.5.1 The Multinomial Probit Model

As we mentioned in section 7.4.1, in this model the stochastic residuals ϵ of (7.2) are distributed multivariate Normal with mean zero and an arbitrary covariance matrix, i.e. in this case the variances may be different and the error terms may be correlated in any fashion. The problem is, of course, that this generality does not allow us to write the model in so simple a form as the MNL (except for the binary case); therefore to solve it numerically we need approximations.

7.5.1.1 The Binary Probit Model

In this case we can write the utility expressions (7.2) as:

$$U_1(\boldsymbol{\theta}, \mathbf{Z}) = V_1(\boldsymbol{\theta}, \mathbf{Z}) + \epsilon_1(\boldsymbol{\theta}, \mathbf{Z})$$
$$U_2(\boldsymbol{\theta}, \mathbf{Z}) = V_2(\boldsymbol{\theta}, \mathbf{Z}) + \epsilon_2(\boldsymbol{\theta}, \mathbf{Z})$$

where $\epsilon(\boldsymbol{\theta}, \mathbf{Z})$ is distributed bivariate $N(0, \boldsymbol{\Sigma})$ with

$$\boldsymbol{\Sigma} = \begin{pmatrix} \sigma_1^2 & \rho\sigma_1\sigma_2 \\ \rho\sigma_1\sigma_2 & \sigma_2^2 \end{pmatrix}$$

where ρ is the correlation coefficient between U_1 and U_2. From (7.6), the probability of choosing option 1 is given by:

$$P_1(\boldsymbol{\theta}, \mathbf{Z}) = \text{Prob}\{\epsilon_2 - \epsilon_1 \leq V_1 - V_2\}$$

but as the Normal distribution is closed to addition and subtraction (as the Gumbel is closed to maximisation) we have that $(\epsilon_2 - \epsilon_1)$ is distributed univariate $N(0, \sigma_\epsilon)$, where:

$$\sigma_\epsilon^2 = \sigma_1^2 + \sigma_2^2 - 2\rho\sigma_1\sigma_2$$

Dividing $(\epsilon_2 - \epsilon_1)$ by σ_ϵ we obtain a standard $N(0, 1)$ variate; therefore we can write the binary probit choice probability concisely as:

$$P_1(\boldsymbol{\theta}, \mathbf{Z}) = \Phi[(V_1 - V_2)/\sigma_\epsilon] \tag{7.21}$$

where $\Phi[x]$ is the accumulated standard Normal distribution function which has tabulated values. Although this is indeed a simple model, it is completely general for binary choice.

7.5.1.2 Multinomial Probit (MNP) and Taste Variations

As we noted in sections 7.3 and 7.4, a potentially important problem of fixed-coefficient random utility models, such as the MNL and HL, is their inability to treat the problem of random taste variations among individuals. In what follows we will first show with an example what is meant by this and then we will proceed to show how the MNP handles the problem.

Example 7.3: Consider a mode choice model with two explanatory variables, cost (*c*) and time (*t*) and the following postulated utility function:

$$U = \alpha t + \beta c + \epsilon$$

Let us suppose, however, that the perception of costs varies with income (I), i.e. poorer individuals are more sensitive to cost changes, such that the true utility function is:

$$U = \alpha t + \phi c / I + \epsilon$$

It can easily be seen, comparing both expressions, that the model will be correct only if β can be considered as a random variable with exactly the same distribution as ϕ/I in the population; in this case then, the model contains random taste variations.

The problem of random taste variations is normally very serious, as has been clearly illustrated by Horowitz (1981a), and may be considered as a special case of one well-known specification error, the omission of a relevant explanatory variable, which we discussed in Chapter 3.

Let us consider again the utility function (7.3) which is linear in the parameters, as discussed in section 7.2. Its most general case considers the parameter set θ to be a random vector distributed across the population; in this case the residuals may be modelled as alternative specific parameters, hence the variables ϵ in (7.2) may be omitted without loss of generality and the equation can be written more concisely as:

$$U_j = \sum_k \theta_k x_{jk} \qquad (7.22)$$

which is the most general linear specification possible as it allows for taste variations across the population. If the vector θ is distributed multivariate Normal, the choice model resulting from (7.22) is of MNP form (see Daganzo 1979). Various procedures for estimating this model have been discussed by Sheffi *et al.* (1982) and Langdon (1984).

7.5.1.3 Panel Data Models

In Chapter 1 we discussed briefly the advantages and problems of having a data panel, particularly in relation to the most popular method of collecting revealed-preference data, which is the cross-sectional approach. At the beginning of the 1980s interest in this issue arose coincidentally in the transport-modelling and more general econometric fields; in the first there was concern about the pervasive influence of habit in choice behaviour, in terms of asymmetries in response such as the phenomenon of *hysteresis* which we will examine in section 7.5.3. The same problem was considered as *state dependence* in the econometric parlance and models were postulated for the analysis of discrete choices made over time (Heckman 1981).

Daganzo and Sheffi (1979) have shown how discrete choice time-series and state-dependent models can be approximated by a time-series model without state dependence. Their method involves the use of a mathematical transformation to enable the application of an MNP model to panel data; because of its interest, in what follows we derive a simple binary-choice/two-period form.

Assume that individuals may choose among J options at periods $n = 1, 2$. The utility of an alternative, U_{jn}, can be expressed as a linear function of a vector of attributes, $\mathbf{X}_{jn}$, of the individual and the alternative in the period:

$$U_{jn} = \boldsymbol{\theta}_n \mathbf{X}_{jn} \tag{7.23}$$

In order to estimate the moments of $\boldsymbol{\theta}_n$ we need to rely on observed choices. Note that if there was only one period n (i.e. the cross-sectional case) and $\boldsymbol{\theta}_n$ was distributed Normal, we would get an ordinary MNP estimation problem.

Let us assume that the $\boldsymbol{\theta}$ vectors are jointly multivariate normally distributed, but independent from period to period; if not, we have the problem of *serial correlation* (see Heckman 1981). We will allow for state dependence by assuming that the utility of alternative A_j in period n depends also on the choice (c_{n-1}) in the previous period:

$$U_{jn} = \boldsymbol{\theta}_n \mathbf{X}_{jn} + \phi U_{j(n-1)} \tag{7.24}$$

where ϕ is a parameter to be estimated. Therefore, if ϕ is very large the choice of the previous period will probably be repeated; on the other hand as $\boldsymbol{\theta}_1$, $\boldsymbol{\theta}_2$ are independent, if ϕ is equal to zero, both decisions are made as if by different people. Thus ϕ tends to capture the inertia of the system. Expression (7.24) also allows for the fact that people exhibiting a high differential of utility in period one will be more reluctant to change, *ceteris paribus*, than people who chose the same option but with a small utility differential.

Let us consider now a binary-choice situation where alternative A_1 is chosen in both periods. The probability of this happening is given by:

$$P(A_1^1, A_1^2) = \text{Prob} \{ U_{11} \geqslant U_{21} \text{ and } U_{12} \geqslant U_{22} \}$$

and replacing the values from equation (7.24) we get:

$$P(A_1^1, A_1^2) = \text{Prob} \{ \boldsymbol{\theta}_1 \mathbf{X}_{11} \geqslant \boldsymbol{\theta}_1 \mathbf{X}_{21} \text{ and } \boldsymbol{\theta}_2 \mathbf{X}_{12} + \phi U_{11} \geqslant \boldsymbol{\theta}_2 \mathbf{X}_{22} + \phi U_{21} \}$$

that is to say,

$$P(A_1^1, A_1^2) = \text{Prob} \{ 0 \geqslant \boldsymbol{\theta}_1 (\mathbf{X}_{21} - \mathbf{X}_{11}) \text{ and } 0 \geqslant \boldsymbol{\theta}_2 (\mathbf{X}_{22} - \mathbf{X}_{12}) + \phi (U_{21} - U_{11}) \} \tag{7.25}$$

This suggests the following transformation for the utilities in (7.24):

$$U_{11} = 0, \; U_{12} = 0$$
$$U_{21} = \boldsymbol{\theta}_1 (\mathbf{X}_{21} - \mathbf{X}_{11}) \tag{7.26}$$
$$U_{22} = \boldsymbol{\theta}_2 (\mathbf{X}_{22} - \mathbf{X}_{12}) + \phi (U_{21} - U_{11})$$
$$= \boldsymbol{\theta}_2 (\mathbf{X}_{22} - \mathbf{X}_{12}) + \phi U_{21}$$

Now denoting by V_{2n} the term $\theta_n(\mathbf{X}_{2n} - \mathbf{X}_{1n})$ this can be written simply as follows:

$$U_{1n} = 0$$
$$U_{2n} = V_{2n} + \phi V_{2(n-1)} \tag{7.27}$$

For a fixed value of ϕ equation (7.27) yields a linear function of θ; if ϕ is not known we have a non-linear specification (where ϕ is just another parameter) which poses no particular estimation problems to currently available software (Daganzo 1979).

Generalising this procedure by defining the utility of the chosen alternative in both periods as zero, an equivalent trinomial model ($N + 1$ in the general binary choice case for N periods) with utilities $\mathbf{W}$ can be formulated as follows (Daganzo and Sheffi 1979):

$$W_0 = 0$$
$$W_1 = \theta_1(\mathbf{X}_{NC1} - \mathbf{X}_{C1}) \tag{7.28}$$
$$W_2 = \theta_2(\mathbf{X}_{NC2} - \mathbf{X}_{C2}) + \phi W_1$$

where NC stands for not chosen and C for chosen. Using this specification the probability of choosing the dummy alternative, Prob $\{W_0 \geqslant W_1 \text{ and } W_0 \geqslant W_2\}$, is given by:

$$\text{Prob } \{0 \geqslant \theta_1(\mathbf{X}_{NC1} - \mathbf{X}_{C1}) \text{ and } 0 \geqslant \theta_2(\mathbf{X}_{NC2} - \mathbf{X}_{C2}) + \phi W_1\}$$

which is essentially (7.25) in general form. Hence it is possible to estimate an MNP model by evaluating the probability of each individual in the sample choosing the dummy alternative. Johnson and Hensher (1982) applied this procedure to Australian data and found that the elasticities derived from the panel data model were very different and lower than those obtained from various optional cross-sectional models.

7.5.2 Choice by Elimination and Satisfaction

In Chapter 8 we discuss the problem of specification and functional form giving particular emphasis to the linear-in-the-parameters form which has accompanied the vast majority of dissaggregate demand (normally of multinomial logit structure) applications. Owing to a growing body of criticism directed at linear-in-the-parameters forms, the early 1980s witnessed an important interest in the specification and estimation of non-linear formulations of varying designs. Commentary on the functional characteristics of these forms was interwined with statements about alternative models of the decision process considered to underpin choice models.

One typical view was that because linear-in-the-parameters forms are associated with a compensatory decision-making process (i.e. a change in one or more of the attributes may be compensated by changes in the others), models cannot be appropriately specified for decision processes characterised by perception of discontinuities which are more plausibly of a non-compensatory nature (i.e. where good aspects of an alternative may not be allowed to compensate for bad aspects which are ranked higher in importance in the selection procedure, simply because that alternative may be eliminated earlier in the search process; see the discussion in Golob and Richardson 1981).

Example 7.4: Let us consider a set of individuals, confronted by a choice, to be endowed with a set of objectives **G** and a set of constraints **B**. A general multi-criterion problem can then be formally stated as:

$$\underset{\text{(options)}}{\text{Max}} \{F_1(Z_1^1) \ldots F_1(Z_N^1)\}$$

$$\vdots$$

$$\underset{\text{(options)}}{\text{Max}} \{F_k(Z_1^k) \ldots F_k(Z_N^k)\} \tag{7.29}$$

$$\vdots$$

$$\underset{\text{(options)}}{\text{Max}} \{F_K(Z_1^K) \ldots F_K(Z_N^K)\}$$

subject to the vector of constraints:

$$\mathbf{f}(\mathbf{Z}) \leq \mathbf{B} \tag{7.30}$$

in which $F_k(Z_j^k)$ is the value of the criterion function associated with attribute Z_j^k of option A_j. For example, we might be interested in finding a mode, in a choice set of size N, which minimises travel time and cost, maximises comfort and safety, and so on. These attributes associated with any particular alternative might, in addition, be required to satisfy absolute constraints such as (7.30).

If a single alternative is found which simultaneously satisfies these optimality criteria (i.e. it optimises the K functions in expression 7.29) and whose attributes are feasible in terms of (7.30), then an unambiguous optimal solution is obtained. In general, however, there will be conflicts between objectives (i.e. options superior in some respects and inferior in others).

A number of important questions can be posed before a choice model based on this multi-criterion problem may be constructed:

- What strategies might be adopted to solve the problem?
- Are there differences in the strategies adopted by different individuals in a given population?
- How can these strategies be formally represented?
- How should the aggregation over the population be conducted to produce a model to be estimated with individual data?

The last point is especially important because choice models are derived by aggregating over the actions of individuals within the population, and while any or all of them may indulge in a non-compensatory decision process, it may or may not be appropriate to characterise the sum total of these decisions and the resultant choice model in these terms (see the discussion in Williams and Ortúzar 1982a).

We will just refer here to the first of these issues, namely how an individual confronted by a hypothetical decision context may resolve the multi-criterion problem. There is of course a wide literature dispersed over several fields, which involves the application of decision theory to problems of this kind. We will mention three methods, starting with the best known, simplest and most widely used approach, the trade-off strategy which forms the basis for compensatory decision models.

7.5.2.1 Compensatory Rule

Here the preferred option is selected by optimising a single objective function $G = G(F_1, F_2, \ldots, F_K)$. If the F_k functions are simply the attributes $\mathbf{Z}^k$, or linear transformations of them, G may be written as:

$$G = G\left(\sum_k \theta_k Z_1^k, \ldots, \sum_k \theta_k Z_j^k, \ldots, \sum_k \theta_k Z_N^k\right) \tag{7.31}$$

and the conventional linear trade-off problem is addressed. The parameters θ are determined from either the stated or revealed preferences of the individual decision maker. One of the characteristics of this trade-off approach is its symmetric treatment of the objective functions.

7.5.2.2 Non-Compensatory Rules

An alternative general approach is to treat the objective functions (7.29) asymmetrically by either ranking them or converting some or all to constraints by introducing norms or thresholds. That is, we might require that any acceptable alternative has, for example, an associated travel cost not exceeding a particular amount; formally the restriction is imposed that:

$$Z_1^k, \ldots, Z_j^k, \ldots, Z_N^k \leqslant Z^k \tag{7.32}$$

in which Z^k is a maximum (or minimum when the inequality sign is reversed) satisfactory value for the attribute. The creation of norms or thresholds restricts the range of feasible alternatives which individuals are considered to impose on their decision process.

Choice by Elimination In this case it is assumed that individuals possess both a ranking of attributes (e.g. cost is more important than waiting time, which in turn is more important than walking time, etc.) and minimum acceptable values or thresholds (7.32) for each. For example, the decision process may solve the multi-criterion problem in the following fashion: first the highest ranked attribute is considered and all alternatives not satisfying the threshold restriction are eliminated (even though they may excel in lesser ranked attributes); the process is repeated until only one option is left, or a group which satisfies all the threshold constraints among which one is selected in a compensatory manner (see Tverski 1972).

Satisficing Behaviour There are, however, a great many ways in which the above search strategy may be considered to be organised; for example, it might be that a complex cyclic process is used by the individual whereby the thresholds become sequentially modified until a unique alternative is found. Equally, a satisficing mechanism might operate in which the individual might be prepared to curtail the search at any point according to a pre-specified rule, in which case some or all of the attributes or alternatives may not be considered. Indeed, when the notion of satisficing (see Eilon 1972) is applied to travel-related decisions involving location, the decision model is closely associated with the acquisition of information in the search process.

As Young and Richardson (1980) remarked, a search may be characterised by an elimination process based on attributes or one based on alternatives. In the former, attributes are selected in turn and options are processed, and maintained or rejected depending on the values of these attributes; in the latter, alternatives are considered in turn and their bundle of attributes examined. At any stage of the process options which do not satisfy norms or other constraints are eliminated. A more detailed consideration of decision strategies is given by Foerster (1979) and Williams and Ortúzar (1982a).

7.5.3 Habit and Hysteresis

At the end of the 1970s there was considerable interest in the relevance and role of habit in travel choice behaviour, particularly in cases of relocation (i.e. migration) or other phenomena granting a fresh look at the individual's choice set. Empirical evidence (Blase 1979) suggested that the effect of habit can be of practical significance and the problem should be treated seriously.

The existence of habit, or what might be considered as inertia accompanying the decision process of an individual, is possibly the most insidious of behavioural aspects which represent divergencies from the traditional assumptions underpinning choice models, for it appears directly in the response context. In order to examine the effects and implications of habit it is appropriate to return to the assumptions behind the conventional cross-sectional approach.

Figure 7.4a reproduces the S-shaped curve relevant to binary choice. For a given difference in utility $(V_2 - V_1)$ there exists a certain unique probability of choice; under conditions of change $(V_2' - V_1')$, the probability will correspond to that observed for that utility difference in the base year, i.e. the response is determined from the cross-sectional dispersion. An implication of this assumption is that response to a particular policy or change will be exactly reversed if the stimulus is removed; the stimulus–response relation is symmetric with respect to the sign and size of the stimulus.

If habit exists it will affect those members of the population who are currently associated with an option experiencing a stimulus to the relative advantage of another alternative. This introduces a basic asymmetry into response behaviour and

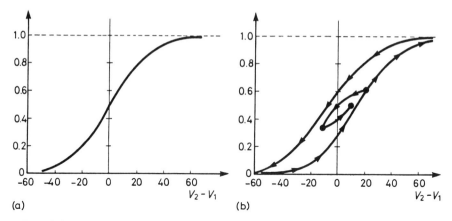

Figure 7.4 Influence of habit in cross-sectional models: (a) logit response curve, (b) hysteresis curve for habit effect

gives rise to the phenomenon of hysteresis (Goodwin 1977), as pictured in Figure 7.4b. In this case the present state of the population identified in terms of the market share of each alternative depends not only on the utility values V_2 and V_1, but on how these variables attained their current value.

Formally, the state of the system P may be expressed as a path integral in the space of utility components $\mathbf{V}$; the value of the integral is path independent when habit is absent but path dependent when it is present (see the discussion in Williams and Ortúzar 1982a).

EXERCISES

7.1 There is interest to study the behaviour of a group of travellers in relation to two transport options A and B, with travel times t_a and t_b respectively. It has been postulated that each traveller experiments the following net utilities from each option:

$$U_a - \alpha t_a + \beta I$$

$$U_b = \alpha t_b$$

where α and β are known parameters and I is the traveller's personal income.

Although there is no reliable data about the income of each traveller, it is known that the variable I has the following distribution in the population:

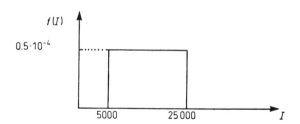

If $\alpha = -0.5$ and $\beta = 2.10^{-4}$, find out the probability function of choosing option A for a given traveller, as a function of the value of $(t_b - t_a)$; sketch the function in appropriate axis.

7.2 Consider a binary logit model for car and bus, where the following representative utility functions have been estimated with a sample of 750 individuals belonging to a particular sector of an urban area:

$$V_c = 3.5 - 0.25t_c - 0.42e_c - 0.1c_c$$
$$V_b = -0.25t_b - 0.42e_b - 0.1c_b$$

where t is in-vehicle travel time (min), e is access time (min) and c is travel cost ($). Assume the following average data is known:

		Variable	
Mode	t	e	c
Car	25	5	140
Bus	40	8	50

If you are informed that the number of individuals choosing each option in the sector and in the complete area are respectively as follows:

	Number of individuals choosing option i	
Option	Sample	Population
Car	283	17 100
Bus	467	68 900

(a) Indicate what correction would be necessary to apply to the model and write its final formulation.
(b) Calculate the percent variation in the probability of choosing car if the bus fares go up by 25%.
(c) Find out what would happen if, on the contrary, the car costs increase by 100%.

7.3 Compute the probabilities of choosing car, bus, shared taxi and underground, according to the following hierarchical logit model:

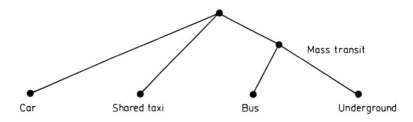

with the following utility functions:

(a) High nest

$$V_c = -0.03t_c - 0.02c_c + 1.25$$
$$V_{st} = -0.03t_{st} - 0.02c_{st} - 0.20$$
$$V_{mt} = 0.60\text{EMU}$$

(b) Mass transit nest

$$V_b = -0.04t_b - 0.03c_b + 0.5$$
$$V_u = -0.04t_u - 0.03c_u$$

and for the average variable values presented in the following table:

Mode	Time (t)	Cost/income (c)
Car	4.5	23.0
Shared taxi	5.5	15.0
Bus	7.5	5.5
Underground	5.5	3.6

7.4 The binary probit model has the following expression:

$$P_1 = \Phi\{(V_1 - V_2)/\sqrt{\sigma_1^2 + \sigma_2^2 - 2\rho\sigma_1\sigma_2}\}$$

Using this result write down the probability of choosing option one in the following binary model:

$$U_i = \theta X_i + \epsilon_i$$

where the ϵ are distributed IID standard Normal, for the following cases:
(a) If the value of θ is fixed and equal to 3
(b) If θ is distributed Normal $N(3, 1)$ and is independent of the ϵ.

8 Specification and Estimation of Discrete Choice Models

8.1 INTRODUCTION

The previous chapter provided an overview of discrete choice modelling and an introduction to different model forms and theoretical frameworks for individual decisions. This chapter is devoted to a discussion of two key issues: how to fully specify a discrete or disaggregate model (DM) and how to estimate such a model once properly specified.

The search for a suitable model specification involves selecting the structure of the model (MNL, HL, probit, etc.), the explanatory variables to consider, the form in which they enter the utility functions (linear, non-linear) and the identification of the individual's choice set (alternatives available). In broad terms the objectives of a specification search include realism, economy, theoretical consistency and policy sensitivity. In other words, we search for a realistic model, which does not require too many data and computer resources, does not produce pathological results and is appropriate to the decision context where we want to use it. Aggregate models such as those we discussed in Chapters 5 and 6 are often critically portrayed as policy insensitive, either because key variables have been completely left out of the model or because important model components have been specified as insensitive to certain policies (e.g. consider the problem of inelastic trip generation). Most of the features of model specification are susceptible to analysis and experimentation (see Leamer 1978) but they are also strongly dependent on study context and data availability.

In this chapter we start by considering how to identify the set of options available to individuals: choice-set determination. This is a key problem as we estimate DM by means of the (generally) observed individual choices between alternatives. These should be the alternatives actually considered, conciously or unconciously, by the individual. The omission of seemingly unimportant options on the grounds of costs may bias results. For example, in the vast majority of aggregate studies only binary choice between car and public transport was considered with the consequence that the multimodal problem could not be treated seriously; in fact, in the best cases the consideration of alternative public-transport options was relegated to the assignment stage employing rather crude all-or-nothing or multipath allocation of trips to submodal network links. In the same vein, the inclusion of alternatives which are actually ignored by certain groups (say walking more than 500 metres for high income individuals), would also bias model estimation.

The chapter then considers the other elements of model specification and in particular functional form and model structure. The criteria of economy, realism, theoretical consistency and decision-making context play a key role in complementing the experience and intuition of the modeller during specification searches. An additional, and often over-riding element, is the availability of specialised software. In fact, one of the reasons behind the immense popularity of the linear-in-the-parameters multinomial logit (MNL) model, is that it can be easily estimated with normally available software; this is not the case for more general structures or functional forms which normally present enormous difficulties (Daganzo 1979; Liem and Gaudry 1987).

The increasing availability of good software to select and estimate these models will certainly alleviate this problem. However, one issue to which we will return is that although we may be able to successfully estimate the parameters of widely different models with a given data set, these (and their implied elasticities) will tend to be different and we often lack the means to discriminate between them, at least with cross-sectional data.

The final specification will then depend heavily on the modeller's experience and theoretical understanding, and context-specific factors such as: time and resources available for the modelling activity (the MNL is considerably simpler and cheaper than its competitors); degree of correlation among alternatives; and required degree of accuracy of the forecasts, among others. It has to be borne in mind that using an inadequate model, such as the MNL when the hypotheses needed to generate it do not hold, may lead to serious errors (Williams and Ortúzar 1982a).

The chapter then discusses in detail model estimation and validation techniques as a function of the sampling method used to generate the observations. Maximum likelihood and the different statistics used to select best fit are presented in section 8.3. Section 8.4 discusses the generation and use of validation samples. The chapter concludes with considerations relevant to model specification and estimation with stated-preferences (as opposed to revealed-preferences) data. Stated-preferences techniques can be used to advantage in order to better specify discrete choice models and they constitute a useful additional element in the modeller's tool-kit.

8.2 CHOICE-SET DETERMINATION

One of the first problems an analyst has to solve, given a typical revealed-preferences cross-sectional data set, is that of deciding which alternatives are available to each individual in the sample. It has been noted that this is one of the most difficult of all the issues to resolve, because it reflects the dilemma the modeller has to tackle in arriving at a suitable trade-off between modelling relevance and modelling complexity; usually however, data availability acts as a yardstick.

8.2.1 Choice-set Size

It is extremely difficult to decide on an individual's choice set unless one asks directly; therefore the problem is closely connected with the dilemma of whether to use reported or measured data, as discussed in Chapter 3. Although in mode choice modelling the number of alternatives is usually small, rendering the problem less severe, in other cases such as destination choice, the identification of options in the choice set is a crucial matter. This is not simply because the total number of alternatives is usually very high, as we will see below, but because we face the added problem of how to measure/represent the attractiveness of each option. Ways of managing a large choice set include:

1. Taking into account only subsets of the options which are effectively chosen in the sample (i.e. in a sampling framework such as the one used by Ben Akiva 1977);
2. Using the *brute force* method, which assumes that everybody has all alternatives available and hence lets the model decide that the choice probabilities of unrealistic options are low or zero.

Both approaches have disadvantages. For example, in case 1 it is possible to miss realistic alternatives which are not chosen owing to the specific sample or sampling technique; in case 2 the inclusion of too many alternatives may affect the discriminatory capacities of the model, in the sense that a model capable of dealing with unrealistic options may not be able to describe adequately the choices among the realistic ones (see Ruijgrok 1979). Other methods to deal with the choice set-size problem are:

3. the aggregation across options, such as in a destination choice model based in zonal data;
4. assuming continuity across alternatives, such as in the work of Ben Akiva and Watanatada (1980).

8.2.2 Choice-set Formation

Another problem in this realm is that the decision maker being modelled may well choose from a relatively limited set; in this sense if the analyst models choices which are actually ignored by the individual, some alternatives will be given a positive probability even if they have no chance of being selected in practice. Moreover, consider the case of modelling the behaviour of a group of individuals who vary a great deal in terms of their knowledge of potential destinations (owing perhaps to varying lengths of residence in the area); because of this, model coefficients which attempt to describe the relationship between predicted utilities and observed choices may be influenced as much by variation in choice sets among individuals (which are not fully accounted for in the model) as by variations in actual preferences (which are accounted for). Because changes in the nature of the destinations may affect both choice set and preferences to different degrees, this confusion may be likely to play havoc with the use of the model in forecasting or with the possibility of transferring it over time and space.

Ways to handle this problem include:

1. The use of heuristic or deterministic choice-set generation rules which allow to exclude certain alternatives (i.e. bus is not available if the nearest stop is more than some distance away) and which may be validated using data from the sample.
2. The obtention of choice-set information directly from the sample, simply by asking respondents about their perception of available options (or better still about unavailable options and why).
3. The use of random choice sets, whereby choice probabilities are considered to be the result of a two-stage process: firstly, a choice-set generating process, in which the probability distribution function over all possible choice sets is defined; and secondly, conditional on a specific choice set, a probability of choice for each alternative is defined (see the discussions by Lerman 1984; and Richardson 1982).

8.3 SPECIFICATION AND FUNCTIONAL FORM

The search for the best model specification is also related to functional form. Although it may be argued that the linear function (7.3) is probably adequate in many contexts, there are others such as destination choice where non-linear functions are deemed more appropriate (Foerster 1981; Daly 1982a). The problems in this case are that, in general, there is no guarantee that the parameter-estimation routine will converge to unique values, and that software is not readily available. Another specification issue related to functional form is how the explanatory variables should enter the utility function, even if this is linear in the parameters.

Three approaches have been proposed in the literature to handle the functional form question:

- The use of conjoint analysis in real or laboratory experiments to determine the most appropriate form of the utility function (Lerman and Louviere 1978); we will briefly come back to this in section 8.5.
- The use of statistical transformations, such as the Box–Cox method, letting the data 'decide' to a certain extent (Gaudry and Wills 1978).
- The constructive use of econometric theory to derive functional form (Train and McFadden 1978; Jara-Díaz and Farah 1987); this is perhaps the most attractive proposition as the final form can be tied up to evaluation measures of user benefit.

As we will see later, it is important to note that in general non-linear forms imply different trade-offs to those normally associated with concepts such as the value of time (Bruzelius 1979); also, it is easy to imagine that model elasticities and explanatory power may vary dramatically with functional form.

8.3.1 Functional Form and Transformations

Linear-in-the-parameters expressions such as (7.3) usually contain a mixture of quantitative and qualitative variables (where the latter are normally specified as dummies, i.e. sex, age, income level), and the problems are how to enter both and where to enter the latter, as we have already discussed. In other words, it would be more appropriate to write (7.3) as:

$$V_{jq} = \sum_k \theta_{kj} f_{kj}(x_{kjq}) \tag{8.1}$$

which is still linear in the parameters, but makes it explicit that the functional form of the x variables is somewhat arbitrary. Usual practice consists in entering them in raw form (i.e. time rather than 1/time or its logarithm) and this would be of no consequence unless the model response were sensitive to functional form.

If we do not have theoretical reasons to back up a given form, it appears interesting to let the data indicate which could be an appropriate one. A class of transformations widely used in econometrics has been successfully adapted for use in transport modelling (see Gaudry and Wills 1978; Liem and Gaudry 1987). We will review two examples, the second one being a generalisation of the first:

8.3.1.1 Basic Box–Cox Transformation

The transformation $x^{(\tau)}$ of a positive variable x, given by:

$$x^{(\tau)} = \begin{cases} (x^\tau - 1)/\tau, & \text{if } \tau \neq 0 \\ \log x, & \text{if } \tau = 0 \end{cases} \tag{8.2}$$

is continuous for all possible τ values. With this we can rewrite equation (8.1) as:

$$V_{jq} = \sum_k \theta_{kj} x_{kjq}^{\tau_k} \tag{8.3}$$

and it is easy to see that if $\tau_1 = \tau_2 = \ldots = \tau_k = 1$, (8.3) reduces to the typical linear form (7.3); furthermore, if all $\tau_k = 0$, we obtain the widely used log-linear form. Therefore both traditional forms are only special cases of (8.3).

8.3.1.2 Box–Tukey Transformation

The basic transformation (8.2) is only defined for $x > 0$; a more general form, for variables that may take negative or zero values, is given by:

$$(x + \mu)^{(\tau)} = \begin{cases} [(x + \mu)^\tau - 1]/\tau, & \text{if } \tau \neq 0 \\ \log (x + \mu), & \text{if } \tau = 0 \end{cases} \tag{8.4}$$

where μ is just a translation constant chosen to ensure that $(x + \mu) > 0$ for all observations.

It can be shown that is an MNL is specified with functional form (8.4) and restricting all τ to be equal, its elasticities are given by:

$$E_{P_J, x_{ki}} = (\delta_{ji} - P_j) x_{ki} \theta_k (x_{ki} + \mu)^{\tau-1} \qquad (8.5)$$

with δ_{ji} equal to 1 if $j = i$ and 0 otherwise. Although it is obvious from (8.5) that the elasticities depend on the values of τ and μ, it is not clear how large the effect might be as the values of $\boldsymbol{\theta}$ also vary.

Using data for Sidney in 1971, Hensher and Johnson (1981) reported 25% differences between train travel time/demand elasticity values for the optimal (τ, μ) combination and the linear case; the incorrect $(\tau = 1)$ specification suggesting the higher values. In Chapter 13 we will discuss the consequences of using Box–Cox models in the derivation of subjective values of time (Gaudry *et al.* 1989).

8.3.2 Theoretical Considerations and Functional Form

Although we have made it clear that in any particular study, data limitations and resource restrictions often play a vital role, it is important to consider the influence of theory in the construction of a demand function. In what follows we will show how the constructive use of economic theory helps to solve the important problem of how to incorporate a key variable, such as income, in a utility function. Throughout we will assume a linear-in-the-parameters form and will not be concerned with model structure, but the analysis may be generalised at a later stage.

The conventional approach to understanding the roles of income, time and cost of travel within the discrete choice framework, is based on the work of Train and McFadden (1978); they established the microeconomic foundations of the theory by considering the case of individuals who choose between leisure (L) and goods consumed (G); the trade-off appears once the link between G and income (I) is formulated: they assume that I depends on the number of hours worked (W). Thus, increasing W allows G to increase, diminishing L. More formally the problem is stated as follows:

$$\text{Max } U(G, L)$$

subject to:

$$\left. \begin{array}{l} G + c_i = wW \\ W + L + t_i = T \end{array} \right\} \quad \forall A_i \in \mathbf{A} \qquad (8.6)$$

where U is the individual utility function, w is the real wage rate (the amount the individual gets paid per hour), c_i and t_i are the money and time spent per trip respectively, $\mathbf{A}$ is the choice set and T is a reference period (hours); the unknowns are G, L and W.

If U in problem (8.6) is given a fairly general form, such as Cobb–Douglas, finding its maximum with respect to $A_i \in \mathbf{A}$, is equivalent to finding the maximum of $(-c_i/w - t_i)$ among other possibilities. This is the origin of the widely used cost/wage rate variable in discrete-mode choice models, which has degenerated into cost/income in certain applications. The possibility of adapting working hours to attain a desired level of income plays a key role in the above derivation; thus, as W is endogenously determined and w is given exogenously, income becomes endogenous. This formulation assumes that the cost of travelling is negligible in relation to income, i.e. that there is no income effect.

However, for many individuals (particularly in less developed countries) both income and working hours are fixed and there may be income effects. In such cases it can be shown that the maximum of U depends on the value of $(-c_i/g - t_i)$ among other possibilities (Jara-Díaz and Farah 1987), where g is an *expenditure rate* defined in general by:

$$g = I/(T - W) \tag{8.7}$$

The presence of such an income variable, reflecting purchasing power in the utility specification, indicates that the marginal utility of income varies with income, i.e. the model allows for an income effect. However, it is interesting to mention that empirical tests have shown that this new specification consistently outperforms the conventional wage-rate specification, even for individuals with no income effect (Jara-Díaz and Ortúzar 1989).

8.3.3 Intrinsic Non-linearities: Destination Choice

Let us treat the singly constrained gravity model (5.13)–(5.14) we examined in Chapter 5 in a disaggregate manner by considering each individual trip maker in zone i as making one of the O_i trips originating in that zone. In this case the probability that a person will make the choice of travelling to zone j is simply:

$$P_j = T_{ij}/O_i = B_j f_{ij} \sum_k B_k f_{ik} \tag{8.8}$$

Now if we define:

$$V_k = \log(B_k f_{ik}) = \log B_k + \log f_{ik} \tag{8.9}$$

the model is seen to be exactly equivalent to the multinomial logit model (7.9). Thus the conventional origin-constrained gravity model may be represented by the disaggregate MNL without any loss of generality (Daly 1982a); note that (8.9) imposes no restrictions on the specification of the separation function f_{ij}. As we saw in Chapter 5, probably the most common function used in practice is the negative exponential of c_{ij}, the generalised cost of travelling between zones i and j; it is interesting to mention that when this form is substituted in (8.9) we obtain:

$$V_k = \log B_k - \beta c_{ij} \tag{8.10}$$

which is in fact linear in the parameter β. The problem of non-linearity arises due to the presence of B_k which may contain variables of the *size* variety that describe not the quality but the number of elementary choices within k and are typical of cases, such as choice of destination, where aggregation of alternatives is required (Daly 1982a).

8.4 STATISTICAL ESTIMATION

This section considers methods for the estimation of DM together with the goodness-of-fit statistics to be used in this task. Model estimation methods need to be adapted to the sampling framework used to generate the observations. This is necessary to improve estimation efficiency and avoid bias.

8.4.1 Estimation of Models from Random Samples

To estimate the coefficients θ_k in (7.3) the maximum likelihood (ML) method is normally used. ML is based on the idea that although a sample could originate from several populations, a particular sample has a higher probability of having been drawn from a certain population than from others. Therefore the ML estimates are the set of parameters which will generate the observed sample most often.

To illustrate this idea consider a sample of n observations of a given variable $\mathbf{Z} = \{Z_1, \ldots, Z_n\}$ drawn from a population characterised by a parameter θ (mean, variance, etc.). As $\mathbf{Z}$ is a random variable it has associated a density function $f(\mathbf{Z}/\theta)$ which depends on the values of θ. The values of $\mathbf{Z}$ in the sample being independent, we can write the joint density function as:

$$f(Z_1, Z_2, \ldots, Z_n/\theta) = f(Z_1/\theta)f(Z_2/\theta) \ldots f(Z_n/\theta)$$

The usual statistical interpretation of this function is with $\mathbf{Z}$ as variables and θ fixed. Inverting the process, the previous equation can be interpreted as a likelihood function $L(\theta)$; if we maximise it with respect to θ, the result is called maximum likelihood estimate because it corresponds to the parameter value which has the greatest probability of having generated the observed sample. Of course the idea may be extended to several parameters (for example, in multiple linear regression it can be shown that the least square coefficients are maximum likelihood estimates).

Let us assume a sample of Q individuals for which we observe their choice (0 or 1) and the values of x_{jkq} for each available alternative, such that for example:

individual 1 selects alternative 2

individual 2 selects alternative 3

individual 3 selects alternative 2

individual 4 selects alternative 1 etc.

As the observations are independent the likelihood function is given by the product of the model probabilities that each individual chooses the option they actually selected:

$$L(\theta) = P_{21} P_{32} P_{23} P_{14} \ldots$$

Defining the following dummy variable:

$$g_{jq} = \begin{cases} 1 & \text{if } A_j \text{ was chosen by } q \\ 0 & \text{otherwise} \end{cases} \tag{8.11}$$

the above expression may be written more generally as:

$$L(\theta) = \prod_{q=1}^{Q} \prod_{Aj \in A(q)} (P_{jq})^{g_{jq}} \tag{8.12}$$

To maximise this function we proceed as usual, differentiating partially with respect to θ and equating it to 0. As in other cases we normally maximise $l(\theta)$, the natural logarithm of $L(\theta)$, which is more manageable and yields the same optima.

Therefore, the function we seek to maximise is (Ortúzar 1982):

$$l(\theta) = \log L(\theta) = \sum_{q=1}^{Q} \sum_{Aj \in A(q)} g_{jq} \log P_{jq} \tag{8.13}$$

When $l(\theta)$ is maximised, a set of estimated parameters θ^* is obtained which is distributed $N(\theta, S^2)$ where:

$$S^2 = \frac{-1}{E\left(\dfrac{\partial^2 l(\theta)}{\partial \theta^2}\right)} \tag{8.14}$$

Also LR $= -2l(\theta)$ is asymptotically distributed χ^2 with Q degrees of freedom (see Ben Akiva and Lerman 1985). All this indicates that even though θ^* may be biased in small samples, the bias is small for large enough samples (normally samples of 500 to 1000 observations are more than adequate).

Now, although we have an explicit expression for the covariance matrix S, determining the parameters θ^* involves an iterative process. In the case of the linear-in-the-parameters MNL the function is well behaved, so the process converges quickly and always to a unique maximum; this explains why software to estimate this model is so easily available. Unfortunately this is not the case for other discrete choice models the estimation processes of which are much more involved; therefore in what follows we will mainly refer to this simpler model.

Substituting the MNL expression (7.9) in (8.13), it can be shown that if the variable set includes an alternative specific constant for option A_j we have:

$$\sum_q g_{jq} = \sum_q P_{jq}$$

and this allows us to deduce that as alternative specific constants tend to capture the effect of variables not considered in the modelling, they ensure that the model always reproduces the aggregate market shares of each alternative. Therefore it is not appropriate to compare, as a goodness-of-fit indicator, the sum of the probabilities of choosing one option with the total number of observations that selected it, because this condition will be satisfied automatically by a model with a full set of constants. As it is also not possible to compare the model probabilities with the g_{jq} values (which are either 0 or 1), a goodness-of-fit measure such as R^2 in ordinary least squares, which is based on estimated residuals, cannot be defined.

Example 8.1: Consider a simple binary-choice case with a sample of just three observations (as proposed by Lerman 1984); let us also assume that there is only one attribute x, such that:

$$P_{1q} = 1/\{1 + \exp[\theta(x_{2q} - x_{1q})]\}; \qquad P_{2q} = 1 - P_{1q}$$

and also that we observed the following choices and values:

Observation (q)	Choice	x_{1q}	x_{2q}
1	1	5	3
2	1	1	2
3	2	3	4

In this case for any given value of θ, the log-likelihood function for the sample is given by:

$$l(\theta) = \log(P_{11}) + \log(P_{12}) + \log(P_{23})$$

and replacing the values we obtain:

$$l(\theta) = 10\theta - \log(e^{5\theta} + e^{3\theta}) - \log(e^{\theta} + e^{2\theta}) - \log(e^{3\theta} + e^{4\theta})$$

Figure 8.1 shows the results of plotting $l(\theta)$ for different values of θ. The optimum, $\theta^* = 0.756$, allows us to predict the following probabilities:

Observation (q)	P_{1q}	P_{2q}
1	0.82	0.18
2	0.32	0.68
3	0.32	0.68

Therefore we would say that the model predicts incorrectly the second observation under a maximum utility criterion.

We mentioned that the ML parameters θ^* are asymptotically distributed Normal with covariance matrix S^2. In general the well-understood properties of the ML method for well-behaved likelihood functions allow, as in multiple regression, a number of statistical tests which are of major importance:

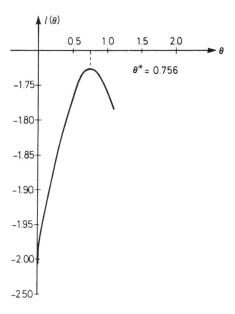

Figure 8.1   Variation of $l(\theta)$ with θ

8.4.1.1 The t-test for Significance of any Component θ_k^* of θ^*

Equation (8.14) implies that θ_k^* has an estimated variance s_{kk}^2, where $\mathbf{S}^2 = \{s_{kk}^2\}$, which is calculated during estimation. Thus if its mean $\theta_k = 0$,

$$t = \theta_k^*/s_{kk} \tag{8.15}$$

has a standard Normal distribution N(0, 1). For this reason it is possible to test whether θ_k^* is significantly different from zero (it is not exactly a t-test as we are taking advantage of a large-sample approximation and t is tested with the Normal distribution). Sufficiently large values of t (typically bigger than 1.96 for 95% confidence levels) lead to the rejection of the null hypothesis $\theta_k = 0$ and hence to accepting that the kth attribute has a significant effect.

The quasi-stepwise variable selection process followed during the specification searches of discrete choice models normally considers both formal statistical tests, such as the above one, and more informal tests such as examining the sign of the estimated coefficient to judge whether it conforms with *a priori* notions or theory. In this sense it is worth noting that rejection of a variable with a proper sign crucially depends on its importance; for example, let us note that the set of available explanatory variables can be usefully divided into two classes:

- highly relevant or policy variables, which have either a solid theoretical backing and/or which are crucial to model forecasting;
- other explanatory variables, which are either not crucial for policy evaluation (for example gender), or for which there are no theoretical reasons to justify or reject their inclusion.

Table 8.1 depicts the cases that might occur when considering the possible interactions in the above framework, and the solutions recommended by current practice. Consider first the case of rejecting a variable of type Other with correct sign; this may depend on its significance level (i.e. it may only be significant at the 85% level) and usual practice is to leave it out if it is not significant at the 80% level.

Current practice also recommends to include a relevant variable with a correct sign even if it fails any significance test. The reason is that the estimated coefficient is the best approximation available for its real value; the lack of significance may just be caused by lack of enough data.

Table 8.1   Variable selection cases

		Policy	Other
		Variable	
Correct sign	Significant	Include	Include
	Not significant	Include	May reject
Wrong sign	Significant	Big problem	Reject
	Not significant	Problem	Reject

Variables of the Other class with a wrong sign are always rejected; however, as variables of the Policy type must be included at almost any cost, current practice dictates in their case model re-estimation, fixing their value to one obtained in a similar study elsewhere. This will be an easy task if the variable is also non-significant, but might be very difficult otherwise as the fixed value will tend to produce important changes in the rest of the model coefficients.

8.4.1.2 The Likelihood Ratio Test

A number of important model properties may be expressed as linear restrictions on a more general linear in the parameters model. Some important examples of properties are:

- Attribute genericity. As mentioned in section 7.3, there are two main types of explanatory variables, generic and specific; the former have the same weight or meaning in all alternatives, whereas the latter have a different, specific, meaning in each of the choice options and therefore can take on a zero value for certain elements of the choice set.
- Sample homogeneity. It is possible to test whether or not the same model coefficients are appropriate for two subpopulations (say living north and south of a river). For this a general model using different coefficients for the two populations is formulated and equality of coefficients may be tested as a set of linear restrictions.

Example 8.2: Let us assume a model with three alternatives, car, bus and rail, and the following choice influencing variables: travel time (TT) and out-of-pocket cost (OPC). Then a general form of the model would be:

$$V_{car} = \theta_1 TT_{car} + \theta_2 OPC_{car}$$
$$V_{bus} = \theta_3 TT_{bus} + \theta_4 OPC_{bus}$$
$$V_{rail} = \theta_5 TT_{rail} + \theta_6 OPC_{rail}$$

However, it might be hypothesised that costs (but not times, say) should be generic. This can be expressed by writing the hypothesis as two linear equations in the parameters:

$$\theta_2 - \theta_4 = 0$$
$$\theta_2 - \theta_6 = 0$$

In general it is possible to express attribute genericity by linear restrictions on a more general model. For extensive use of this type of test, refer to Dehghani and Talvitie (1980).

Because of the properties of ML it is very easy to test any such hypotheses, expressed as linear restrictions, by means of the well-known likelihood ratio test (LR). To perform the test the estimation program is first run for the more general case to produce estimates θ^* and the log-likelihood at convergence $l^*(\theta)$. It is then run again to attain estimates θ_r^* of θ and the new log-likelihood at maximum $l^*(\theta_r)$ for the restricted case. Then if the restricted model under consideration is a correct specification, the LR statistic,

$$-2\{l^*(\theta_r) - l^*(\theta)\}$$

is asymptotically distributed χ^2 with r degrees of freedom, where r is the number of linear restrictions; rejection of the null hypothesis implies that the restricted model is erroneous. It is important to note that to carry out the test we require one model to be a restricted or nested version of the other. Train (1977) offers examples of use of this test to study questions of non-linearity, non-genericity and non-homogeneity. Horowitz (1982) has discussed the power and properties of the test in great detail and should be consulted for further reference.

8.4.1.3 The Overall Test of Fit

A special case of likelihood ratio test is to verify whether all the components of θ are equal to 0. This model is known as the equally likely (EL) model and satisfies:

$$P_{jq}(A(q), \theta) = 1/N_q$$

with N_q the choice set size of individual q. The test is not very helpful in general because we know that a model with alternative-specific constants will reproduce the data better than a purely random function. For this reason a more rigorous test of this class is to verify whether all variables, except the specific constants, are 0. This preferable reference or null model is the market share (MS) model, where all the explanatory variables are 0 but the model has a full set of alternative specific constants; if all individuals have the same choice set in this case we get:

$$P_{jq}(\mathbf{A}(q), \boldsymbol{\theta}) = \mathrm{MS}_j$$

with MS_j the market share of option A_j.

Let us first look at the test for the EL model because it is simpler. Consider a model with k parameters and with, as usual, a log-likelihood value at convergence of $l^*(\theta)$, and denote by $l^*(0)$ the log-likelihood value of the associated EL model; then under the null hypothesis $\boldsymbol{\theta} = 0$ we have that the LR statistic:

$$-2\{l^*(0) - l^*(\theta)\}$$

is distributed χ^2 with k degrees of freedom; therefore we can choose a significance level (say 95%) and check whether LR is less than or equal to the critical value of χ^2 $(k, 95\%)$, in which case the null hypothesis would be accepted. However, we already hinted that the test is weak becase if it is rejected (as always happens) it only means that the parameters $\boldsymbol{\theta}$ explain the data better than a model with no significant explanatory power. Actually the best feature of this test is its low cost as $l^*(0)$ does not require a special program run since it is usually computed as the initial log-likelihood value by most search algorithms.

To carry out the test with the market share model we require to compute $l^*(C)$ its log-likelihood value at convergence; if there are $(k - c)$ parameters which are not specific constants, the appropriate value of LR is compared with $\chi^2(k - c, 95\%)$ in this case. In general an extra run of the estimation routine is required to calculate $l^*(C)$ except for models where all individuals face the same choice set, in which case it has the following close-form equation:

$$l^*(C) = \sum_j Q_j \log (Q_j/Q) \tag{8.16}$$

where Q_j is the number of individuals choosing option A_j.

Figure 8.2 shows the notional relation between the values of the log-likelihood function, for the set of parameters that maximise it, $l^*(\theta)$, for the two previous models, $l^*(0)$ and $l^*(C)$ respectively, and for a fully saturated (perfect) model with an obvious value $l(*) = 0$.

8.4.1.4 The ρ^2 Index

Although it is not possible to build an index such as R^2 in this case, it is always interesting to have an index which varies between 0 (no fit) and 1 (perfect fit) in order to compare alternative models. An index that satisfies some of the above

Figure 8.2   Notional relation between log-likelihood values

characteristics was initially defined as:

$$\rho^2 = 1 - \frac{l^*(\theta)}{l^*(0)} \tag{8.17}$$

However, although its meaning is clear in the limits (0 and 1) it does not have an intuitive interpretation for intermediate values; in fact, values around 0.4 may be excellent fits.

Because a ρ^2 index may in principle be computed relative to any null hypothesis, it is important to choose an appropriate one. For example, it can be shown that the minimum values of ρ^2 in (8.17), in models with specific constants, vary with the proportion of individuals choosing each alternative. Taking a simple binary case, Table 8.2 shows the minimum values of ρ^2 for different proportions choosing option 1 (Tardiff 1976). It can be seen that ρ^2 is only appropriate when both options are chosen in the same proportion.

Table 8.2   Minimum ρ^2 for various relative frequencies

Sample proportion selecting the first alternative	Minimum value of ρ^2
0.50	0.00
0.60	0.03
0.70	0.12
0.80	0.28
0.90	0.53
0.95	0.71

These values mean, for example, that a model estimated with a 0.9/0.1 sample yielding a ρ^2 value of 0.55, would be undoubtedly much weaker than a model yielding a value of 0.25 from a sample with an equal split. Fortunately, there is a simple adjustment that allows us to solve this difficulty; it consists of calculating the index with respect to the market share model:

$$\rho^2 = 1 - l^*(\theta)/l^*(C) \tag{8.18}$$

This statistic lies between 0 and 1, is comparable across different samples and is related to the χ^2 distribution. Finally, it is worth noting that in the case of the nested logit model the index takes the form:

$$\rho^2 = 1 - \frac{l_1^*(\theta) + l_2^*(\theta) + \ldots + l_S^*(\theta)}{l_1^*(C) + l_2^*(C) + \ldots + l_S^*(C)} \tag{8.19}$$

the subscripts 1 to S referring to the MNL models at each nest.

8.4.1.5 The Percentage Right or First Preference Recovery (FPR)

This is an aggregate measure that simply computes the proportion of individuals effectively choosing the option with the highest modelled utility. FPR is easy to understand and can readily by compared with the chance recovery (CR) given by the equally likely model:

$$CR = \sum_q (1/N_q)/Q$$

Note that if all individuals have a choice set of equal size N, then $CR = 1/N$. FPR can also be compared with the market share recovery (MSR) predicted by the best null model (Hauser 1978):

$$MSR = \sum_{Aj} (MS_j)^2$$

Disadvantages of the index are exemplified by the fact that although an FPR of 55% may be good in general, it is certainly not so in a binary market; also an FPR of 90% is normally good in the binary case, but not if one of the options has a market share of 95%. Another problem of the index, worth noting in the sense of not being an unambiguous indicator of model reliability, is that too high a value of FPR should lead to model rejection as well as a too low value; to understand this point it is necessary to define the expected value of FPR for a specific model as:

$$ER = \sum_q P_q \tag{8.20}$$

where P_q is the calculated (maximum) probability associated with the best option for individual q. Also, because FPR is an independent binomial random event for individual q, occurring with probability $1/N_q$ in the CR case and P_q in the ER case, their variances are given respectively by:

$$Var\,(CR) = (1/N_q)(1 - 1/N_q) \tag{8.21}$$

and

$$Var\,(ER) = P_q(1 - P_q) \tag{8.22}$$

Thus, a computed value of FPR for a given model can be compared with CR and ER; if the three measures are relatively close (given their estimated variances) the model is *reasonable but uninformative*; if FPR and ER are similar and larger than CR, the model is *reasonable and informative*; finally, if FPR and ER are not similar, the model does not explain the variation in the data and should be rejected whether FPR is larger or smaller than ER (see Gunn and Bates 1982).

8.4.2 Estimation of Models from Choice-based Samples

As mentioned in Chapter 3, estimating a model from a choice-based sample may be of great interest because the data-collection costs are often considerably lower than those for typical random or stratified samples. The problem of finding a tractable estimation procedure possessing desirable statistical properties is not an easy one, and the state of the art is provided by the excellent papers of Coslett (1981) and Manski and McFadden (1981).

It has been found in general that maximum likelihood estimators specific to choice-based sampling are impractical, except in very restricted circumstances, due to computational intractability. However, if it can be assumed that the analyst knows the fraction of the decision-making population selecting each alternative, then a tractable method can be introduced. The approach modifies the familiar maximum likelihood estimator of random sampling by weighting the contribution of each observation to the log-likelihood by the ratio Q_i/S_i, where the numerator is the fraction of the population selecting option A_i and the denominator the analogous fraction for the choice-based sample.

Manski and Lerman (1977) have shown that the unweighted random-sample ML estimator is generally inconsistent when applied to choice-based samples and in most choice models this inconsistency affects all parameter estimates. However, as we saw in section 7.3.2, for simple MNL models with a full set of alternative-specific constants the inconsistency is fully confined to the estimates of these dummy variables if all individuals have the same choice set.

8.4.3 Comparison of Non-nested Models

The likelihood ratio test outlined above requires testing a model against a parametric generalisation of itself, that is, it requires the model to be *nested*. Models whose utility functions have significantly different functional forms, or models based on different behavioural paradigms, cannot be compared by this test.

It is easy to conceive of situations in which it would be useful to test a given model against another which is not a parametric generalisation of itself. The following example, provided by Horowitz (1982), is very illustrative.

Example 8.3: Consider one model with a representative utility function specified as:

$$V = \theta_1 Z_1 + \theta_2 Z_2$$

and another with a representative utility function given by:

$$W = \theta_3 Z_3 Z_4$$

and assume we want to test both models to determine which explains the data best. Clearly, there is no value of θ_3 that causes V and W to coincide for all values of θ_1 and θ_2 and the attributes $\mathbf{Z}$. If both models belong to the same general family, however, it is possible to construct a hybrid function; for example, in our case we could form a model whose measured utility X contains both V and W as special cases:

$$X = \theta_1 Z_1 + \theta_2 Z_2 + \theta_3 Z_3 Z_4$$

and using log-likelihood ratio tests both models could be compared against the hybrid; the first one would correspond to the hypothesis $\theta_3 = 0$ and the second to the hypotheses $\theta_1 = \theta_2 = 0$.

Several other tests, including cases where the competing models do not belong to the same general family, are discussed at length in the excellent paper by Horowitz (1982). We will briefly come back to this issue in section 8.5.

8.5 VALIDATION SAMPLES

As we already mentioned in Chapter 5, the performance of any model should be judged against data other than that being used to specify it and, ideally, taken at another point in time (perhaps after the introduction of a policy in order to assess the model response properties). This is true for any model. We will define a subsample of the data, or preferably, another sample *not used* during estimation, as a *validation sample*.

We will first briefly describe a procedure to estimate the minimum size of such a validation sample (ideally to be subtracted from the total sample available for the study) conditional on allowing us to detect a difference between the performance of two or more models, when there is a true difference between them. The method, which is based on the FPR concept, was devised by Hugh Gunn and first applied by Ortúzar (1983).

| | | Model 2 | |
		Not FPR	FPR
Model 1	Not FPR	n_{11}	n_{12}
	FPR	n_{21}	n_{22}

Consider the 2×2 table layout shown above, where n_{ij} is the number of individuals assigned to cell (i, j). For all individuals in a validation sample, choice probabilities and FPR are calculated for each of two models under investigation and the cells of the table are filled appropriately (e.g. assigning to cell $(1, 1)$ if not FPR in both models, and so on). We are interested in the null hypothesis that the probabilities with which individuals fall into cells $(1, 2)$ and $(2, 1)$ are equal, for in that case the implication on simple FPR is that the two models are equivalent; on this null hypothesis the following statistic M is distributed χ^2 with one degree of freedom (see Foerster 1979):

$$M = \frac{n_{12} - n_{21}}{n_{12} + n_{21}} \tag{8.23}$$

Thus, a test of the equivalence of the two models in terms of FPR is simply given by computing M and comparing the result with $\chi^2(1, 95\%)$; if M is less than the appropriate critical value of χ^2 (3.84 for the usual 95% confidence level) we cannot reject the null hypothesis and we conclude that the models are equivalent on these terms.

Given this procedure we can select whichever level of confidence seems appropriate for the assertion that the two models under comparison differ in respect of the expected number of FPR. This gives us control over the fraction of times that we will incorrectly assert a difference between similar models. As usual, the aim of choosing a particular sample size is to ensure a corresponding control over the proportion of times we will make the other type of error, namely incorrectly concluding that there is no difference between different models.

Now, to calculate the probability of an error of the second type we need to decide what it the minimum difference we should like to be able to detect; with this we can calculate the sample size needed to reduce the chance of errors of the second kind to an acceptable level for models which differ by exactly this minimum amount, or more.

Example 8.4: Consider the case of two models such that, on average, model 2 produces 10 extra FPR per 100 individuals modelled as compared to model 1. Note that here it does not matter whether this arises as a result of model 1 having 20% FPR and model 2 having 30% FPR, or the first 80% and the second 90%; in other words, both models can be inadequate.

In this simple case n_{21} is zero and M simply becomes n_{12}. If we are ensuring 95% confidence that any difference we establish could not have arisen by chance from equivalent models, we will compare n_{12} with the value 3.84; for any given sample size n, the probability that r individuals will be assigned to cell $(1, 2)$ is simply the binomial probability $\binom{n}{r}p^r(1 - p)^{(n-r)}$ where p denotes the probability of an individual chosen at random being assigned to cell $(1, 2)$ i.e. the minimum difference we wish to detect. Given n and taking $p = 0.05$ as usual, we can calculate the probabilties of 0, 1, 2, and 3 individuals being assigned and sum these to give the total probability of accepting the null hypothesis (i.e. committing an error of the second kind). Table 8.3 gives the resulting probabilities for different sample sizes.

It is clear that the required validation sample size needs to be relatively large given that typical estimation data sets have only a few hundred observations. Also recall that Table 8.3 is for the simple case of one model being better than or equal to the

Table 8.3 Probability of an error of the second kind for given sample size and models as defined

Sample size	Minimum difference 5% Prob (error II)
50	0.75
100	0.26
150	0.05
200	0.01
250	0.00

other in each observation; the method of course may easily be extended to cases where both the (1, 2) and (2, 1) cells have non-zero probability.

An especially helpful feature of validation samples is that provided their size is adequate the issue of ranking non-nested models is particularly easily resolved, as likelihood ratio tests can be performed on that sample regardless of difference in model structure or parameters. This is because the condition of one model being a parametric generalisation of the other is only required for tests with the same data used for estimation (see Gunn and Bates 1982; Ortúzar 1983).

8.6 MODELLING WITH STATED-PREFERENCE DATA

In Chapter 3 we discussed the experimental design and SP data-collection process in some detail; in section 8.3 we noted that SP experiments could be instrumental in helping to decide which is the most appropriate functional form to model a given choice situation. In this section we will first briefly review how this can be done and then we will proceed to discuss what changes are introduced to the general discrete choice modelling approach by the use of stated-preference data.

8.6.1 Identifying Functional Form

The travel-demand model estimation literature is heavily oriented towards the problem of estimating a set of model parameters given a functional specification; only occasionally are alternative model structures tested. The favoured functional forms are those which can be deduced from (economic) first principles and also satisfy the condition of being easily estimable; for this reason the vast majority of studies has considered linear (in the parameters) utility functions. A notable exception to this rule is the increasing use of transformations to search for functional form but, as we saw in section 8.3, in these cases the computational problem of model estimation is greatly increased; in fact, estimation methods have only been developed for the simpler MNL model in this case.

In contrast, the literature in the area of psychological measurement procedures that use laboratory or interview data, has been deeply concerned with questions of functional form for a long time (see Louviere 1988). In these studies subjects are asked to make judgements about hypothetical alternatives; for example, in a mode choice context they may be asked to select the preferred alternative from a hypothetical set, or to rank the options, or to associate a level of utility to each of them.

Because an individual can be asked to make a fairly large number of judgements in a single interview, the experiment designer can explore, for example, the effects on response of changes to one variable while keeping all the others constant. This allows a much more detailed assessment of functional form, since the analyst can almost trace the shape of response along each variable. A very interesting finding of such studies is that for any particular decision, functional forms tend to be fairly stable across the population even though the values of their parameters can vary widely (see Meyer *et al.* 1978).

Let us assume that travel behaviour is influenced by a set of independent factors which may be quantitative or qualitative in nature. Following Lerman and Louviere (1978), let us denote the set of J quantitative factors for option A_i by $\mathbf{S}_i = \{S_{ji}\}$ and the set of L qualitative factors by $\mathbf{Q}_i = \{Q_{ji}\}$. The total number of factors is $K = J + L$, and the entire attribute vector $\mathbf{X}_i = \{X_{ki}\}$ is simply $\mathbf{S}_i$ and $\mathbf{Q}_i$.

Let us also assume that each factor has associated a certain value (which may be obtained by some or other measurement process) and that the utility of this quantity as perceived by the individual is $u_{ki} = f_{ki}(X_{ki})$, where f is a perception function.

Consider now an experimental context where we observe the response to a combination of $(S_{1i}, \ldots, S_{Ji}; Q_{1i}, \ldots, Q_{Li})$ on a psychological measurement scale. If we assume that this response measure is connected to the utility U_i of option A_i by some algebraic combination rule, we can write:

$$U_i = g_i(u_{1i}, \ldots, u_{Ki}) \tag{8.24}$$

Finally if we postulate that the vector of responses $\mathbf{U} = \{U_i\}$ is connected to observed (i.e. non-experimental) behaviour B by another algebraic function, we can write:

$$B = h(\mathbf{U}) \tag{8.25}$$

and by substituting, we get:

$$B = h\{g[f(\mathbf{S}, \mathbf{Q})]\} \tag{8.26}$$

As this is too general a formulation for modelling purposes, in practical applications one must make explicit assumptions about the functions f, g and h, and deduce their consequences.

Now, for the purposes of developing an appropriate functional form, the critical component of this approach is the specification of equation (8.24). Alternative forms, such as multiplicative or linear cases, may be tested and selected by means of analysis of variance; however, in order to successfully apply it two conditions must be satisfied: first, the pattern of statistical significance of the utility responses to various combinations of the independent variables must be of a specific nature in order to permit diagnosis or testing of model form; second, corresponding graphical evidence must support the diagnosis or test.

Example 8.5: Consider a residential location model where individuals are assumed to trade off the total cost of travel (including travel times) with house price, independently of one another, i.e. it is assumed that they combine the effects of the two variables linearly. This hypothesis may be tested directly by an analysis of variance. Suppressing the option index i for simplicity, we can write:

$$U_{mn} = U_m^1 + U_n^2 + \epsilon_{mn}$$

where U_l^k are utility values assigned to the lth level of the kth factor in a factorial design, U_{mn} stands for the overall utility assigned by individuals to combinations of levels of both factors, and ϵ_{mn} is a random term with zero mean.

A test for independence of the two effects corresponds to a test of the significance of the interaction effect $U_m^1 U_n^2$. As Lerman and Louviere (1978) point out, in an analysis of variance this is a global test for any and all interactions between both variables; thus if the interaction effect is not significant, the hypothesis of linear form cannot be rejected. If the interaction is significant, on the other hand, it implies that a simple linear combination is not appropriate.

This test should be accompanied by a graphical plot of the interaction. If the linear hypothesis (no interaction) is correct, the data should plot as a series of parallel lines when plotted against either utility value. It can be shown that this is true regardless of the form assumed for the marginal relationships (8.24); it can also be shown that this is true for any multilinear utility model and for any forms less restrictive than simple addition or multiplication.

8.6.2 SP Data and Discrete Choice Modelling

There are two particular features of SP data that lend it to different analysis methods, *vis à vis* other sources of disaggregate data: first, the fact that each respondent may contribute with more than one observation and, second, the different forms in which preferences can be expressed. In Chapter 3 we mentioned that there are three main types of SP responses: ratings, rankings and choices. In the first case, the subject is asked to rate each option using a number between 1 and 5 or 10. The result of this exercise may be interpreted as the strength of the individual preference for each alternative. Therefore, normal algebraic operations can be carried out on them, for example extracting a ratio or subtracting one from another. However, this is now believed to be a weak element in SP work as there is no evidence to support the assertion that individual preferences can be elicited and translated into cardinal scales of this kind.

Simpler, and more reliable, tasks are to ask individuals to rank alternatives in order of preference, or simply to make several choices between hypothetical alternatives. In the case of *ranking* experiments the individual is asked to rank a set of N alternatives in order of preference. If r_i denotes the alternative ranked in the ith position, the response implies that:

$$U(r_1) \geqslant U(r_2) \geqslant \ldots \geqslant U(r_N)$$
(8.27)

In the case of *choice* exercises the individual is only asked to choose his preferred option from the alternatives (two or more) in the choice set; therefore in this case the response corresponds with the usual discrete choice RP approach, except for the fact that both alternatives and choices are hypothetical. This type of exercise can be extended by allowing respondents to express their degree of confidence in the stated choice. To this end, the respondent is offered a semantic scale, the most typical having five points (*1: Definitively prefer first option; 2: Probably prefer first option; 3: Indifferent; 4: Probably prefer second option; 5: Definitively prefer second option*). This exercise is sometimes also called *rating* in the specialised literature although it is actually a generalization of a choice experiment (see for example Ortúzar and Garrido 1991; Pearmain *et al.*, 1991). This generalisation offers advantages and disadvantages: on the one hand it permits a richer range of modelling techniques to be applied to the data; on the other hand, it may weaken the specificity of the choice and that of the response, increasing the difference between experiment and behaviour.

Taking advantage of the special features of SP data there are four broad groups of techniques for analysis:

(i) Naive or graphical methods
(ii) Least square fitting, including linear regression
(iii) Non-metric scaling
(iv) Logit and probit analysis

These methods can be used to provide different levels of analysis of the SP experiments. In general, all of them seek to establish the weights attached to each of the attributes in an (indirect) utility function estimated for each alternative. These weights are sometimes referred to as 'preference weights', 'part utilities' or simply 'coefficients' associated to each attribute. Once these 'part utilities' have been estimated they can be used for various purposes:

(a) To determine the relative importance of the attributes included in the experiment;
(b) An extension of this is the estimate of the rate at which one attribute is traded-off with another (a typical example is the estimation of 'values-of-time' when both time and cost attributes have been included in the experiment); it is also possible to estimate the value of more qualitative attributes like reliability, security levels, and so on;
(c) To specify utility functions for forecasting models, including questions of model structure.

The nature of SP data and the objective of the analysis will be determinant factors in the choice of model estimation techniques.

8.6.2.1 Naive Methods

The naive or graphical methods utilise a simple approach based on the fact that in most designs each level of each attribute appears the same number of times. Therefore, some indication of the relative utility of that attribute-level pair can be obtained by computing the mean average rank, rating or choice score for each option in which it was included and comparing that with similar mean averages for other levels and attributes. In effect, just plotting these means on a graph often gives very useful indications about the relative importance of the various attributes included in the experiment. This model does not make use of any statistical theory and therefore fails to give us an indication of the statistical significance of the results.

Example 8.6: Consider an SP exercise comparing three alternative modes of transport, a traditional diesel bus (DB), a modern mini bus (MB) and an electric light rail vehicle (LRT). The attributes included in the SP experiments are in-vehicle travel time IVT, the headway, the fare, and, of course, the vehicle type. The following table shows the different levels to be tested for each attribute:

	Level 1	Level 2	Level 3
Travel Time (min)	25	15	35
Fare (£)	£1.30	£1.00	£1.50
Headway (min)	5	10	20
Vehicle type	DB	LRT	MB

A fractional factorial design is used, and the respondents are asked to rate, or score, the alternatives (10 is the highest or best service). The results are as follows:

Travel time	Fare	Headway	Vehicle type	Score
25	1.30	5	DB	8
25	1.00	10	MB	9
25	1.50	20	LRT	4
15	1.30	10	LRT	10
15	1.00	20	DB	7
15	1.50	5	MB	8
35	1.30	20	MB	4
35	1.00	5	LRT	4
35	1.50	10	DB	1

It is now possible to calculate a 'naive' value for each attribute by calculating the average score for that level and attribute and comparing it with the difference in values. For instance, in the case of travel time the following table can be constructed:

Travel Time Level	Value (min)	Difference in values	Average rating	Difference in rating	Rating per minute
1	25	—	21/3	—	—
2	15	−10 (2 − 1)	25/3	1 (2 − 1)	−1/10
3	35	+20 (3 − 2)	9/3	−5 (3 − 2)	−5/20

and in the case of fares:

Fare Level	Value (£)	Difference in value	Average rating	Differences in rating	Rating per £
1	1.3	—	22/3	—	—
2	1.00	−0.3 (2 − 1)	20/3	−2/3	2.2
3	1.50	0.5 (3 − 2)	13/3	−7/3	−14/3

From this we can estimate the subjective value of time (SVT) as follows: SVT = $(-5/20)/(-14/3) = 0.054$, that is the ratio of ratings per minute over ratings per £. The reader can calculate the values of headway and vehicle type in the same way. Two interesting reflections can follow this very simple example: the values of time or other attributes do depend on the 'difference' being considered, for instance moving from 15 to 25 minutes does not produce the same SVT as moving from 25 to 35 minutes. The second comment is that we have estimated the values of these coefficients using the scores produced by a single respondent; that is, because each interview generates several observations in many cases we can estimate individual rather than sample based models.

The naive method is seldom used in practice, except as a quick way of estimating indicators like the value of time to provide an initial, 'in the field' validation of an experiment. However, this example has served to illustrate some of the ideas behind SP data analysis.

8.6.2.2 Discrete Choice Modelling with Rating Data

The rating data analyst sets to find a quantitative relation between the set of attributes and the response expressed in the semantic scale. For this he needs first to associate a numerical value R_m to each sentence m ($m = 1, ..., M$) of the scale and postulate a linear model such as:

$$\theta_0 + \theta_1 X_1 + \theta_2 X_2 + \ldots + \theta_K X_k = r_j \tag{8.28}$$

where θ_0 is a constant, X_k is typically the difference between the kth attributes of the two competing options in the situation considered; θ_k is the coefficient of X_k and r_j represents a transformation of the response of individual j (i.e. it defines a unique correspondence between the semantic scale and the numerical scale R_m). Thus, when the questionnaire is completed the analyst obtains the chosen values of the dependent variable R_m and knowing the attribute values X_k he can perform a *multiple regression analysis* to estimate the values of θ_k.

Ordinary least squares or weighted and generalised least squares have been used to this end. One of the advantages of using these techniques is the ability to obtain goodness-of-fit indicators and measures of the significance of the model parameters. The main problem with this approach is that there are innumerable numerical scales that could be associated to the response scale. It may occur therefore, that the results of the analysis (estimated coefficients, their ratios and model goodness of fit) will depend on the definition of R_m; this hints at the importance of choosing the scale correctly. This issue will be discussed in greater detail when considering the analysis of extended choice data.

8.6.2.3 Discrete Choice Modelling with Rank Data

Rank data is arguably simpler and more reliable than rating data. Individuals are expected to be able to say that they prefer A to C and C to B with greater confidence and consistency than they can have in assigning scores to each alternative. There are several ways of exploiting rank data.

Monotonic Analysis of Variance or MONANOVA (Kruskal 1965) has been used for many years as a method for non-metric scaling. MONANOVA is a decomposition technique specifically developed to analyse rank order data. The method estimates part utilities iteratively thus estimating 'utility values' corresponding to each alternative. The first of these part utility estimates is generated using the naive method just discussed. These utilities permit the modelling of a ranking of alternatives; MONANOVA uses a 'stress' measure to indicate how much the modelled ranking differs from the ranking actually elicited from each individual. MONANOVA then seeks to improve the estimates of the 'part utilities' in order to reduce the stress (or badness-of-fit) indicator. MONANOVA, as the naive method, is also capable of generating one model for each individual. Despite its uses, MONANOVA lacks a robust statistical grounding and fails to provide global goodness-of-fit and measures of significance indicators; MONANOVA also restricts the type of utility function that can be specified and it is less well suited to the development of forecasting models.

A more interesting form of analysing rank data is to convert it into implicit choices. In the case above the rank ACB would be converted into the choices A better than C, C better than B and A also better than B. The data thus transformed can now be analysed using logit or probit discrete choice modelling software. For the multinomial logit model this can be done using the following theorem (Luce and Suppes 1965):

$$\text{Prob}\,(r_1, r_2, r_3, \ldots) = \text{Prob}\,(r_1/\mathbf{C})\,\text{Prob}\,(r_2, r_3, \ldots)$$

where $\text{Prob}\,(r_1, r_2, r_3, \ldots)$ is the probability of observing that the ranking indicates that r_1 is preferred to r_2 and so on, and $\text{Prob}\,(r_1/\mathbf{C})$ is the probability of r_1 being chosen from the choice set $\mathbf{C} = \{r_1, r_2, r_3, \ldots\}$.

If the theorem is applied recursively, an expression for the probability of the ranking in terms of $N - 1$ probabilities of choice is obtained:

$$\text{Prob}\,(r_1, r_2, r_3, \ldots) = \text{Prob}\,(r_1/\mathbf{C})\,\text{Prob}\,(r_2/\mathbf{C} - \{r_1\}) \ldots$$

where, for instance, $\mathbf{C} - \{r_1\}$ indicates the choice set excluding alternative r_1. Using this theory, Chapman and Staelin (1982) proposed that the content of a ranking of choices (8.27) can be exploded into $N - 1$ statistically independent choices as:

$$(U_1 \geqslant U_n, n = 1, 2, \ldots, N)(U_2 \geqslant U_n, n = 2, 3, \ldots, N) \ldots (U_{N-1} \geqslant U_N) \quad (8.29)$$

and these data can simply be estimated by a multinomial logit routine. However, care must be taken with the following potential problems.

1. As the ranking considers hypothetical options it is likely that the information will contain some noise. This may be particularly serious in the case of less attractive alternatives which are often treated with less care by respondents and bunched together at the bottom of the ranking. This type of behaviour is not consistent

with the independence of irrelevant alternatives axiom of the logit model, so its occurrence must be statistically tested.

2. The rankings must be constructed in decreasing order of preference (i.e. from the best to the worse alternative) by each respondent; failure to do this might generate noisy data which can invalidate the modelling results.

Problems with this approach have been reported by Ben Akiva *et al.* (1992). They found that the response data from different depths of the ranking (i.e. not exploding the full rank) are not equally reliable in the sense of producing statistically significantly different utility estimates. For further discussion of these problems the reader is referred to the excellent review by Hensher (1993a). Finally, a comparison with rating and choice data for similar samples of the same population, on the basis of model accuracy versus production costs is reported by Ortúzar and Garrido (1991).

8.6.2.4 Modelling with SP Choice Data

The use of simple choice SP data enables us to use the whole range of analysis tools available for RP discrete choice modelling; this includes Nested Logit, because we are not restricted to only two options nor do we require the IIA property to hold in order to exploit the data fully.

An interesting difference between RP and SP choice data, is that the latter, by design, lacks some sources of error. In particular, there is no measurement error since all attribute values are *presented* to respondents (although there may be some perception problems). However, we have already discussed other features of SP surveys that weaken the behavioural value of the data: lack of realism in the decision context, artificiality of the alternatives and the difference between ranking or rating and actually choosing a particular option.

Apart from specification error, which clearly does still apply, there is another potentially serious source of error related to the response itself. Although practical results are generally encouraging, in terms of suggesting that most respondents do understand what it is expected of them, there is no guarantee that they are able to complete an SP experiment with complete accuracy. In fact, a good review by Bates (1988a) discusses the following types of potential error applying to all types of SP data:

- Respondent fatigue, which obviously increases with the complexity of the experimental design (see the discussion in Chapter 3);
- Policy response bias, which might occur if the respondent is interested in affecting the outcome of the analysis;
- Self-selectivity bias, when respondents either inadvertently or on purpose, cast their existing behaviour in a better light.

The outcome of all this is that we may have measurement error in the dependent variable, i.e. instead of getting a true estimate of the utility U, we are obtaining some pseudo utility $\ddot{U}$ which can be linked to our general formulation (7.2) by:

$$U_i = V_i + \epsilon_i = \ddot{U}_i + \tau_i \tag{8.30}$$

Assuming homoscedastic τ_i (although it is quite possible that their variance varies across experiments either due to fatigue or learning), the estimation of the parameters of $\mathbf{V}$ presents no problems as (8.30) can be rewritten as:

$$\ddot{U}_i = V_i + (\epsilon_i - \tau_i) \tag{8.31}$$

and the normal estimation methodology may be employed. The problem comes in forecasting, because in that case we are interested in making estimates of $\mathbf{U}$, and what we would get from applying this model are estimates of $\ddot{\mathbf{U}}$ provided the same distribution of errors apply in the design year. In other words:

> ... we are making estimates of relative preferences *as expressed in a Stated Preference experiment* rather than of what would occur in the market (Bates, 1988a).

The only way to get round this problem is to apportion the error between ϵ_i and τ_i, using both SP and RP data to estimate the models, and this is somewhat similar to the problem of using aggregate data in model estimation which we discuss in Chapter 9. Bates (1988a) notes that an understanding of the magnitude of τ_i is of crucial importance to the use of SP in forecasting. Only if it is insignificant in relation to ϵ_i, could the estimated model be used directly to give forecasts. This calls for special care in the design of the SP experiments to reduce respondent fatigue, enhance realism, prevent policy-response bias and minimise self-selectivity bias. However the problem remains normally serious and so current practice recommends mixed estimation with RP data whenever possible (see Bradley and Daly 1991).

8.6.2.5 Model Estimation with Generalised Choice Data

In the case of generalised or extended choice surveys the respondent is allowed to express degrees of confidence in her choices. If conventional Logit modelling is used two models can be estimated, one including only the 'definitely choose A' and another including also the 'probably choose A' and the results compared for goodness-of-fit and parameter significance.

Alternatively, one can research more closely what is the best transformation of the semantic scale into a numerical one, in the sense of producing the best possible models. Several practitioners have used the following symmetric scale: $R_1 = 2.197$, $R_2 = 0.847$, $R_3 = 0.000$, $R_4 = -0.847$, $R_5 = -2.197$, which corresponds to the Berkson–Theil transformation of the following choice probabilities: 0.1, 0.3, 0.5, 0.7, 0.9 (see for example the review in Bates and Roberts 1983) and it has become almost standard practice in transport circles. However, this is not necessarily the most 'appropriate' scale for any given study and it may be important to investigate if scale selection has really a significant effect on the analysis' results. In fact, Garrido (1992) shows that depending on the values of the option attributes and socioeconomic characteristics of the individuals, it is possible to generate large differences in coefficient ratios (such as the subjective value of time) using different scales.

Example 8.7: A group of staff and students participated in a rating SP experiment comparing two options in the following context: a morning trip from home to work (about 10 km away), involving a choice between bus and light rail (an option which does not exist today). For simplicity the experimental design considered only four attributes:

- Travel Cost (varying at three levels)
- Travel Time (varying at two levels)
- Walking Distance (varying at three levels)
- Waiting time, estimated as half of the Public Transport Headway (varying at two levels)

Thus we had a $3^2 \times 2^2$ factorial design and since we were looking for main effects only, we just required nine options (Master Plan 3, columns 1, 2, 7, 8; see Kocur *et al.*, 1982). The following table shows the attribute differences (instead of their absolute values) between the two options; the design (in terms of the options offered) was based on combinations of such differences. This implicitly assumes the resulting model will be generic (e.g. same coefficient for in-vehicle time for each mode) but reduces the size of the design.

Bus attribute minus LRT Attribute	Attribute Level Difference		
	Low	Medium	High
Travel cost ($)	−10	60	80
In-vehicle time (min)	15	25	na
Walking distance (blocks)	−7	−3	0
Headway (min)	−3	2	na

Consider now the four probability scales defined in the following table:

	Scale 1	Scale 2	Scale 3	Scale 4
R_1	0.100	0.010	0.300	0.200
R_2	0.300	0.400	0.450	0.400
R_3	0.500	0.500	0.500	0.500
R_4	0.700	0.600	0.850	0.880
R_5	0.900	0.990	0.950	0.970

The next table presents SVT (i.e. coefficient ratios of the variables time and cost) derived from models estimated after applying the Berkson–Theil transformation to the four probability scales:

Value of Time	Scale 1	Scale 2	Scale 3	Scale 4
In-vehicle travel	4.01	1.73	3.98	4.11
Waiting	20.68	18.67	23.89	23.24
Walking	23.68	21.63	24.91	24.74
R^2	0.48	0.44	0.46	0.45

As can be seen, scale selection does indeed influence the modelling results. The SVT values do not only differ but belong to models with different goodness of fit to the data. Furthermore, the differences do not seem to depend on whether the scale is symmetrical or not; that is, although one could expect a symmetric scale (like scales 1 and 2) to produce more reasonable results, the fitted models and estimated SVT values reject this notion.

One way of avoiding the problem described above is to consider an approach not requiring the analyst to specify the numerical scale *a priori* in order to estimate the model. McKelvey and Zavoina (1975) developed an approach with this feature, called 'ordinal probit'; its theoretical underpinnings are summarised below.

Let Y_j be a variable representing the degree of preference of one option over another for individual j ($j = 1, \ldots, n$) in a situation presented on a rating experiment (i.e. if A and B are the pair of options compared, Y_j would represent how much is A preferable to B or vice-versa). Let us assume also that this variable satisfies the following linear model:

$$Y_j = \mathbf{X}_j \boldsymbol{\theta} + \epsilon_j \qquad (8.32)$$

where $\mathbf{X}_j$ is the K-dimensional vector of independent variables (e.g. attributes of the individuals and alternatives) for individual j; θ is a K-dimensional vector of model coefficients and ϵ_j is an error term distributed Normal with zero mean and covariance matrix $\sigma^2 \mathbf{I}$, where $\mathbf{I}$ represents the identity matrix.

However, the actual value of Y_j cannot be measured with the measurement techniques available; it is only possible to observe an ordinal version of this variable, say Z_j, which *does not* satisfy the linear model (8.32). Now, let us assume that the variable Z_j has M response categories, $R_1, R_2, \ldots, R_M$, and that these are related to the unobservable variable Y_j as follows. Firstly, let us assume that $\mu_0, \mu_1, \ldots, \mu_M$, are $M + 1$ real numbers where:

$$\mu_0 = -\infty$$

$$\mu_M = +\infty$$

Besides, these numbers must also satisfy the following condition:

$$\mu_0 \leqslant \mu_1 \leqslant \ldots \leqslant \mu_{M-1} \leqslant \mu_M$$

such that:

$$Z_j \in R_m \Leftrightarrow \mu_{m-1} < Y_j \leqslant \mu_m \qquad \text{for } 1 \leqslant j \leqslant n \qquad (8.33)$$

Now, as Z_j is ordinal it can be represented by a set of dummy variables:

$$Z_{jm} = \begin{cases} 1 & \text{if } Z_j \in R_m \quad \text{for } 1 \leqslant j \leqslant n, 1 \leqslant m \leqslant M \\ 0 & \text{otherwise} \end{cases}$$

On the other hand, expressions (8.29) and (8.30) allow us to write the probability function of the observed dependent variable Z_j directly:

$$\text{for } 1 \leqslant m \leqslant M \text{ y } 1 \leqslant j \leqslant n$$

$$\mu_{m-1} < Y_j < \mu_m \Leftrightarrow \mu_{m-1} < \sum_{k=0}^{K} \theta_k X_{kj} + \epsilon_j < \mu_m \tag{8.34}$$

$$\Leftrightarrow \frac{\mu_{m-1} - \sum_{k=0}^{K} \theta_k X_{kj}}{\sigma} < \frac{\epsilon_j}{\sigma} < \frac{\mu_m - \sum_{k=0}^{K} \theta_k X_{kj}}{\sigma}$$

where X_{0j} takes the value of 1 (i.e. θ_0 is a specific constant). Now, as ϵ_j is distributed Normal, we can write:

$$\Pr(Z_{jm} = 1) = \Pr(Z_j \in R_m) = \phi\left(\frac{\mu_m - \sum_{k=0}^{K} \theta_k X_{kj}}{\sigma}\right) - \phi\left(\frac{\mu_{m-1} - \sum_{k=0}^{K} \theta_k X_{kj}}{\sigma}\right) \tag{8.35}$$

where $\phi(\cdot)$ is the standard cumulative Normal function. Finally, it is important to note that any linear transformation of the unobservable variable Y_j applied to the series μ_m, yields exactly model (8.35). Thus, without loss of generality we can assume that $\mu_1 = 0$ and $\sigma = 1$ in order to identify the final model coefficients:

$$\Pr(Z_{jm} = 1) = \phi\left(\mu_m - \sum_{k=0}^{K} \theta_k X_{kj}\right) - \phi\left(\mu_{m-1} - \sum_{k=0}^{K} \theta_k X_{kj}\right) \tag{8.36}$$

The calibration problem consists of estimating the $M + K - 1$ coefficients $\mu_2, \ldots, \mu_{M-1}, \theta_0, \ldots, \theta_K$. More information about the model, its calibration process and some statistics associated with its development are given by McKelvey and Zavoina (1975), Terza (1985) and Johnson (1990), while Hensher (1991) and Ortúzar and Garrido (1993a) report applications of ordinal probit in the transport field.

If an ordinal probit estimation routine is not at hand, one can estimate the response scale during the model fitting process by effectively considering each value of the scale as an additional variable. In this case a coordinate search method is used, starting with the typical symmetric scale 1 in Example 8.7. The procedure consists simply of changing in turn each point of the scale (say R_i) by a small amount and estimating a linear regression model with the new values. The search continues until R^2 is maximised and the value of R_i is fixed. The procedure is repeated for each point of the scale (save for R_3 which is always kept as 0.5) in an iterative routine until a best fit is found in each case (that with the highest R^2). This process is repeated again to check for differences. Ortúzar and Garrido (1993a) found that the search never involved more than two iterations before convergence, but they have no proof that a global optimal solution may be found.

Example 8.8: The following table shows the original symmetric scale and the scales found after performing the above 'optimal scale linear regression approach' on two samples for the rating experiment of Example 8.7.

Scale	Initial	Students	Staff
R_1	0.1	0.284	0.228
R_2	0.3	0.286	0.278
R_3	0.5	0.500	0.500
R_4	0.7	0.714	0.722
R_5	0.9	0.900	0.842

The results suggest the possibility of testing if the original number of points in the semantic scale is appropriate. If only one value was used for the first two points of the scale in the optimal scale models (which appear strikingly close) it would be interesting to see what consequences this apparent loss of information brings about. On the plus side a four-point scale would have one parameter fewer to be estimated. The next table shows the optimal values of the new scale obtained when R_1 and R_2 are replaced by a single point R_1'. In this scale, as in the previous one, the probability value of R_3 was fixed to 0.5 as it corresponds to the point of indifference between both modes.

Scale	Students	Staff
R_1	0.277	0.121
R_2	0.500	0.500
R_4	0.716	0.776
R_5	0.899	0.922

As can be seen, the scale values in both samples are further apart than in the previous table which suggests that no other point fusion would be necessary. Also, all values appear reasonable in relative terms, i.e. they correspond to increasing probability values from R_1' to R_5.

8.6.2.6 Model Estimation with Mixed SP and RP Data

Consider the MNL model (7.9) and the inverse relation that its parameter β has with the single standard deviation σ of the Gumbel residuals ϵ. This relation explains why it is not correct to postulate the same error distribution for estimation and forecasting as mentioned above; the near and extreme right hand side expressions in (8.35) should yield different values for β. This produces 'scale' differences on the parameters and if such equality is improperly assumed we might finish estimating pseudo utilities instead of 'true' utilities. To avoid this problem we need to adjust the SP data to actual behaviour, exploiting the advantages of the RP data in this sense, and estimating the parameters $\boldsymbol{\theta}$ jointly.

In econometrics the estimation of models with different data sources is called 'mixed estimation'. Often these data are divided into two sets: *primary* and *secondary* data. The primary data provide direct information about the main modelling parameters. The secondary data provide additional (indirect) information about the parameters. For example, in discrete choice modelling the primary data could be information coming from a survey at the disaggregate level, and the secondary one could be data coming from an aggregate survey. In our case RP data constitute the primary set, since these data capture the actual behaviour of the individuals, and SP data constitute the secondary set. We will use the framework developed by Ben-Akiva and Morikawa (1990). This framework postulates that the difference between the errors in RP and SP may be represented as a function of the variances of each type of error ϵ and η. This can be written as follows:

$$\sigma_\epsilon^2 = \mu^2 \sigma_\eta^2 \tag{8.37}$$

where μ is an unknown *scale coefficient*. This leads to the following utility functions for a certain alternative A_i:

$$U_i^{RP} = \boldsymbol{\theta}\mathbf{X}_i^{RP} + \boldsymbol{\alpha}\mathbf{Y}_i^{RP} + \epsilon_i$$
$$\mu U_i^{SP} = \mu(\boldsymbol{\theta}\mathbf{X}_i^{SP} + \boldsymbol{\phi}\mathbf{Z}_i^{SP} + \tau_i) \tag{8.38}$$

where $\boldsymbol{\alpha}$, $\boldsymbol{\phi}$ and $\boldsymbol{\theta}$ are parameters to be estimated; $\mathbf{X}^{RP}$ and $\mathbf{X}^{SP}$ are attributes (of both alternatives and individuals) at the RP and SP levels respectively. $\mathbf{Y}^{RP}$ and $\mathbf{Z}^{SP}$ are attributes which only belong to the RP or SP sets respectively (notice that vector $\mathbf{X}$ is present in both types of data).

The consideration of the utility functions (8.38) allows to homogenise the type of error as multiplying the SP parameters by μ makes the associated stochastic error to have the same variance as the corresponding RP error. We would normally expect the SP data to have more noise than the RP data; therefore it is sensible to expect that μ should take values between zero and one. If this is not the case, it may be assumed that the RP data is more noisy and this can be tested re-estimating a model with the converse structure. Thus, assuming that both stochastic errors have a Gumbel distribution with zero mean but with a different variance, the choice probabilities would be given by the following expression (Morikawa *et al.* 1992):

$$P_i^{RP} = \frac{e^{(\boldsymbol{\theta}\mathbf{X}_i^{RP} + \boldsymbol{\alpha}\mathbf{Y}_i^{RP})}}{\sum_j e^{(\boldsymbol{\theta}\mathbf{X}_j^{RP} + \boldsymbol{\alpha}\mathbf{Y}_j^{RP})}}$$

$$\tag{8.39}$$

$$P_i^{RP} = \frac{e^{\mu(\boldsymbol{\theta}\mathbf{X}_i^{SP} + \boldsymbol{\phi}\mathbf{Z}_i^{SP})}}{\sum_j e^{\mu(\boldsymbol{\theta}\mathbf{X}_j^{SP} + \boldsymbol{\phi}\mathbf{Z}_j^{SP})}}$$

from these expressions it is possible to obtain the following joint likelihood function:

$$L(\boldsymbol{\theta}, \mu, \boldsymbol{\alpha}, \boldsymbol{\phi}) = \left(\prod_{n=1}^{N^{RP}} \prod_{A_i \in A(q)} P_{iq}^{RP}\right) * \left(\prod_{n=1}^{N^{SP}} \prod_{A_i \in A(q)} P_{iq}^{SP}\right) \tag{8.40}$$

which should be maximised to yield the parameter estimates.

Expression (8.40) is a non-linear function because μ is multiplying not only the attributes but also the SP parameters. So, the estimation procedure is simply an operational research problem. One approach to solve this problem is to apply a software especially designed to deal with non-linear likelihood functions directly. However, this type of computational package is often scarce; therefore it would be very useful to find a technique that allows us to estimate these models through the more easily available programs developed for discrete choice analysis. We can mention at least two well founded methods to do this task: simultaneous estimation (Bradley and Daly 1991) and sequential estimation (Ben-Akiva and Morikawa 1990).

Sequential estimation method This method has the advantage of allowing the use of ordinary logit or probit estimation software. The algorithm is as follows:

Step 1 Estimate the SP model according to (8.38) in order to obtain the estimators of $\mu\boldsymbol{\theta}$ and $\mu\boldsymbol{\phi}$. Then, define a new variable:

$$\hat{V}_i^{RP} = \mu\boldsymbol{\theta}\mathbf{X}_i^{RP} \tag{8.41}$$

Step 2 Estimate the following RP model with the new variable included, in order to estimate the parameters λ and $\boldsymbol{\alpha}$:

$$U_i^{RP} = \lambda\hat{V}_i^{RP} + \boldsymbol{\alpha}\mathbf{Y}_i^{RP} + \epsilon_i \tag{8.42}$$

where $\lambda = 1/\mu$.

Step 3 Multiply $\mathbf{X}^{SP}$ and $\mathbf{Z}^{SP}$ by μ to obtain a modified SP data set. Pool the RP data and the modified SP data and then estimate the two models jointly.

Simultaneous estimation method This method, developed by Bradley and Daly (1991), consists of constructing an artificial tree which has twice as many alternatives as there are in reality. Half of these are labelled RP alternatives, the other half are SP alternatives. The utility functions are $\mathbf{U}^{RP}$ and $\mathbf{U}^{SP}$ (see (8.38)). As indicated in Figure 8.3, the RP alternatives are placed just below the root of the tree; however, the SP alternatives are each placed in a single-alternative nest. For an RP observation, the SP alternatives are set unavailable and the choice is modelled as in a standard logit model. For an SP observation, the RP alternatives are set unavailable and the choice is modelled by a nested (tree) logit structure. For the SP observations, the mean utility of each of the dummy composite alternatives is computed as usual (see Daly 1987):

$$V^{COMP} = \mu\log\sum e^{V^{SP}} \tag{8.43}$$

where the sum is taken over all of the alternatives in the nest corresponding to the composite alternative and

$$V^{SP} = U^{SP} - \tau = \boldsymbol{\theta}\mathbf{X}^{SP} + \boldsymbol{\phi}\mathbf{Z}^{SP} \tag{8.44}$$

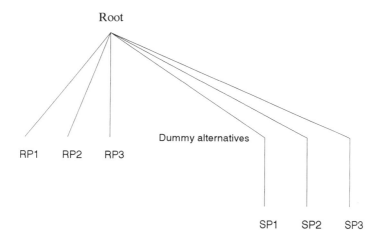

Figure 8.3   Artificial tree structure for joint RP and SP estimation

is simply the measured part of the SP utility. Then, because each nest contains only one alternative in this specification,

$$V^{\text{COMP}} = \mu V^{\text{SP}} = \mu\boldsymbol{\theta}\mathbf{X}^{\text{SP}} + \mu\boldsymbol{\phi}\mathbf{Z}^{\text{SP}} \qquad (8.45)$$

we have exactly the form required as long as the value of μ is constrained to be the same for each of the dummy alternatives. Since the dummy composite alternatives are placed just below the root of the tree, as are the RP alternatives, a standard estimation procedure will ensure that μ is estimated to obtain uniform variance at this level. This artificial construction does not require the usual consistency assumption for nested logit models that μ should not exceed one, because the individuals are not modelled as choosing from the whole choice set. However, as noted before, the value of μ may be taken as providing an indication of which data set is more accurate.

Example 8.9: Consider again the experiment described in Example 8.7. Apart from the SP data, information was also collected on the choices actually made by the participants, their socioeconomic characteristics and the LOS of the RP alternatives in their choice set. Mixed models were estimated with both approaches just described. Their final specification considered the variables *Travel cost, In-vehicle time* and *Waiting time* as common attributes in both data sets (i.e. the vector **X** defined above); the variables *Walking time, Age* and an *LRT-specific constant* were considered as exclusive SP attributes, and a *Bus-specific constant* as an exclusive RP attribute (i.e. vectors **Y** and **Z** defined before). Table 8.4 shows the parameters obtained for each pure model, that is, taking the RP and SP data sets separately. The table also presents subjective values (SV) of time computed in the standard fashion (Gaudry et al. 1989).

We can observe that all parameters have the correct sign and that most of them are statistically significant (*t*-statistics are shown in brackets) at the 95% level. Although the SP parameters have higher *t*-ratios than the corresponding RP

Table 8.4 Parameters of the pure models

Attribute	SP Model	RP Model
Travel cost	−0.0026	−0.0037
	(−7.7)	(−3.8)
In-vehicle travel time	−0.0647	−0.0879
	(−6.2)	(−3.2)
Walking time	−0.2940	na
	(−12.6)	
Age	−0.0434	—
	(−5.4)	
Waiting time	−0.0926	−0.530
	(−2.0)	(−2.4)
PT constant	−0.5410	1.1100
	(−1.7)	(2.8)
SV of travel time ($/min)	23.90	25.20
	(4.9)	(2.4)
SV of walking time ($/min)	60.88	na
	(6.6)	
SV of waiting time ($/min)	41.70	6.00
	(1.9)	(2.0)
ρ^2	0.17	0.12
Sample size	1589	201

parameters, this does not necessarily imply that they actually are more significant because the sample size is different (1589 SP versus 201 RP observations); remember also that the value of t in SP just gives a threshold for significance but not its exact value. The RP model has no walking time parameter because the RP data set did not contain this attribute. The age parameter turned out to be not significant. Regarding the general goodness of fit of each model, we can observe that the ρ^2 of both models is relatively high. The value of travel time is similar (and significant) in both models; the value of waiting time, however, presents higher differences.

Table 8.5 shows the parameters obtained for the mixed models using both approaches. Again all the parameters have a correct sign and most are significantly different from zero at the 95% level. We can observe that the estimated scale coefficient μ is lower than one (as expected) and highly significant, confirming the hypothesis that the SP data has more noise than the RP data. Notice also that the ratio between the common parameters of SP and RP in Table 8.4 (i.e. vector $\mathbf{X}$) fluctuates between 0.6 and 0.74, which is consistent with the value of μ stemming from both approaches (Ortúzar and Garrido 1993b).

There is a great similarity in the values of each corresponding parameter in Table 8.5. This confirms empirically that both mixed estimation approaches produce consistent estimates. However, the sequential method yields parameters with higher t-statistics. This may be due to the fact that the simultaneous estimation method uses

Table 8.5 Parameters of the mixed models

Attribute	Sequential method	Simultaneous method
Travel cost	−0.0037	−0.0037
	(−8.4)	(−4.3)
In-vehicle travel time	−0.0912	−0.0921
	(−7.8)	(−4.2)
Walking time	−0.4286	−0.4239
	(−12.9)	(−3.8)
Waiting time	−0.1353	−0.1423
	(−3.1)	(−3.0)
Age	−0.0596	−0.0627
	(−6.9)	(−3.3)
LRT constant	−0.4377	−0.7377
	(−1.6)	(−1.6)
Bus constant	−	1.143
		(3.3)
μ	0.686	0.694
	(4.9)	(4.0)
SV of travel time ($/min)	24.65	24.89
	(5.7)	(3.0)
SV of walking time ($/min)	61.78	61.10
	(7.0)	(2.8)
SV of waiting time ($/min)	36.57	38.46
	(2.9)	(2.5)
ρ^2	0.175	0.165
Sample size	1790	1790

the same sample size to estimate (jointly) more parameters. The general goodness of fit is also higher in the sequential approach, which is slightly surprising.

All the subjective values of time, in both tables, are significantly different from zero at the 95% confidence level. Besides, they are quite similar between themselves (including the corresponding values in the SP and RP models, which is not usual). In fact, large differences have been observed in previous studies; for example Bradley and Daly (1991) obtained differences between SP–SV versus mixed model–SV and RP–SV versus mixed model–SV of up to 48.6% and 57.8% respectively.

Stated-preference analysis is entering maturity. It is currently accepted by academics and practitioners as a much needed approach to extend the tool-kit of travel behaviour analysts. The interested reader is advised of the existence of at least one monograph (Pearmain *et al.* 1991) and three special issues of international journals which are devoted to the subject (Bates 1988b, Louviere 1992, Hensher 1993b); these should be consulted for further information as the only worrying problem of such a big success is the growing number of badly designed applications which can only ruin the reputation of even the best methodologies.

EXERCISES

8.1 Consider the following mode choice model:

$$V_1 = \theta_1 t_1 + \theta_3 c_1 + \theta_4 Nc + \theta_7$$
$$V_2 = \theta_1 t_2 + \theta_2 e_2 + \theta_5 c_2 + \theta_8$$
$$V_3 = \theta_1 t_3 + \theta_2 e_3 + \theta_6 c_3$$

where t_k is in-vehicle ravel time, e_k is access time, c_k is cost divided by income and Nc is the number of cars in the household.

(a) Indicate which variables are generic, which are specific and what is the real meaning of θ_7 and θ_8.

(b) Discuss the implications of having obtained the following values during model estimation:

$$\theta_1 = -0.115 \qquad \theta_2 = -0.207 \qquad \theta_3 = -0.301$$
$$\theta_4 = 1.730 \qquad \theta_5 = 0.476 \qquad \theta_6 = -0.301$$
$$\theta_7 = -1.250 \qquad \theta_8 = 2.513$$

8.2 During specification searches you obtained the set of mode choice models for car (1), bus (2) and underground (3), shown in the table below; the units of time and cost/income are minutes, sex is a dummy variable which takes the value of 1 for males and 0 for females; EMU is the expected maximum utility of the transit nest (bus–underground).

(a) Indicate which model you prefer explaining very clearly why.

(b) The sample you used for estimation comprised 1000 individuals having all alternatives available. If 250 choose car, 600 choose bus and the rest underground, compute $l^*(0)$, the log-likelihood value for the equally likely model, and $l^*(C)$, the log-likelihood for the constants only model.

Variable (option entered)	Coefficient (t-ratio)			
	MNL-1	MNL-2	HL-1	HL-2
Car time (1)	−0.112	—	−0.114	—
	(−6.10)		(−6.00)	
Transit time (2, 3)	0.006	—	−0.001	—
	(1.25)		(−0.94)	
Travel time (1–3)	—	−0.071	—	−0.083
		(−3.34)		(−3.60)
Cost/income (1–3)	−0.031	−0.040	−0.035	−0.033
	(−2.56)	(−3.52)	(−2.83)	(−3.10)
No. of cars (1)	1.671	1.823	1.764	1.965
	(4.21)	(4.80)	(4.12)	(5.14)
Sex (2,3)	−0.752	−0.776	−0.739	−0.701
	(−1.87)	(−1.98)	(−2.01)	(−1.83)
EMU	—	—	0.875	0.800
			(5.12)	(13.4)
ρ^2	0.412	0.284	0.376	0.315

8.3 You were asked to estimate a multinomial logit (MNL) model and an independent probit (IP) model with the same data set; you obtained the values shown in the following table:

Parameters	MNL	IP
θ_1	2.242	1.698
θ_2	−0.045	−0.034
θ_3	−0.213	−0.162
σ^2	Not applicable	2.870

Indicate whether these results appear to be consistent; if your answer is affirmative explain which is the cause of the differences. If your answer is negative, explain why.

8.4 While conducting an SP survey you asked three individuals to rank the three options whose attributes are given below:

Option	Travel time (min)	Fare ($)
1. High speed train	30	10
2. Express train	40	8
3. Luxury Coach	60	5

After completing the survey you obtained the following results:

Individual	Ranking
1	1, 2, 3
2	2, 3, 1
3	2, 1, 3

You are interested in estimating a MNL model with linear in the parameters utility function given by:

$$V_i = \theta_1 t_i + \theta_2 c_i$$

If you are told that $\theta_1 = -0.03$, find a maximum likelihood estimate for θ_2. Discuss your results.

9 Model Aggregation and Transferability

9.1 INTRODUCTION

The planning and evaluation of transport improvement projects requires models both to make forecasts and to examine their sensitivity with respect to changes in the values of key variables under the control of the analyst. The forecasts themselves normally need to be aggregate, i.e. to represent the behaviour of an entire population or market segment.

In many practical studies until the late 1980s, the models used have been of the classical aggregate four-stage form despite many (and often justifed) criticisms about their inflexibility, inaccuracy and cost. One important reason for this persistence, apart from the familiarity (e.g. they have been considered accepted practice for many years) is that they offer a tool for the complete modelling process, from data collection through to the provision of forecasts of flows on links. This has not often been the case with disaggregate model approaches perhaps because the data necessary to make aggregate forecasts with them is not readily available (see the discussion by Daly and Ortúzar 1990).

In an econometric interpretation of demand models, the aggregation over *unobservable* factors (either attributes or personal characteristics) results in a probabilistic decision model and the aggregation over the *distribution* of observables results in the conventional aggregate or macro relations (Williams and Ortúzar 1982b). Cast in these terms, the difficulty of the aggregation problem depends on how the components of the system are described within the frame of reference employed by the modeller; it is this framework which will determine the degree of variability to be accounted for in a *causal* relation. To give an example, if the framework used by the analyst is that provided by the entropy-maximising approach we saw in Chapter 5, the explanation of the statistical dispersion in a given data set will be very different to that provided by another modeller using a random utility approach, even if they both finish with identical model functions; this *equifinality issue* is discussed by Williams (1981).

In the case of disaggregate random utility models the aggregation problem is how to obtain from data at the level of the individual, aggregate measures such as market shares of different modes, flows on links, and so on. This can be achieved in one of two ways, by having the process of aggregating individual data either before or after model estimation, as shown in Figure 9.1.

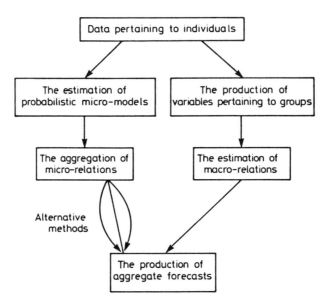

Figure 9.1 Alternative aggregation strategies

In the first case we have variations of the classical aggregate approach, which can be easily criticised for being inefficient in the use of the data, not accounting for their full variability and for risking statistical distortion such as the ecological fallacy discussed in section 7.1. The second approach answers most of the above criticisms; the question that remains is how exactly to perform the aggregation operation over the micro relations.

Daly and Ortúzar (1990) have studied the problem of aggregation of exogenous data in some depth. They concluded that in the case of models representing the behaviour of more than one individual (as is the case with the classical aggregate model) some degree of aggregation of the exogenous data is inevitable and the issue becomes one of to what extent greater accuracy (i.e. smaller zones) is desirable. However, when the model represents the behaviour of a single individual it is conceivable that exogenous data can be obtained and used separately for each traveller; therefore the issue is whether it is preferable on cost or other grounds to use less accurate data; their findings support the notion that the cost/accuracy trade-off is heavily dependent on context. For example, it is clear that for model choice modelling and short-term forecasting the use of highly disaggregate data is desirable; however, the plot thickens considerably for other choice contexts and longer-term forecasting. The next two sections will consider aggregation bias and forecasting methods in greater detail.

9.2 AGGREGATION BIAS AND FORECASTING

Let us consider the multinomial logit (MNL) model (7.9) we derived in section 7.3 and the inverse relation (7.10) that its parameter β has with the single standard deviation σ of the residuals ϵ. If we also consider the typical linear form (7.3) for the measurable utilities $\mathbf{V}$ it is easy to see that it is not possible to estimate β separately from the parameters $\boldsymbol{\theta}$; in fact the calibration process will yield estimates

$$\phi = \beta\boldsymbol{\theta} \tag{9.1}$$

which correspond to the marginal utilities $\boldsymbol{\theta}$ deflated by σ.

We are interested in examining the effect of the manner in which the attributes $\mathbf{x}$ (or at least some of them) are calculated, measured or codified, on the estimated demand functions. As usual we will assume that the MNL model (7.9) is well specified (i.e. there are no taste variations or correlation problems).

Let us assume now that we replace one of the attributes, for example x_1, by an aggregate estimate z_1, where:

$$x_{1i} = z_{1i} + \tau_1 \tag{9.2}$$

and the τ_i are distributed Gumbel $(0, \sigma_\tau)$; then replacing (7.3) and (9.2) in (7.2) we get:

$$U_i = \theta_1 z_{1i} + \sum_k \theta_k x_{ki} + \delta_i \tag{9.3}$$

where the error term δ_i has variance $(\theta_1^2 \sigma_\tau^2 + \sigma^2)$. In this case the coefficient estimates will not be

$$\phi_k = \frac{K\theta_k}{\sigma} \tag{9.4}$$

as before, where $K = \pi\sqrt{6}$, but

$$\psi_k = \frac{K\theta_k}{\sqrt{(\theta_1^2\sigma_\tau^2 + \sigma^2)}} \tag{9.5}$$

that is to say, $\psi_k \leq \phi_k$, $\forall k$. This is normally known as *aggregation bias* and has led to the recommendation that use of average zonal variables for estimating disaggregate demand models should be avoided whenever possible (see for example Horowitz 1981a). The previous analysis may be extended to examine the consequences of this bias in forecasting, as in the following example taken from Gunn (1985a).

Example 9.1: Consider a choice situation modelled by an MNL model such as (7.9) and assume that attribute x_{1j} is doubled, *ceteris paribus*, for each option (including the distribution of the stochastic residuals). It is clear that neither $\boldsymbol{\theta}$ nor σ would be affected if the model was re-estimated with a new data bank containing a consistent choice set; that is to say, the values ϕ from the original context would predict satisfactorily in the new context.

Consider now what would happen if after doubling x_{1j}, each of these values was replaced by its aggregate estimate z_{1j} (for example, the zonal average). In fact we would obtain equation (9.3) again, but the variance of δ_i would now be $(\theta_1^2 4\sigma_\tau^2 + \sigma^2)$; in other words, if the model was re-estimated with the new data it would yield coefficients with expected value given by

$$\psi_k' = \frac{K\theta_k}{\sqrt{(\theta_1^2 4\sigma_\tau^2 + \sigma^2)}} \tag{9.6}$$

that is $\psi_k > \psi_k'$ and the ψ would produce greater than normal predictions in these conditions. Alternatively, attribute reduction policies would imply underpredictions of the model calibrated with aggregate data (see Ortúzar and Ivelic 1987).

9.3 AGGREGATION METHODS

While a disaggregate model allows us to estimate individual choice probabilities, we are normally more interested in the prediction of aggregate travel behaviour. If the choice model was linear the aggregation process would be trivial, amounting only to replacing the average of the explanatory variables for the group in the disaggregate model equation; see for example the aggregation of household-based trip generation models in Chapter 4. However, if the model is non-linear, this method, called *naive aggregation*, will generally produce bias as shown in Figure 9.2. The correct aggregate probability for a group of two individuals A and B is $(P_A + P_B)/2$; the naive method yields the probability $P_C = P[(V_A + V_B)/2]$. As can be seen, if the model was linear both values would coincide.

Discrete choice models such as those we have discussed can be represented in general by:

$$P_{jq} = f_j(\mathbf{x}_q)$$

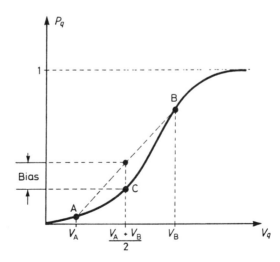

Figure 9.2 Bias of the naive aggregation method

where P_{jq} is the probability that individual q selects option A_j, $\mathbf{x}_q$ is the set of variables influencing her decision, and f_j is the choice function for A_j (for example, the MNL).

For a population of Q individuals the aggregate proportion choosing A_j, according to the model, is the expected value (or enumeration) of the probabilities of each individual:

$$P_{jQ} = \frac{1}{Q}\sum_q f_j(\mathbf{x}_q) \qquad (9.7)$$

Unfortunately this method requires a large data set and computational work. However, if we accept that the sample used to estimate the model is representative of the population, we can use (9.7) and refer to *sample enumeration*. This is a good practical method for moderate size choice sets and is excellent for mode choice models in short-term predictions. However, the method is not so useful in the long term because it does not allow us to address overall contexts which are very different to that of the base year (it assumes that the distribution of the attributes will not differ from that of the sample in the future); it is also unable to produce aggregate zone-to-zone flows necessary for the estimation of demand at the link level.

To cope with these problems the *artificial sample* enumeration approach may be used (see Daly and Gunn 1986). An artificial sample is one in which personal characteristics of members of existing households, believed to be representative of the population of the study area, are matched with locational characteristics of a number of locations also believed to be typical of the area. Thus, the marginal distributions of both personal and locational characteristics are by construction typical of the study area; the approximation made is that the joint distribution of these characteristics can be represented by the product of the two marginals.

Given suitable networks, zoning systems and planning data, the marginal distributions of locations can be those of actual locations in the study area (details of their accessibility to available destinations are needed); if the locations are distributed over the whole of the study area, we can be reasonably confident of overall representativity in large samples (see Gunn 1985).

For personal characteristics the following steps are needed to achieve realism:

1. Actual households' members are drawn at random from a large nationally representative data set (e.g. a census).
2. For each zone of the study area, different expansion factors are found for each of these households such that the expanded sample corresponds as closely as possible to known or forecast aggregate totals (i.e. of variables such as numbers of workers, numbers of individuals by sex and age grouping, etc.).
3. The expansion factors, or more commonly the number of households in each group, are chosen such that the overall distribution of households in terms of a given stratification (say size, number of workers and age of head of household) is not too different from the overall national average (note that when classifying data in this form there are several impossible strata, e.g. households of size 1 with more than 1 worker). To achieve this it may be appropriate to minimise an

objective function such as the following:

$$S(N_i) = \sum_k W_k \left[\sum_i (X_{ik} N_i - Y_k)^2 \right] + \sum_i (N_i - R_i)^2 \qquad (9.8)$$

where N_i is the required number of households in stratum i; W_k is a weight chosen to increase or decrease the importance of the fit to the kth variable (e.g. number of workers, number of males between 18 and 65, etc.); X_{ik} is the average value of variable k for household stratum i; Y_k is the average (observed) value of variable k for (each zone of) the study area; and R_i the number of households in stratum i in the base-year sample. Various other constraints can be put on the process, as discussed by Gunn (1985b).

The artificial sample replicates the population of each zone of the study area; thus aggregate forecasts can simply be obtained by applying the enumeration method to these data.

Another practical method is known as the *classification approach*, which consists in approximating (9.7) by finite number of relatively homogeneous classes, as in:

$$P_{jQ} = \sum_c f_j(\mathbf{X}_c) Q_c / Q \qquad (9.9)$$

where $\mathbf{X}_c$ is the mean of the variable set vector for subgroup c and Q_c/Q the proportion of individuals in the subgroup.

The accuracy of the method depends on the number of classes c and their selection criteria (in the limit it equals the naive method, when the number of subgroups $c = 1$, and the enumeration method, when $c = Q$). Good methods to define the classes have been reported (McFadden and Reid 1975) and the approach is recommended for cases where sample enumeration is not appropriate (Koppelman 1976).

An obvious method to define classes is to use as market segmenting variables those that present the greatest variance, or those which limit in some way the available choice set of each individual. Thus in the mode choice case good variables are the number of cars per household and family income.

9.4 MODEL UPDATING OR TRANSFERANCE

9.4.1 Introduction

During the 1980s a substantial body of literature emerged with empirical evidence about the stability (or, in most cases, lack of it) of parameters of disaggregate travel demand models, across space, cultures and time (see for example Gunn *et al.* 1985; Koppelman and Wilmott 1982; Koppelman *et al.* 1985a,b). The reasons were simple: firstly, evidence of stable values of estimated parameters could provide a direct indication of model validity; secondly, a model that is not stable over time is likely to produce inaccurate predictions; finally, and not less importantly, transferable models should allow for more cost-effective analyses of transport plans and policies.

Because it is unrealistic to expect an operational model in the social sciences to be perfectly specified, it is quite obvious that any estimated model is in principle context dependent. For this reason, it is not very useful to look for perfect model stability and to consider model transferability in terms of equality of parameter values in different contexts (although many studies initially took this view; see Galbraith and Hensher 1982; Ortúzar 1986).

A more appropriate modern view considers model transference as a practical approach to the problem of estimating a model for a study area with little resources or a small available sample. In this sense the model-transfer approach is based on the idea that estimated parameters from a previous study may provide useful information for estimating the same model in a new area, even when their true parameter values are not expected to remain the same. Now, as transferred models cannot be expected to be perfectly applicable in a new context, updating procedures to modify their parameters are needed so that they represent behaviour in the application context more accurately. Depending on the information available in the new environment, different updating procedures may be applied (see Ben Akiva and Bolduc 1987).

9.4.2 Methods to Evaluate Model Transferability

If we define transferability as the usefulness of a transferred model, information or theory in a new context, we can attempt to measure it by comparing the model parameters and, more interestingly, its performance in the two contexts. For this we will assume that we have estimated the parameters independently in the two contexts and that we would like to measure the errors involved in using the first model in the second context. The following tests and measures were used in such analyses in many practical studies (Galbraith and Hensher 1982; Koppelman and Wilmott 1982; Ortúzar *et al*. 1986).

9.4.2.1 Test of Model Parameter for Equality

To evaluate the absolute difference between coefficients of a given model estimated in two different contexts, the t^*-statistics may be used; thus, if (9.10) holds, the null hypothesis that this difference is 0 cannot be rejected at the 95% level:

$$t^* = \frac{\theta_i - \theta_j}{\sqrt{[(\theta_i/t_i)^2 + (\theta_j/t_j)^2]}} < 1.96 \tag{9.10}$$

where θ denotes coefficients, t their t-ratios, i stands for the original context and j for the new context. Galbraith and Hensher (1982) recommend the application of this test only to parameters with low standard error (high t-ratio); otherwise, the t^*-statistic may reject the alternative hypothesis (i.e. the parameters are different) even if they exhibit substantial differences.

9.4.2.2 Disaggregate Transferability Measures

These are based on the ability of a transferred model to describe individual observed choices in the new context and rely on measures of log-likelihood as those that were depicted in Figure 8.2. In addition we need to define $l_j^*(\theta_i)$ as the log of the likelihood that the observed data in the application context j were generated by the transferred model estimated in context i; note that we need to denote the previous measures as $l_j^*(\theta_j)$, $l_j^*(C)$ and $l_j^*(0)$ respectively. Figure 9.3 shows the expected relation among these values.

A natural measure of the transferability of a model estimated in context i for the application in context j, is the difference in log-likelihood (i.e. likelihood ratio) between this model and one originally estimated in context j: $-\{l_j^*(\theta_i) - l_j^*(\theta_j)\}$. This measure has been used to build two specific indices of transferability:

1. Transferability test statistics (TTS), defined by Atherton and Ben Akiva (1976) as twice the difference in log-likelihood identified above:

$$\text{TTS} = -2\{l_j^*(\theta_i) - l_j^*(\theta_j)\} \tag{9.11}$$

   This statistic is distributed χ^2 with degrees of freedom equal to the number of model parameters, under the assumption that the parameter vector of the transferred model is fixed. The test is not symmetric; therefore it is both possible and reasonable to accept transferability in one direction, between a pair of contexts, but reject it in the other direction.
2. Transfer index (TI), which describes the degree to which the log-likelihood of the transferred model exceeds a null or reference model (such as the market shares model), relative to the improvement provided by a model developed in the new context. It is defined by Koppelman and Wilmott (1982) as:

$$\text{TI}_j(\theta_i) = \frac{l_j^*(\theta_i) - l_j^*(C)}{l_j^*(\theta_j) - l_j^*(C)} \tag{9.12}$$

TI has an upper bound of one (which is obtained when the transferred model is as accurate as the local one), but does not have a lower bound; negative values imply only that the transferred model is worse than the local reference model.

The two measures defined above are interrelated by their dependence on the difference in log-likelihood between transferred and local models. However, they offer different perspectives on model transferability: TI provides a relative measure and TTS a statistical test measure (Koppelman and Wilmott 1982).

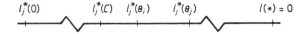

Figure 9.3   Expected relation between log-likelihood values

9.4.3 Updating with Disaggregate Data

The most general presentation of the MNL model (7.9) with linear utility functions $\mathbf{V}$ given by (7.3), considers not only the explicit inclusion of relation (7.10) — as we saw in section 9.2 — but also the explicit inclusion of a set of location parameters w_i as in:

$$P_{iq} = \frac{\exp\left[(w_i + \boldsymbol{\theta}\mathbf{X}_{iq})/\sigma\right]}{\sum_j \exp\left[(w_j + \boldsymbol{\theta}\mathbf{X}_{jq})/\sigma\right]} \tag{9.13}$$

where the location parameters represent the mode of the distribution of errors for each alternative, the scale parameter σ is the standard deviation of the distribution of the error term, and the parameters $\boldsymbol{\theta}$ the attribute weightings employed by the individual in evaluating alternatives.

In his analysis of model mis-specification, Tardiff (1979) shows that the omission of explanatory variables should have the following effects:

- shift the mean of the error distribution, represented in the model by w_i, and increase its variance reflected by σ;
- bias the estimates of the parameters associated with the included variables.

When comparing models which are incompletely specified, in different contexts, it is expected that the differences in the mean values of the error distribution will be relatively large, the differences in the error standard deviation will be smaller, and the differences in the parameters estimates the smallest. Thus, efforts to improve model transfer to a specific application environment should emphasise adjustment of constants first, parameter scale second and relative values of the parameter last; this has been confirmed by several practical studies using both aggregate and disaggregate data (Gur 1982; Dehghani and Talvitie 1983; Koppelman *et al*. 1985b; Gunn and Pol 1986).

The parameters in equation (9.13) are of course not uniquely identifiable and therefore cannot all be estimated; as we have seen, in the case of the alternative specific constants one is arbitrarily (and with no loss of generality) set to 0. Also, it is not possible to estimate σ but only the ratios $\mathbf{w}/\sigma$ and $\boldsymbol{\theta}/\sigma$; defining these ratios by $\boldsymbol{\mu} = \mathbf{w}/\sigma$ and $\boldsymbol{\phi} = \boldsymbol{\theta}/\sigma$ we obtain the more familiar version of the MNL as:

$$P_{iq} = \frac{\exp\left(\mu_i + \boldsymbol{\phi}\mathbf{X}_{iq}\right)}{\sum_{Aj \in A(q)} \exp\left(\mu_j + \boldsymbol{\phi}\mathbf{X}_{jq}\right)} \tag{9.14}$$

where one of the μ_i is constrained to zero.

9.4.3.1 Updating the Constants

Parameter estimates for a choice model are obtained by maximising a log-likelihood expression such as (8.13), where embedded in the probability function P_{iq} are expressions for the representative utility of each option formulated as:

$$V_{iq} = \mu_i + \phi \mathbf{X}_{iq} \tag{9.15}$$

Let us denote as ϕ_T a set of parameters estimated in one context to be transferred to a new application context; in this case the transferred portion of the utility function can be defined as (Koppelman *et al.* 1985b):

$$Z_{iq}^A = \phi_T \mathbf{X}_{iq}^A \tag{9.16}$$

where $\mathbf{X}_{iq}^A$ is a vector of attributes of alternative A_i for individual q in the application (A) context. The updating of the alternative specific constants is accomplished by modifying the utility function in equation (9.15) for the application context to:

$$V_{iq}^A = \mu_i^A + Z_{iq}^A \tag{9.17}$$

where V_{iq}^A is the representative utility of option A_i in the application context and μ_i^A its updated alternative specific constant. To estimate the updated value of the constants it is necessary to maximise the log-likelihood function:

$$l(\mu^A) = \sum_q \sum_{Aj \in A(q)} g_{jq} \log P_{jq}(\mathbf{Z}_q^A, \mu^A) \tag{9.18}$$

where as before, g_{jq} is defined by:

$$g_{jq} = \begin{cases} 1 & \text{if } A_j \text{ was chosen by } q \\ 0 & \text{otherwise} \end{cases}$$

9.4.3.2 Updating of Constants and Scale

The methodology just outlined can be trivially extended to adjust the scale of the transferred parameters as well as the constants. The coefficient of Z_{iq}^A in equation (9.17) was restricted to one in the preceding approach; to update the parameter scale that restriction is relaxed yielding the following representative utility (Koppelman *et al.* 1985b):

$$V_{iq}^A = \mu_i^A + \lambda^A Z_{iq}^A \tag{9.19}$$

where λ^A is the scaling parameter for the application context relative to the estimation or original context. In this case the log-likelihood function to be maximised is as (9.18) but including the extra parameter λ^A. Note that this adjusts the scale of the explanatory variables but does not affect their relative importance. Practical applications of this method have been reported by Gunn *et al.* (1985) and a discussion of further refinements to this problem can be found in Ben Akiva and Bolduc (1987).

9.4.4 Updating with Aggregate Data

Consider the same problem as before with the exception that no disaggregate data are available in the application context; however, assume we possess data on observed market shares P_{jq}^*, and also average values for the explanatory variables $\bar{X}_{jz}$, for certain groups z (say residents of a given zone) in both contexts.

Consider a naive aggregation in the original context, where the measured utility of option A_j for group z is given by:

$$\bar{V}_{jz} = \mu_j + \phi\bar{X}_{jz} \tag{9.20}$$

Updating both alternative constants and scale in this case, requires first to compute non-constant utility for the application context as:

$$\bar{Z}_{jz}^A = \phi_T \bar{X}_{jz}^A \tag{9.21}$$

then postulate an expression for the representative utility of group z in the application context as:

$$\bar{V}_{jz}^A = \mu_j^A + \tau^A \bar{Z}_{jz}^A \tag{9.22}$$

where μ^A and τ^A are chosen so as to maximise the following log-likelihood function (Koppelman *et al.* 1985a):

$$l(\mu^A, \tau^A) = \sum_z W_z \sum_j P_{jz}^* \log P_{jz}(\bar{Z}_{jz}^A, \mu^A, \tau^A) \tag{9.23}$$

with W_z a weight, usually the number of observations, which indicates the relative importance of the group in the data set. Other (more suspect) methods to update the constants only have been proposed by Dehghani and Talvitie (1983) and Gur (1982).

The aggregation issue in the presentation above is not trivial as it is well known that the naive method may introduce severe bias. In this sense it is interesting to mention that the methodology just discussed is wholly consistent with the aggregation approach implicit in most aggregate transport studies (recall Figure 9.1 and the discussion in Chapter 5). There, disaggregate model parameters have been traditionally used as fixed coefficients of generalised cost functions, and later *scale* and *bias* parameters have been fitted using aggregate data (Williams and Ortúzar 1982b).

It is also of interest to note that a more elaborate version of this approach has also been used in practice; for example, in the Greater Santiago Strategic Transport Study (ESTRAUS 1989) disaggregate mode choice parameters were firstly estimated with a mixture of data for 1983 to 1986 (Ortúzar and Ivelic 1988); these were used to build generalised cost functions whose scale and bias parameters were then calibrated using 1977 network and survey data. Finally, the resulting aggregate distribution and modal-split models were validated using volume counts and other aggregate data for 1986.

An interesting alternative, if available, is the use of purposely designed synthetic samples in an enumeration approach (Gunn *et al.* 1982). An important advantage of this method, as we discussed in section 9.3, is that no major adjustments need to be made to the disaggregate models if the artificial sample provides unbiased information to the model system.

EXERCISES

9.1 A group of 800 heads of household with different income levels and located in various parts of an urban area, are confronted with choice between two transport services A and B, for travelling to the central business district. The first, which is more oriented to the population segment with higher income, has a cost C_a and the second a cost C_b.

It has been estimated that the utilities of each alternative are given by the following linear functions:

$$U_a = -0.30C_a + 3.23I$$
$$U_b = -0.30C_b$$

where I is family income (1000$/week).

Estimate the number of households that would choose service A using the following information:

Family income (100$/week)	Number of households	C_a ($)	C_b ($)
Between 1 and 2	450	150	120
Between 2 and 3	250	175	145
Between 3 and 4	100	160	130

9.2 Consider the urban corridor depicted in the figure

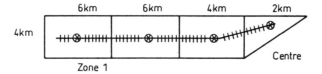

which has the following characteristics:

— Underground and highway run parallel to each other
— There are underground stations at each zone
— The households in the corridor have different income levels, different car ownership and different access to the underground, as shown in Table 1.

We are interested in the trips between zone 1 and the centre of town. We are informed that a binary logit model has been estimated yielding the following representative utilities:

Table 1   Distribution of households with trips between zone 1 and the centre

CO	Access	Income 5000	10000	15000	Total
	$U(DA)$	0	0	350	350
1.0	$U(CA)$	0	50	150	200
	Total	0	50	500	550
	$U(DA)$	150	100	0	250
0.5	$U(CA)$	200	0	0	200
	Total	350	100	0	450
	$U(DA)$	150	100	350	600
Total	$U(CA)$	200	50	150	400
	Total	350	150	500	1000

$$V_c = -2.0 + 9 \times 10^{-5}I + 2.84CO - 0.03t_c - 0.68e_c/d - 50.0c_c/I$$
$$V_u = -0.03t_u - 0.68e_u/d - 50.0c_u/I$$

where t is in-vehicle travel time (min), e is access time (min), c is cost ($\$$), d is distance (km), I is income ($\$$/month) and CO is the number of cars divided by the number of licences in the household.

Underground trips are divided according to access into $U(DA)$, underground with direct access (i.e. on foot), and $U(CA)$, underground with car access. The levels of service by individuals travelling between zone 1 and the centre are summarised in Table 2.

Find out, using an appropriate method, the aggregate probability (i.e for the whole population) of choosing underground.

9.3 Consider a binary logit model for car and bus with the following representative utility functions:

$$V_c = 1.35 - 0.03t_c - 0.15c_c$$
$$V_b = -0.03t_b - 0.15c_b$$

where t is total travel time (min) and c is travel cost divided by income (min). Assume the data in the table on the next page is known about individuals from zone A travelling to work at zone C:

Table 2   Levels of service

	t_c	e_c	c_c	t_u	e_u	c_u	d
$U(DA)$	11.3	5	122.5	14	8	50	14.5
$U(CA)$	14.2	5	131.3	22	15	75	16.3

(a) Find out the aggregate proportion choosing car by the naive aggregation method and by the sample enumeration method. Compute the naive aggregation error in this case.

(b) Find now the aggregate proportion using car by the classification method (using income as stratification variable). Plot your results and those of the naive aggregation method; discuss your graph.

(c) Compare all your results and discuss them critically.

Individual	Chosen option	Income level	t_c (min)	t_b (min)	c_c (min)	c_b (min)
1	Car	High	47.5	83.2	14.8	7.0
2	Car	High	30.2	45.0	10.4	5.0
3	Car	High	22.2	30.4	12.6	4.0
4	Bus	High	45.0	50.6	8.2	5.0
5	Bus	Low	15.3	20.5	50.0	17.0
6	Car	Low	34.8	50.2	55.0	35.0
7	Bus	Low	65.5	100.5	200.3	53.5
8	Bus	Low	12.0	14.0	44.6	17.0

9.4 You are interested in transferring the model of Exercise 9.3 to a new context, where you have taken a small sample of five individuals whose characteristics are presented in the following table:

Individual	Chosen Option	t_c (min)	t_b (min)	c_c (min)	c_b (min)
1	Car	37.5	70.2	16.8	10.0
2	Car	20.2	30.0	16.4	8.0
3	Car	12.0	15.4	18.6	7.0
4	Bus	35.0	35.6	14.2	8.0
5	Bus	5.3	6.5	56.0	20.0

Assuming there are no mode specific constants, estimate the value of τ, the transfer scale parameter, using the data above. Discuss your result.

10 Assignment

10.1 BASIC CONCEPTS

10.1.1 Introduction

The last six chapters have dealt in detail with the key models currently in use to represent the demand for travel in a study area. This chapter deals mainly with the supply side of transport modelling and Chapter 11 with the equilibrium between supply and demand. The supply side of the transportation system was outlined in Chapter 3 where the zoning and network system were introduced. The network system, and in the case of public transport the characteristics of the services offered such as frequency and capacity, represent the main elements of the supply side in transport.

In conventional economic thinking the actual exchanges of goods and services take place as a result of combining the demand for them with their supply. The equilibrium point resulting from this combination defines the price at which the goods will be exchanged and their respective flows (quantities exchanged) in the market. The equilibrium point is found when the marginal cost of producing and selling the goods equals the marginal revenue obtained from selling them. Economic theory admits that this equilibrium may never actually happen in practice as the system of prices and production levels is under permanent adjustment to cope with changes in purchasing power, tastes, technology and production techniques. However, the concept of equilibrium is still valuable in understanding the movement of the economy and to forecast its future states.

It is useful to consider the transport system within that context. The supply side is made up of a road network $S(L, C)$ represented by links L (and their associated nodes) and their costs C. The costs are a function of a number of attributes associated to the links, e.g. *distance, free-flow speed, capacity* and a *speed–flow relationship*. The demand side is made up of an indication of the number of trips by O–D pair and mode that would be made for a given level of service, i.e. that assumed in their estimation. One of the main elements defining levels of service is, in this context, travel time, but often monetary costs (fares, fuel) and features like comfort for the public may be relevant too. If the actual level of service offered by the transport network turns out to be lower than estimated, then a reduction in the demand and perhaps a shift to other destinations, modes and/or times of day would be expected. The speed–flow (or generalised cost–flow) relationship is important as it relates the use of the network to the level of service it can offer.

The public-transport network must also be defined in similar terms to the private network. However, it should contain additional specification of the services offered in terms of their routes, capacities, frequency and ideally, though seldom in practice, their quality, reliability and regularity.

In the case of a transport system one can see equilibrium taking place at several levels. The simplest one is equilibrium in the road network where travellers from a fixed trip matrix seek routes to minimise their travel costs (times). This results in their trying alternative routes, exploring new ones and perhaps settling into a relatively stable pattern after much trial and error. This allocation of trips to routes yields a pattern of path and link flows which could be said to be in equilibrium when travellers can no longer find better routes to their destinations: they are already travelling on the best routes available. This is the *road network* equilibrium. A similar, but perhaps less dramatic, phenomenon takes place in public-transport networks where passengers may seek routes (i.e. combinations of services) to reduce their generalised journey costs as affected by overcrowding, waiting and walking times, and in-vehicle times.

There are, however, other (higher) levels of interaction. As car congestion increases, buses operating on the same roads will have their journey times increased as well. This may induce some public-transport users (and bus operators) to change their routes to avoid these delays. These choices interact with those of car drivers as the new arrangements may provide additional capacity in some links and therefore new equilibrium points. These are *multimode network* equilibrium problems and are discussed in Chapter 11.

At an even higher level, the resulting flow pattern may affect choices of mode, destination and time of day for travel. Each of these shifts in demand will induce in turn changes in the corresponding equilibrium points. In modelling terms, the new flow pattern produces levels of service for routes and modes which may or may not be consistent with those assumed in estimating the (presumed) fixed trip matrix. This requires re-estimating the matrix and therefore feeding back the new levels of service into the estimation process to obtain a new one. The process may need to be repeated in a systematic way until the trip matrices (and therefore trip time, destination and mode) are obtained with values for travel costs which are consistent with the flows estimated for each network. This higher level we shall call *system equilibrium* as opposed to *network equilibrium*.

This chapter is organised as follows. We consider first the problem of assigning a fixed trip matrix to a road network. In order to treat this problem we consider typical characteristics of speed– or cost–flow curves. The assignment problem is split into a route choice model and the loading of the trip matrix onto the identified routes. Different conditions require different loading methods. Stochastic methods allow for variability in drivers' perception or route costs; these methods are discussed in section 10.4. The most interesting deterministic assignment methods try to include consistently the effect of congestion on route choice. This chapter considers only pragmatic methods under the general title of congested assignment in section 10.5; we leave a more rigorous treatment of equilibrium assignment for Chapter 11. Section 10.6 considers the problems and approaches required to model public-transport assignment.

10.1.2 Definitions and Notation

Some further notation will be introduced as required but the basic elements used in this chapter are:

T_{ijr} is the number of trips between i and j via route r,
V_a is the flow on link a,
$C(V_a)$ is the cost–flow relationship for link a,
$c(V_a)$ is the actual cost for a particular level of flow V_a; the cost when $V_a = 0$ is referred to as *free-flow* cost,
c_{ijr} is the cost of travelling from i to j via route r,

$$\delta_{ijr}^{a} = \begin{cases} 1 & \text{if link } a \text{ is on path (or route) } r \text{ from } i \text{ to } j \\ 0 & \text{otherwise} \end{cases}$$

A superscript n will be used to indicate a particular iteration in iterative methods. A superscript $*$ will be used to indicate an optimum value, e.g. c_{ij}^{*} is the minimum cost of travelling between i and j.

10.1.3 Speed–Flow and Cost–Flow Curves

A familiar relationship in traffic engineering is that relating the speed on a link to its flow. This concept was originally developed for long links in motorways, tunnels or trunk roads. A speed–flow relationship is usually presented as in Figure 10.1; as flow increases, speed tends to decrease after an initial period of little change; when flow approaches *capacity* the rate of reduction in speed increases. Maximum flow is obtained at capacity and when attempts are made to force traffic volumes beyond this value an unstable region with low flows and low speeds is reached.

For practical reasons, in traffic assignment this type of relationship is handled in terms of travel time per unit distance versus flow, or more generally, as a cost–flow relationship, as also shown in Figure 10.1. Traffic assignment methods taking into

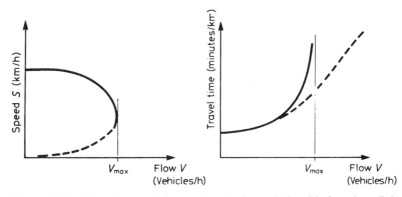

Figure 10.1 Typical speed–flow and cost–flow relationship for a long link

account congestion effects need a set of suitable functions relating link attributes (capacity, free flow speed) and flow on the network with the resulting speeds or costs. This can be written in general terms as:

$$C_a = C_a(\{\mathbf{V}\}) \tag{10.1}$$

that is the cost on a link a is a function of all the flows $\mathbf{V}$ in the network, i.e. not just the flow on the link itself. This general formulation is relevant in urban areas where there is a good deal of interaction between flows on different links and their corresponding delays, for example at priority junctions or roundabouts. However, this can be simplified if one considers long links, that is links where most of the travel time takes place on the link rather than at the end junctions. In this case the function is said to be *separable* and we can write:

$$C_a = C_a(V_a) \tag{10.2}$$

that is, the cost on the link depends just on its flow and the link characteristics. This assumption simplifies the estimation of these functions and the development and use of suitable trip assignment techniques. It must be recognised, however, that it becomes much less realistic as one works with denser and more congested urban areas.

A number of general functional forms have been proposed to embody the general relationship in equation (10.2). The fact that our main concern in this section is traffic assignment permits us to concentrate on a smaller set of these functions, in particular those with good mathematical properties. The following are desirable properties from the point of view of traffic assignment:

- Realism; the modelled travel times should be realistic enough.
- The function should be non-decreasing and monotone; increasing flow should not reduce travel time. This is not only reasonable but also desirable, as we shall see below.
- The function should be continuous and differentiable.
- The function should allow the existence of an overload region, i.e. it should not generate infinite travel time, even when flow is equal or greater than capacity. This may happen as part of an iterative process when more traffic is assigned to a link than its capacity; a high positive value for travel time should be produced but infinity will generate overflow in computer programs, an undesirable occurrence. Moreover, short-term overload can certainly happen in practice without generating anything approaching infinite delay! The dotted line in the cost–flow curve in Figure 10.1 simulates this.
- For practical reasons the cost–flow relationship should be easy to transfer from one context to another; the use of engineering parameters like free-flow speed, capacity, and number of junctions per kilometre is therefore desirable.

One would expect the cost–flow relationship to be an increasing function with flow, except perhaps at very low flow levels when travel times may remain constant despite small increases in traffic volume. The total operating cost on a link will then be given by $V_a C_a(V_a)$; it is interesting to consider the corresponding marginal cost, that is the contribution to total cost made by the marginal addition of a vehicle to the stream:

$$C_{ma} = \frac{\partial[V_a C_a(V_a)]}{\partial V_a} = C_a(V_a) + V_a \frac{\partial C_a(V_a)}{\partial V_a} \tag{10.3}$$

On the right-hand side we have two terms, the first one corresponding to the average cost on the link and the second to the contribution to delay to other traffic made by the marginal vehicle. This is an external effect and corresponds to the additional costs incurred by other users of the link when a new car is added to it. As the cost–flow curve is an increasing one this contribution is always greater than zero. It is also clear that in economic terms the average and marginal costs will only be the same in the flat part of the cost–flow curve, if any.

A number of authors have suggested functional forms for cost–flow relationships. These usually rely on the assumption that one is trying to model steady-state conditions and some kind of average behaviour. Branston (1976) has produced a good review of the practical problems encountered when trying to calibrate these cost–flow functions:

- There are problems with the length of the observation period in particular in congested areas and where an upstream junction acts as bottleneck; the exact location of flow and delay measuring areas plays a critical role in determining the quality of the results obtained.
- The assumption that delays depend only on flow on the link itself is unrealistic in most dense urban networks and this is particularly critical in trying to estimate cost–flow functions.

Branston (1976) also reviews cost–flow curves proposed by other authors. Some of the most used are the following:

1. Smock (1962) for the Detroit Study:

$$t = t_0 \exp(V/Q_s) \tag{10.4}$$

   where t is travel time per unit distance, t_0 is travel time per unit distance under tree flow conditions, and Q_s is the steady-state capacity of the link.
2. Overgaard (1967) generalised (10.4) as follows:

$$t = t_0 \alpha^{\beta(V/Q_p)} \tag{10.5}$$

   where Q_p is the practical capacity of the link, and α and β are parameters for calibration.
3. The Bureau of Public Roads (1964) in the USA proposed what is probably the most common function of this type:

$$t = t_0[1 + \alpha(V/Q_p)^\beta] \tag{10.6}$$

4. Finally, the Department of Transport in the UK has produced a large number of cost–flow curves for a variety of link types in urban, sub-urban and inter-urban

roads. Some have a general form which considers first the speed–flow $s(V)$ curve:

$$s(V) = \begin{cases} S_0 & V < F_1 & (10.7a) \\ S_0 - \dfrac{S_0 - S_1}{F_2 - F_1}(V - F_1) & F_1 \leqslant V \leqslant F_2 & (10.7b) \\ S_1/(1 + (S_1/8d)(V/F_2 - 1) & V > F_2 & (10.7c) \end{cases}$$

where

S_0 is the free flow speed,
S_1 is the speed at capacity flow F_2,
F_1 is the maximum flow at which free-flow conditions prevail, and
d is the distance or length of the link.

Then the time-flow $T(V)$ relationship becomes:

$$T(V) = \begin{cases} d/S_o & V < F_1 & (10.8a) \\ d/S(V) = \dfrac{d}{S_0 + SS_{01}F_1 - SS_{01}V} & F_1 \leqslant V \leqslant F_2 & (10.8b) \\ d/S_1 + (V/F_2 - 1)/8 & V > F_2 & (10.8c) \end{cases}$$

with SS_{01} given by:

$$SS_{01} = \frac{S_0 - S_1}{F_1 - F_2} \qquad (10.9)$$

Typical values for these coefficients (Department of Transport 1985) are given in Table 10.1.

Table 10.1   Typical speed–flow curve coefficients in the UK

Type	S_0 km/h	S_1 km/h	F_1 pcu/h/lane	F_2 pcu/h/lane
Single 2 lane, rural	63	55	400	1400
Dual 2 lane, rural	79	70	1600	2400
Single 2 lane, urban, outer area	45	25	500	1000

In some cases a cut-off point in speed reductions is assumed; for example the speed may be assumed to remain at F_2 for $V > F_2$. All the above speed or cost–flow curves produce information about travel time on a link. However, it is recognised that most users might wish to minimise a combination of link attributes including time and distance. Conventional practice recommends the use of a simplified version of the generalised cost concept, namely a linear weighted combination of time and distance:

$$C_a = \alpha(\text{travel time})_a + \beta(\text{link distance})_a \qquad (10.10)$$

This cost could be measured in generalised time or generalised money units. It is also possible to include an out-of-pocket expenditure element, for example a toll to be applied on a given link.

The calibration of cost–flow relationships is time consuming and requires a good deal of high-quality data: observations of travel times on links under different flow levels. For this reason, this is rarely attempted and many countries have developed their own functions for use in their studies. See also the limitations of link-based cost–flow functions in urban areas as discussed in section 11.3.

Suh *et al.* (1990) have put forward an innovative approach to estimate cost–flow curves based on traffic counts; they use a bi-level optimisation method that, in essence, seeks to establish the parameters for the cost–flow curves minimising a measure of difference between assigned and observed flows. The value of this approach is limited by the errors in the assignment process as discussed, again, in section 11.3: e.g. errors in the network, trip matrix, in the assumption of perfect information and that all users perceive link costs in the same way. The cost–flow curves estimated by this bi-level optimisation will then incorporate these errors and will be, therefore, difficult to transfer to other areas or even schemes.

10.2 TRAFFIC ASSIGNMENT METHODS

10.2.1 Introduction

During the classic traffic assignment stage a set of rules or principles is used to load a fixed trip matrix onto the network and thus produce a set of links flows. This is not, however, the only relevant output from the assignment stage; this has several objectives which are useful to consider in detail. Not all of them receive the same emphasis in all situations nor can all be achieved with the same level of accuracy. The main objectives are:

1. Primary:
   - to obtain good *aggregate* network measures, e.g. total motorway flows, total revenue by bus service;
   - to estimate zone-to-zone travel costs (times) for a given level of demand;
   - to obtain *reasonable* link flows and to identify heavily congested links.
2. Secondary:
   - to estimate the routes used between each O–D pair;
   - to analyse which O–D pairs use a particular link or route;
   - to obtain turning movements for the design of future junctions.

In general terms we shall attain the primary objectives more accurately than the secondary ones. Even within objectives we are likely to be more accurate with those earlier in the list. This is essentially because our models are more likely to estimate correctly aggregate than disaggregate values.

The basic inputs required for assignment models are:

- A trip matrix expressing estimated demand. This will normally be a peak-hour matrix in urban congested areas, and perhaps other matrices for other peak and off-peak periods. A 24-hour matrix is sometimes used for assignment of uncongested networks. The conversion of 24-hour matrices into single hours is seldom satisfactory in terms of congestion, as these matrices are symmetric and single-hour trips seldom are. The matrices themselves may be available in terms of person trips; therefore, they should be converted into vehicle trips as capacity- and speed–flow relationships are described in these terms.
- A network, namely links and their properties, including speed–flow curves.
- Principles or route selection rules thought to be relevant to the problem in question.

10.2.2 Route Choice

The basic premise in assignment is the assumption of a rational traveller, i.e. one choosing the route which offers the least perceived (and anticipated) individual costs. A number of factors are thought to influence the choice of route when driving between two points; these include journey time, distance, monetary cost (fuel and others), congestion and queues, type of manoeuvres required, type of road (motorway, trunk road, secondary road), scenery, signposting, road works, reliability of travel time and habit. The production of a *generalised* cost expression incorporating all these elements is a difficult task. Furthermore, it is not practical to try to model all of them in a traffic assignment model, and therefore approximations are inevitable.

The most common approximation is to consider only two factors in route choice: *time* and *monetary cost*; further, monetary cost is often deemed proportional to travel distance. The majority of traffic assignment programs allow the user to allocate weights to travel time and distance in order to represent drivers' perceptions of these two factors. The weighted sum of these two values then becomes a generalised cost used to estimate route choice. There is evidence to suggest that, at least for urban car traffic, time is the dominant factor in route choice. Outram and Thompson (1978) compared drivers' stated objective with their actual performance in route choice. They found that the proportion of drivers being successful in achieving their objectives was relatively low. They also found that the combination of time and distance gave the best explanation of route choice. However, even if we allow the combination of time and distance in a generalised cost function, we can only explain something of the order of 60 to 80% of the routes actually observed in practice. As the marginal contribution of other factors in untangling route choice is very small, the unexplained part must be attributed to factors like differences in perception, imperfect information on route costs or simply errors.

The fact that different drivers often choose different routes when travelling between the same two points may be ascribed to two different types of reasons:

1. Differences in individual perceptions of what constitutes the 'best route'; different individuals may not only incorporate different features in their generalised cost function but perceive them in different ways.
2. Congestion effects affecting shorter routes first and making their generalised costs comparable to initially less attractive routes.

Example 10.1: Consider an idealised town with a low-capacity through route (1000 vehicles per hour) and a high-capacity bypass, as in Figure 10.2. The bypass is a longer but faster route with a capacity of 3000 vph. Assume that during the morning peak 3500 drivers approach the town and that everyone would like to use the shortest route, i.e. via the town centre. It is clear that it would not be possible for all of them to do so as the route would become too congested even before its ultimate capacity is reached. Many would opt then for second choice to avoid long queues and delays. Presumably drivers would experiment with the two routes until they find a more or less stable arrangement when none can improve their time by switching to the other route. This is a typical case of Wardrop's equilibrium, which is discussed in greater detail below. Diversion across routes in this case is due to *capacity restraint*.

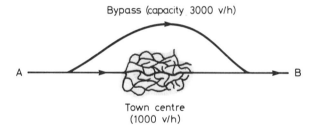

Figure 10.2   Town served by a bypass and a town centre route

However, not all 3500 drivers will think alike; some would always prefer the bypass because of its uninterrupted flow conditions or its scenery, where as others would value other features of the town-centre route. These differences in objectives and perceptions would also lead to a spread of routes and such effect is customarily referred to as the *stochastic* element in route choice.

Particular types of models are more suited to representing one or more of the above influences. A possible classification of traffic assignment methods is given in Table 10.2. The details and characteristics of each method are discussed below.

Table 10.2   Classification scheme for traffic assignment

| | | Stochastic effects included? | |
		No	Yes
Is capacity restraint included?	No	All-or-nothing	Pure stochastic Dial's, Burrell's
	Yes	Wardrop's equilibrium	Stochastic user equilibrium

Each assignment method has several steps which must be treated in turn. Their basic functions are:

- To identify a set of routes which might be considered attractive to drivers; these routes are stored in a particular data structure called a *tree* and therefore this task is often called the *tree-building stage*.
- To assign suitable proportions of the trip matrix to these routes or trees; this results in flows on the links in the network.
- To search for convergence; many techniques follow an iterative pattern of successive approximations to an ideal solution, e.g. Wardrop's equilibrium; convergence to this solution must be monitored to decide when to stop the iterative process.

10.2.3 Tree Building

Tree building is an important stage in any assignment method for two related reasons. First, it is performed many times in most algorithms, at least once per iteration. Second, a good tree-building algorithm can save a great deal of computer time and costs. By a good algorithm we mean an efficient one which is also well programmed in a suitable language. Van Vliet (1978) has produced a good discussion of the most widely used algorithms for tree building and this section is based on his paper.

There are two basic algorithms in general use for finding the shortest (cheapest) paths in road networks, one due to Moore (1957) and one due to Dijkstra (1959). The two will be discussed using a more convenient node-oriented notation: the length (cost) of a link between A and B in the network is denoted by $d_{A,B}$. The path or route is defined by a series of connected nodes, A-C-D-H, etc., whilst the length of the path is the arithmetic sum of the corresponding link lengths in the path. Let d_A denote the minimum distance from the origin of the tree S to the node or centroid A; P_A is the *predecessor* or *backnode* of A so that the link (P_A, A) is part of the shortest path from S to A.

The procedure for building a minimum path tree from S to all other nodes may be described as follows:

Initialisation Set all $d_A = \infty$ (a suitable large number depending on computer and compiler) except d_S which is set equal to 0; set up a *loose-end table L* to contain nodes already reached by the algorithm but not fully explored as predecessors for further nodes. They are the tip of the tree as branches grow to reach all nodes. Initialise all entries L_i in L to zero, and all P_A to a suitable default value.

Procedure Starting with the origin S as the 'current' node $= A$:

1. examine each link (A, B) from the current node A in turn and, if $d_A + d_{A,B} < d_B$ then set $d_B = d_A + d_{A,B}$, $P_B = A$ and add B to L;
2. remove A from L, if the loose-end table is empty, stop; otherwise,

3. select another node from the loose-end table and return to step 1 with it as the current node.

Three comments should be made at this stage. First, in general routes are not allowed to use centroids, therefore in step 1, B would not be added to L if it was a centroid. Second, the essential difference between Moore's and Dijkstra's algorithms lies in the procedure for selecting a node from L. Moore selects the top entry, that is the oldest entry in the table; Dijkstra selects the node nearest to the origin, i.e. the node L_i such that d_{L_i} is a minimum. This requires some additional calculations (including sorting of nodes) but ensures that each link is examined once and only once. It is well known that Dijkstra's algorithm is superior to Moore's, in particular for larger networks; it is however, more difficult to program. Finally, trees are often stored in the computer in one of two forms: as a set of ordered *backnodes* in which A is the backnode of B if link (A, B) forms part of the tree; or as a set of *backlinks* with a similar definition.

Van Vliet (1977) also identified a lesser known algorithm which performs very well even in large networks: D'Esopo's algorithm, as described and tested by Pape (1974). D'Esopo's uses a 'two-ended' loose-end table so that node B is entered at one or other end depending on its 'status'. If B had not been previously reached by the tree then it is entered at the bottom of L; if it is currently on the table no entry is made; but, if it has already been entered to L, examined and removed from the table then it is entered at the top. A simple array can be used to record the status with three potential values ($+1, 0$ or -1) representing each case for each node. As shown by Van Vliet (1977), D'Esopo's algorithm can reduce CPU times by 50% relative to Moore's. Furthermore, its performance is very close and often better compared with that of the best implementations of Dijkstra's; it has the added advantage of being much simpler to program.

Trees have two important additional uses in transport planning. They are often employed to extract cost information in a network. For example, the total travel time between two zones can be obtained by following the sequence of links in the tree connecting them and accumulating their travel times. This operation is often referred to as 'skimming' a tree. Trees built for, say, travel time can be skimmed for other attributes, for example generalised cost, distance, number of nodes, etc. Trees can also be used to produce information on which O–D pairs are likely to use a particular link. This facility, often called a 'selected link analysis', permits the identification of who is likely to be affected by a network change. Moreover, it can also be used to cordon a trip matrix for a smaller study area; in this case the selected links are used to identify entry and exit points to the small study area and the trees to combine the original zones into single external ones for the new sub-area.

10.3 ALL-OR-NOTHING ASSIGNMENT

The simplest route choice and assignment method is 'all-or-nothing' assignment. This method assumes that there are no congestion effects, that all drivers consider the same attributes for route choice and that they perceive and weigh them in the same way. The absence of congestion effects means that link costs are fixed; the assumption that all drivers perceive the same costs means that every driver from i to j must choose the same route. Therefore, all drivers are assigned to one route between i and j and no driver is assigned to other, less attractive, routes. These assumptions are probably reasonable in sparse and uncongested networks where there are few alternative routes and they are very different in cost.

The assignment algorithm itself is the procedure that *loads* the matrix **T** to the shortest path trees and produces the flows $V_{A,B}$ on links (between nodes A and B). All load algorithms start with an initialisation stage, in this case making all $V_{A,B} = 0$ and then apply one of two basic variations: pair-by-pair methods and once-through approaches.

Pair-by-pair This is probably the simplest but not necessarily the most efficient method. In this case we start from an origin and take each destination in turn. First, we initialise all $V_{A,B} = 0$. Then for each pair (i, j):

1. set B to the destination j;
2. if (A, B) is the backlink of B then increment $V_{A,B}$ by T_{ij}, i.e. make $V_{A,B} = V_{A,B} + T_{ij}$;
3. set B to A;
4. if $A = i$ terminate (i.e. process the next (i, j) pair), otherwise return to step 2.

Once-through This is sometimes called a 'cascade' method as it loads accumulated flow from nodes to links following the minimum cost trees from an origin i. Let V_A be the cumulative flow at node A:

1. set all $V_A = 0$ except for the destinations j for which $V_j = T_{ij}$;
2. set B equal to the most distant node from i;
3. increment V_A by V_B where A is the backnode of B, i.e. make $V_A = V_A + V_B$;
4. increment $V_{A,B}$ by V_B, i.e. make $V_{A,B} = V_{A,B} + V_B$;
5 set B equal to the next most distant node; if $B = i$ then the origin has been reached, begin processing the next origin, otherwise proceed with step 3.

In this form V_B represents the total number of trips from i passing through node B en route to destinations further away from i. By selecting nodes in reverse order of distance, each node is processed once only. This algorithm requires the trees to be stored in terms of backnodes ordered by distance from the origin.

Example 10.2: Consider the simple network in Figure 10.3 and its associated trip matrix: A-C = 400, A-D = 200, B-C = 300 and B-D = 100. Section (a) shows the travel costs (times) on each link; section (b) the corresponding trees based on these costs together with the contributions to the total flow after assignment; these are shown in section (c).

All-or-nothing assignment is generally of limited interest to the planner; it may be used to represent some sort of 'desire line', i.e. what drivers would like to do in the absence of congestion. However, its most important practical feature is as a basic building block for other types of assignment techniques, e.g. equilibrium and stochastic methods.

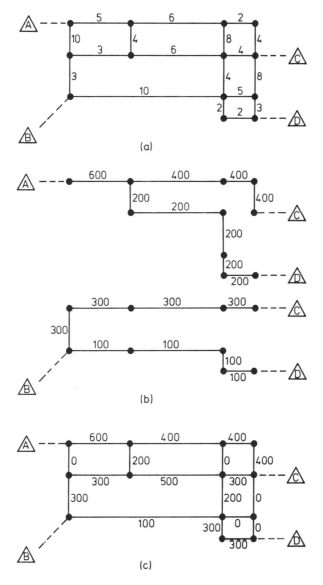

Figure 10.3 A simple network, its trees and flows from loading a trip matrix

10.4 STOCHASTIC METHODS

Stochastic methods of traffic assignment emphasise the variability in drivers' perceptions of costs and the composite measure they seek to minimise (distance, travel time, generalised costs). Stochastic methods need to consider second-best routes (in terms of engineering or modelled costs); this generates additional problems as the number of alternative second-best routes between each O–D pair may be extremely large. Several methods have been proposed to incorporate these aspects but only two have relatively widespread acceptance: *simulation-based* and *proportion-based* methods. The first use ideas from stochastic (Monte Carlo) simulation to introduce variability in perceived costs. The proportion-based methods, on the other hand, allocate flows to alternative routes from proportions calculated using logit-like expressions.

10.4.1 Simulation-Based Methods

A number of techniques use Monte Carlo simulation to represent the variability in drivers' perceptions of link costs; in particular, the technique developed by Burrell (1968) has been widely used for many years. These methods usually rely on the following assumptions:

- For each link in a network one should distinguish objective or engineering costs as measured/estimated by an observer (modeller) and subjective costs as perceived by each driver. It is further assumed that there is a distribution of perceived costs for each link with the engineering costs as the mean, as shown in Figure 10.4.

  The various implementations of these ideas differ in their assumptions about the shape of these distributions: while Burrell's assumes a uniform distribution, other models hypothesise a normal distribution. In either case one also needs to assume or calibrate a standard deviation or range for the distribution of perceived costs.

- The distributions of perceived costs are assumed to be independent;
- Drivers are assumed to choose the route that minimises their perceived route costs, which are obtained as the sum of the individual link costs.

A general description of these algorithms would be as follows. Select a distribution (and spread parameter σ) for the perceived costs on each link. Split the population travelling along each O–D pair into N segments, each assumed to perceive the same costs.

1. Make $n = 0$.
2. Make $n = n + 1$.
3. For each $i - j$ pair:

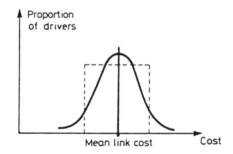

Figure 10.4   Distribution of perceived costs on a link

- compute perceived costs for each link by sampling from the corresponding distributions of costs by means of random numbers;
- Build the minimum perceived cost path from i to j and assign T_{ij}/N trips to it accumulating the resulting flows on the network.

4. If $n = N$ stop, otherwise go to step 2.

In practice many short-cuts are taken to reduce computation times, for example:

- generate new sets of random costs per origin and not per O–D pair;
- use N equal to just 3 or 5 and generate one set of random costs for each matrix and not for each O–D pair or origin;
- use small values for N, even 1 in some circumstances.

This type of approach uses simulation in order to reduce the number of second-best routes to be considered. If a wider range of routes is thought necessary, one can increase the value of N and/or the spread parameter in the distribution of link costs. Burrell's approach has the advantage of generating cheap routes more often than more expensive ones: if a route is expensive it is much less likely to appear as the cheapest as a result of the stochastic variations in link costs. Although the uniform distribution is efficient in computer time, it is not very realistic. A better function, but more expensive in terms of CPU time, is the normal distribution with variance proportional to the mean engineering costs.

As in all Monte Carlo methods, the final results are dependent on the series of random numbers used in the simulation. Increasing the value of N reduces this problem. There are, however, more serious difficulties with this approach:

- The link perceived costs are not independent, as drivers usually have preferences, for example, for motorway links or to avoid priority junctions or minor roads. The assumption of independence in perceived costs may lead to unrealistic switching between parallel routes connected by minor roads.
- No explicit allowance is made for congestion effects.

In compensation, these methods often produce a reasonable spread of trips, are relatively simple to program and do not require the choice or estimation of speed–flow relationships (which may turn out to be a problem in some cases).

10.4.2 Proportional Stochastic Methods

Virtually all these methods are based on a loading algorithm which splits trips arriving at a node between all possible exit nodes, as opposed to the all-or-nothing method which assigns all trips to a single exit node. Very often the implementation of these methods reverses the problem so that the division of trips at a node is actually based upon where the trips are coming from rather than where they are going to. Consider node B in Figure 10.5; there are a number of possible entry points denoted by A_1, S_2, A_3, A_4 and A_5 for trips from I to J.

The 'splitting factors' f_i are defined by:

$$
\begin{aligned}
f_i &= 0 && \text{if } d_{A_i} \geq d_B \\
0 &\leq f_i \leq 1 && \text{if } d_{A_i} < d_{B_i}
\end{aligned}
$$

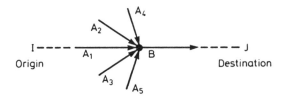

Figure 10.5   A node (B) and links feeding trips into it

where d_{A_i} represents the minimum cost of travel from the origin i to node A_i. The first condition requires that f_i should be zero if an entry node A_i is further from the origin than B, therefore ensuring that trips are allocated to routes which take them efficiently away from the origin. The trips T_B that pass through B are divided according to the equation:

$$F(A_i, B) = \frac{T_B f_i}{\sum_i f_i} \tag{10.11}$$

The assignment procedure is now equivalent to the cascade method for all-or-nothing assignment. Implementations of these ideas differ mainly in the way in which they define the splitting function f_i. The single-path method due to Dial (1971) requires that:

$$f_i = \exp\left(-\Omega \delta d_i\right) \tag{10.12}$$

where δd_i is the extra cost incurred in travelling from the origin to node B via node A_i rather than via the minimum cost route. In this way, if A_i is in the minimum-cost route, δd_i is equal to zero and $f_i = 1$. Nodes that lie on more expensive routes have $\delta d_i > 0$ and their f_i values are less than 1. In this way shorter routes are favoured over more expensive ones.

Dial originally described a double-pass algorithm which effectively uses a logit-type formulation to split trips from i to j among alternative routes r:

$$T_{ijr} = \frac{T_{ij} \exp\left(-\Omega C_{ijr}\right)}{\sum_r \exp\left(-\Omega C_{ijr}\right)} \tag{10.13}$$

The parameter Ω can be used to control the spread of trips among routes. The algorithm involves a forward and a backward pass:

1. The forward pass: take each node A in ascending order of d_A and define a weight for each exit link (A, B) such that:

$w_{(A,B)} = W_A \exp\left(-\Omega \delta d_{(A,B)}\right)$ if $d_A < d_B$ or zero otherwise; W_A is the accumulated weight at A defined as:

$W_A = \sum_{A'} w_{(A',A)}$ and $W_I = 1$ [A' is a predecessor of A]

2. The backward pass: identical to the single-pass algorithm with the exception that the weights $w_{(A,B)}$ are used to work out the split of trips rather than the splitting fators F_i.

Example 10.3: A practical problem with Dial's assignment is that it is biased against trunk routes as opposed to secondary links. Consider the problem of a town served by a bypass and a town-centre route with three small variations as illustrated in Figure 10.6. Assume also that there are 4000 trips from A to B and that all routes have approximately the same cost.

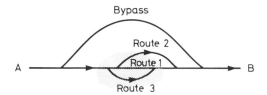

Figure 10.6   Town served by a bypass and three city-centre routes

In this case Dial's algorithm would split the 4000 trips as follows: 1000 via the bypass and 1000 via each of the town-centre routes. However, most users would regard this problem as one with only two alternatives: bypass or town centre. Recall the discussion about the independence of irrelevant alternatives property of the logit model in Chapter 5. Dial's runs into trouble when it considers every possible route even if some permutations or combinations of links may differ just in a few percentage points of their total cost. In behavioural terms Dial ignores the correlation between similar routes. In practice, Dial tends to allocate more traffic to dense sections of the network with short links, compared with sparser parts of the network with relatively longer links. In fact, coding strategies for networks can affect the allocation of flows.

10.5 CONGESTED ASSIGNMENT

10.5.1 Wardrop's equilibrium

If one ignores stochastic effects and concentrates on capacity restraint as a generator of a spread of trips on a network, one should consider a different set of models. For a start, capacity restraint models have to make use of functions relating flow to the cost (time) of travel on a link. These models usually attempt, with different degrees of success, to approximate to the equilibrium conditions as formally enunciated by Wardrop 1952):

Under equilibrium conditions traffic arranges itself in congested networks in such a way that no individual trip maker can reduce his path costs by switching routes.

If all trip makers perceive costs in the same way (no stochastic effects):

Under equilibrium conditions traffic arranges itself in congested networks such that all used routes between an O–D pair have equal and minimum costs while all unused routes have greater or equal costs.

This is usually referred to as Wardrop's first principle, or simply Wardrop's equilibrium. It is easy to see that if these conditions did not hold, at least some drivers would be able to reduce their costs by switching to other routes.

Example 10.4: Consider again the case of a bypass and a single town-centre route as discussed in section 10.2.2 (Figure 10.2). Assume now that the absolute capacity restriction for each route is replaced with two corresponding time–flow relationships as illustrated in Figure 10.7.

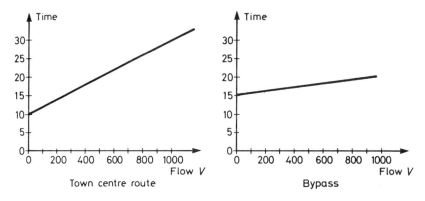

Figure 10.7   Time-flow relationships for Figure 10.2

The flows on the two routes will satisfy Wardrop's equilibrium when the corresponding costs are identical. In this case it is relatively simple to write two equations for travel time versus flow and equate them to find the equilibrium solution, for example:

$$C_b = 15 + 0.005 V_b \qquad (10.14a)$$

$$C_t = 10 + 0.02 V_t \qquad (10.14b)$$

where C_b and C_t are travel costs via the bypass and the town-centre routes respectively, and V_b and V_t are their corresponding flows.

By equating C_b to C_t it is possible to find, in this simple case, the direct solution to Wardrop's equilibrium as a function of the total flow $V_b + V_t = V$:

$$15 + 0.005 V_b = 10 + 0.02(V - V_b)$$

that is:

$$V_b = 0.8V - 200 \qquad (10.15)$$

Expression (10.15) has meaning only for non-negative flows, i.e. for V greater than or equal to $200/0.8 = 250$. For $V < 250$, $C_t < C_b$, $V_b = 0$ and $V_t = V$, i.e. all traffic chooses the town-centre route. For situations where $V > 250$ the two routes will be used; for example, the reader can verify that for $V = 2000$ the equilibrium flows are $V_b = 1400$ and $V_t = 600$ and the costs by each route are 22 minutes.

The same idea would apply to flows on networks where the costs of travel by each of the routes used between two points will be the same under Wardrop's equilibrium. The problem is, of course, that in anything but the simplest cases it is not possible to solve the equilibrium flows algebraically; rather an algorithmic solution method is required.

Several techniques have been proposed as reasonable approximations to Wardrop's equilibrium: some of them are simple heuristic approaches and the most interesting ones follow a more rigorous mathematical programming framework. In order to compare these algorithms against each other the following properties are of interest:

- Is the solution stable?
- Does it converge to the correct solution (Wardrop's equilibrium)?
- Is it efficient in terms of computational requirements?

The indicator δ, defined in the following equation, is often used to measure how close a solution is to Wardrop's equilibrium:

$$\delta = \frac{\sum\limits_{ijr} T_{ijr}(C_{ijr} - C_{ij}^*)}{\sum\limits_{ij} T_{ij} C_{ij}^*} \tag{10.16}$$

where $C_{ijr} - C_{ij}^*$ is the excess cost of travel over a particular route relative to the minimum cost of travel for that (i, j) pair. These costs are calculated after the last iteration has been performed and total flows obtained for each link. Therefore δ is a measure of the total cost of excess travel via less than optimal routes, with denominator introduced so that the measure is recorded in relative rather than absolute terms.

Wardrop (1952) proposed an alternative way of assigning traffic onto a network and this is usually referred to as his second principle:

> *Under social equilibrium conditions traffic should be arranged in congested networks in such a way that the average (or total) travel cost is minimised.*

This is a *design* principle, in contrast with his first principle which endeavours to model the behaviour of individual drivers trying to minimise their own trip costs. The second principle is oriented towards transport planners and engineers trying to manage traffic to minimise travel costs and therefore achieve an optimum *social equilibrium*. In general the flows resulting from the two principles are not the same but one can only expect, in practice, traffic to arrange itself following an approximation to Wardrop's first principle, i.e. *selfish* or *users' equilibrium*.

10.5.2 Hard and Soft Speed-Change Methods

Some of the first heuristic methods still maintained the idea of assigning all trips per O–D pair to a single route (all-or-nothing assignment), but acknowledged the fact that speeds, and therefore travel times, responded to flow levels. The simplest of these methods involves just recalculating link travel times after an all-or-nothing assignment so that they are consistent with the current flow levels. A new all-or-nothing assignment is then performed with the new costs and trees. It is easy to see that in general this is a poor approach as the chosen routes will oscillate and the flow pattern will, in general, never converge. In the case of the town-centre bypass problem of Example 10.4 with, say, $V > 250$, the flows would oscillate between all via the town centre in one iteration and all via the bypass in the next one. This phenomenon will be repeated in larger networks although in some cases it may be more difficult to identify.

In an attempt to dampen these route and flow oscillations it has been proposed to use an average speed of two or more all-or-nothing assignments to perform the next iteration. This if often called a *soft* speed change as opposed to the *hard* speed change of the original method. However, this may only provide an apparent improvement as the main weakness of these two approaches is that they still assign all traffic to a single route for each O–D pair, therefore contradicting Wardrop's principle. Taking again the case of Example 10.4, it can easily be seen that the soft speed-change method will still load all traffic alternatively via one route and then the other in the next iteration. Both methods produce unstable solutions, are inherently non-convergent and the use of soft speed changes will only disguise this fact (to some extent) in larger networks.

In algorithmic terms the two variants can be described as follows:

1. Select an initial set of current link costs, usually free-flow travel times; make $n = 0$.
2. Build the set of minimum cost trees with the current costs and assign (all-or-nothing) the matrix to them obtaining a new set of flows; increment n by 1 (make $n = n + 1$).
3. (a) Hard speed change: recalculate the link costs corresponding to the new flows;
   (b) Soft speed change: calculate the current link costs as the arithmetic average of the costs in the previous iteration and those calculated in (a).
4. If the flows or current link costs have not changed significantly in two consecutive iterations, stop; otherwise, proceed to step 2.

As suggested above, this algorithm will probably never terminate at step 4 unless an arbitrary limit to the number of iterations n is imposed.

10.5.3 Incremental Assignment

This is a more interesting and realistic approach. In this case the modeller divides the total trip matrix **T** into a number of fractional matrices by applying a set of proportional factors p_n such that $\sum_n p_n = 1$. The fractional matrices are then loaded, incrementally, onto successive trees, each calculated using link costs from the last accumulated flows. Typical values for p_n are: 0.4, 0.3, 0.2 and 0.1. The algorithm can be written as follows:

1. Select an initial set of current link costs, usually free-flow travel times. Initialise all flows $V_a = 0$; select a set of fractions p_n of the trip matrix $\mathbf{T}$ such that $\sum_n p_n = 1$; make $n = 0$.
2. Build the set of minimum cost trees (one for each origin) using the current costs; make $n = n + 1$.
3. Load $\mathbf{T}_n = p_n \mathbf{T}$ all-or-nothing to these trees, obtaining a set of auxiliary flows F_a; accumulate flows on each link:

$$V_a^n = V_a^{n-1} + F_a$$

4. Calculate a new set of current link costs based on the flows V_a^n; if not all fractions of T have been assigned proceed to step 2; otherwise stop.

This algorithm does not necessarily converge to Wardrop's equilibrium solution even if the number of fractions p is large and the size of the increments $(p_n\mathbf{T})$ is small. Incremental loading techniques suffer from the limitation that once a flow has been assigned to a link it is not removed and loaded onto another one; therefore, if one of the initial iterations assigns too much flow on a link for Wardrop's equilibrium conditions to be met (for example, because the link is short but has very low capacity), then the algorithm will not converge to the correct solution.

However, incremental loading has two advantages:

- it is very easy to program;
- its results may be interpreted as the build-up of congestion for the peak period.

Example 10.5: Consider again the problem of the two routes, town centre and bypass, of Example 10.4. We split the demand of 2000 trips into four increments of 0.4, 0.3, 0.2 and 0.1 of this demand, i.e. 800, 600, 400 and 200 trips. At each increment we calculate the new travel costs using equations (10.14). The following table summarises the results of this algorithm:

N	Increment	Flow town	Cost town	Flow bypass	Cost bypass
0	0	0	10	0	15
1	800	800	26	0	15
2	600	800	26	600	18
3	400	800	26	1000	20
4	200	800	26	1200	21

It can be seen that the algorithm does not converge, in this case, to the correct equilibrium solution. This is because once the wrong flow (800) has been loaded onto the town-centre route, this method cannot reduce it; therefore the flow and cost via the town centre remain overestimated. As a matter of interest, the value of the δ indicator for the solution above is:

$$\delta = [800(26 - 21) + 1200(21 - 21)]/(2000 \times 21) = 0.095$$

The reader can verify that the use of more and smaller increments would produce closer solutions to true equilibrium. Note that if one starts with an increment of 0.3 times the total demand, the solution is true equilibrium; however, this is just a chance occurrence in this case.

10.5.4 Method of Successive Averages

Iterative algorithms were developed, at least partially, to overcome the problem of allocating too much traffic to low-capacity links. In an iterative assignment algorithm the 'current' flow on a link is calculated as a linear combination of the current flow on the previous iteration and an auxiliary flow resulting from an all-or-nothing assignment in the present iteration. The algorithm can be described by the following steps:

1. Select a suitable initial set of current link costs, usually free-flow travel times. Initialise all flows $V_a = 0$; make $n = 0$.
2. Build the set of minimum cost trees with the current costs; make $n = n + 1$.
3. Load the whole of the matrix **T** all-or-nothing to these trees obtaining a set of auxiliary flows F_a.
4. Calculate the current flows as:

$$V_a^n = (1 - \phi)V_a^{n-1} + \phi F_a \qquad (10.17)$$

$$\text{with } 0 \leqslant \phi \leqslant 1$$

5. Calculate a new set of current link costs based on the flows V_a^n. If the flows (or current link costs) have not changed significantly in two consecutive iterations, stop; otherwise proceed to step 2. Alternatively, the indicator δ in (10.16) could be used to decide whether to stop or not. Another, less good but quite common, criterion for stopping is simply to fix the maximum number of iterations; δ should be calculated in this case as well to know how close the solution is to Wardrop's equilibrium.

Iterative assignment algorithms differ in the method used to give a value to ϕ. A simple rule is to make it constant, for example $\phi = 0.5$. A much better approach due to Smock (1962), is to make $\phi = 1/n$. The reader may verify that equal weight is given to each auxiliary flow F_a in this case; for this reason, the algorithm is also known as the method of successive averages (MSA). It has been shown (see, for example, Sheffi 1985) that making $\phi = 1/n$ produces a solution convergent to Wardrop's equilibrium, albeit not a very efficient one. As we shall see in Chapter 11, the Frank–Wolfe algorithm estimates optimal values for ϕ in order to guarantee and speed up convergence.

Example 10.6: Consider the same bypass versus town-centre problem of Example 10.5 and use $\phi = 1/n$. The following table summarises the steps in the MSA algorithm:

Iteration		ϕ	Flow town	Cost town	Flow bypass	Cost bypass
1	F		2000		0	
	V^n	1	2000	50	0	15
2	F		0		2000	
	V^n	1/2	1000	40	1000	20
3	F		0		2000	
	V^n	1/3	667	23.3	1333	21.7
4	F		0		2000	
	V^n	1/4	500	20	1500	22.5
5	F		2000		0	
	V^n	1/5	800	26	1200	21
6	F		0		2000	
	V^n	1/6	667	23.3	1333	21.7
7	F		0		2000	
	V^n	1/7	572	21.4	1428	22.1
8	F		2000		0	
	V^n	1/8	750	25	1250	21.25
9	F		0		2000	
	V^n	1/9	667	23.3	1333	21.7
10	F		0		2000	
	V^n	0.1	600	22	1400	22

It can be seen that it takes a number of iterations to approximate to the right solution. Of course, the value of δ after iteration 10 is zero in this case. However, the reader will note that the algorithm was close to the correct equilibrium solutions in iterations 3, 6 and 9 but only reached it in iteration 10. This is due to the rigid nature of the rule to calculate ϕ. For more realistic networks the number of iterations needed to reach satisfactory convergence may be very high.

Another lesson from this simple example is that fixing the maximum number of iterations is not a good approach from the point of view of evaluation. Link and total costs can vary considerably in successive iterations and this may affect the feasibility of a scheme.

10.6 PUBLIC-TRANSPORT ASSIGNMENT

10.6.1 Introduction

In this section the problems associated with route choice and assignment for passengers using public-transport networks will be discussed. These problems are, in many ways, more difficult than those encountered by private-transport assignment; computer requirements tend to be heavier and even the best methods require important simplifying assumptions. These difficulties, coupled with a lower profile in public policy during the 1980s, resulted in fewer research resources being allocated to public transport than to car route choice. The trend may be reversed as more attention is being paid to improvements in public-transport service provision and operational efficiency.

We shall discuss first the issues that make public-transport assignment different from private vehicle route choice; then, we will outline some of the approaches that have been implemented to tackle them in practice.

10.6.2 Issues in Public-Transport Assignment

10.6.2.1 Supply

The network of public-transport services is different from that of private cars. It includes, as links, sections of the bus or rail services running between two stops or stations. The concept of link capacity is associated to the capacity of each unit (bus, train) and its corresponding frequency. The travel time has an in-vehicle component as well as components for waiting at stops and walking to and from them. Many of the public-transport sections will use road links, e.g. most buses and some light rail-transit (LRT) services with street running. There will be other public-transport sections or services which will use completely different links, e.g. busways, segregated rail track, etc. The nature of these links generally produces a more complex network, an example of which is given in Figure 10.8.

10.6.2.2 Passengers

In public-transport route choice we are dealing with the movement of passengers and not of vehicles. Passengers can walk to a stop, interchange between two services and even drive part of the way to board a public-transport service later. This calls for the need to provide and specify walk and transfer links between different services, different public-transport modes (bus, rail) and between public- and private-transport facilities (e.g. 'Park and Ride').

10.6.2.3 Monetary Costs

In private car networks it is usually assumed that the monetary cost is directly associated to fuel consumption, which in turn is directly proportional to travel distance. These are both approximations but they are usually accepted as drivers do not perceive these costs in such a direct way as a passenger buying a ticket when

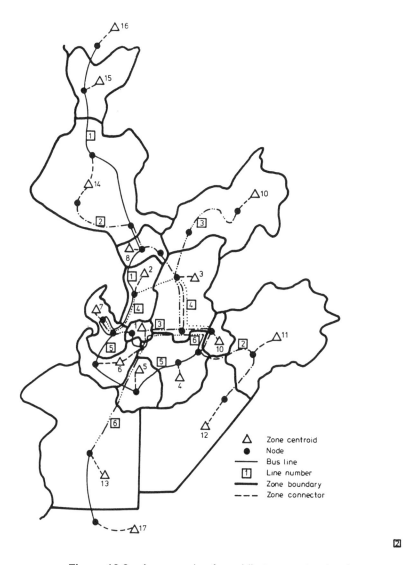

Figure 10.8   An example of a public-transport network

starting a bus journey. Recent years have seen a wide variety of fare structures being introduced in most public-transport operations: fares variable with distance, flat fares (independent of distance travelled), zonal fares (for one or more specific geographic zones), combination and transfer tickets (valid for two or more services), time limit fares (e.g. valid for any number of boardings in an hour), daily, weekly and other season tickets for a fixed service or covering one or more zones and modes. This wide range of fares places difficult requirements on route choice and assignment models, as monetary costs do not depend directly on distance but in general on the location of the origin and destination, and on the route chosen.

10.6.2.4 The Definition of Generalised Costs

In the case of public-transport assignment the generalised cost of travelling may be defined as follows:

$$C_{ij} = a_1 t_{ij}^v + a_2 t_{ij}^w + a_3 t_{ij}^t + a_4 t_{ij}^n + a_1 \delta^n + a_5 F_{ij} \qquad (10.18)$$

where

t_{ij}^v is the in-vehicle travel time between i and j,
t_{ij}^w is the walking time to and from stops (stations),
t_{ij}^t is the waiting time at stops,
t_{ij}^n is the interchange time,
δ^n is an intrinsic 'penalty' or resistance to interchange, measured in time units (typically 2 to 5 minutes),
F_{ij} is fare charged to travel between i and j
a_1 to a_5 are coefficients associated to the elements of cost above.

Usually either a_1 or a_5 is equal to 1.0 in order to measure generalised costs in time or monetary units respectively. Again, it is usual to find that a_2, a_3 and a_4 are taken to be two to three times the value of a_1 as passengers dislike a minute spent walking or waiting more than if spent travelling in-vehicle.

In modelling terms, the software should be able to handle these variables and produce good estimates of each of the component times (in-vehicle, walking, waiting, transfer) if they are not provided externally. In-vehicle travel time depends on the speed attainable and the number and duration of stops en route; walking time, which depends on proximity to the best stop, is in some cases approximated by an average value for a whole zone; interchange time depends on station/stop configuration and separation; waiting time depends essentially on the frequency of the service and its reliability. A general formulation for waiting time is:

$$t^w - \frac{(h^2 + \sigma^2)}{2h} \qquad (10.19)$$

where h is the expected headway of the service and σ its standard deviation (the less regular a service, the greater the expected waiting time). This formulation assumes that passengers arrive at random at the stop and that no passenger fails to board the next bus because of lack of space in it. This 'bus congestion' problem is difficult to solve but algorithms incapable of handling it will tend to produce unrealistic loadings in terms of actual service capacity, see De Cea and Fernández (1989). If the service is perfectly regular, i.e. $\sigma = 0$, then the expected waiting time is half of the headway. It is known, however, that if the frequency of the service is low, passengers will try to arrive just a few minutes before the next departure, thus setting an upper limit to the expected waiting time of perhaps 5 to 10 minutes; how close to the timetabled departure are passengers aiming to come will depend, of course, on the reliability of the service.

10.6.2.5 The Common Lines Problem

This is probably one of the most difficult and typical problems of public-transport assignment. The problem arises when for at least some O–D pairs there are sections in a path which have more than one parallel service offered and passengers can choose the one suiting them better. This choice is often not trivial for passengers ('I wish I had known that an express service was going to come three minutes after the slow one I have taken!'), nor simple from from a modelling point of view. We are used to the idea that a driver chooses a single path from a choice set of all possible paths. In the case of public-transport passengers, they may choose a *set of paths* and let the vehicle that arrives first determine which of the paths they will actually use. The choice is therefore more complex and calls for a more detailed treatment.

A full review of the most suitable algorithms for public-transport assignment is outside the scope of this book. Instead, we shall discuss the main approaches to modelling route choice first and then assignment; not surprisingly, these different approaches result from the treatment they give to some of the issues above, in particular to the parallel or common lines problem, and to the choice of all-or-nothing, stochastic or capacity restraint-assignment methods.

10.6.3 Modelling Public-Transport Route Choice

It is worthwhile defining some terms such as route, line and section in a bit more detail before embarking on a discussion of the route choice problem in the presence of common lines.

A *public-transport (or transit) line*, or simply a *line*, is a fleet of vehicles that run between two points (terminals) on a network. They generally have the same characteristics of size, capacity, speed, etc. Vehicles stop at each node in their path to allow passengers to alight and board. Therefore, each transit line is defined by the vehicle characteristics, the sequence of nodes it serves and its frequency.

A *line section* is any portion of a public-transport line between two, not necessarily consecutive, nodes.

A *public-transport route* is any path a user can follow on the transit network in order to travel between two nodes. The portion of a route between two consecutive transfer nodes is called a *route section*, and each route section has associated a set of *attractive* or *common lines*.

A *strategy* is a set of rules that allows the traveller to reach his destination.

Example 10.9: Consider the public-transport network of Figure 10.9; a simple strategy could be:

- Take line 2 to stop H; transfer to line 3 and then exit at stop J;

   A more complex one may take the form:

- Wait up to 3 minutes for a line 5 vehicle or up to 4 minutes for a line 2 vehicle; otherwise take line 1; if line 5 is taken and you see a line 4 vehicle at stop F then

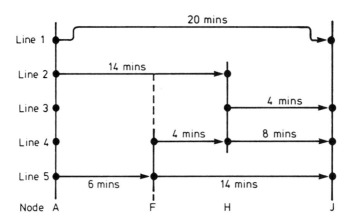

Figure 10.9   A simple public-transport network showing running times

board it and alight at J; if no line 4 vehicle at F continue to J; if line 2 vehicle was taken then transfer at H to line 4 if about to depart, otherwise wait for line 3 to reach J; etc.

In general terms a good flexible strategy will produce shorter expected travel times than the choice of the single path that minimises travel time; the choice of this single minimum path has been for many years the conventional approach to the problem. In contrast, a more realistic flexible strategy allows the passenger to take advantage of the variability of waiting times and the opportunistic choice of a good, but low-frequency, service. This is well illustrated in Spiess and Florian (1989).

One can then define, for each node, the set of attractive lines that would be part of a good strategy to reach a given destination j. Given a strategy, an actual trip is then carried out according to a mechanism like:

1. set i to origin node;
2. board the first arriving vehicle from the set of attractive lines at i;
3. alight at a predetermined node;
4. if not yet at destination, set i to the current node and return to step 2; otherwise the trip is completed.

Note that although this mechanism has a well-defined destination node, the origin is not part of the strategy. A strategy is the set of rules that enables travellers to reach their destination starting from any node in the network. This treatment is helped by the following additional notation:

S_{jk} = set of line sections connecting directly nodes j and k;

L_j^+ = set of outgoing (ingoing if − instead of + is used) line sections from node j;

v_s = flow on line section s;

t_s = in-vehicle travel time on line section s;

f_s = frequency associated to line section s;

g_j = number of trips going to destination node j;

V_{jk} = total flow on route section jk.

We can now identify the set of attractive routes emanating from node j using the $(0, 1)$ variable X_s; X_s is equal to 1 if the line section s, belonging to the set of sections from j to k, is attractive, and zero otherwise. Then, for a given pair of nodes jk the associated values X_s ($s \in S_{jk}$) define the optimum or attractive set of lines towards k.

The total waiting time for users travelling from j to k can be written as:

$$w_{jk} = \frac{V_{jk}}{\sum\limits_{s \in S_{jk}} f_s X_s} \tag{10.20}$$

The problem of finding an optimum strategy for travelling from all origins to a destination can now be written as:

$$\text{Minimise} \sum_s v_s t_s + \sum_{jk} w_{jk} \tag{10.21}$$

subject to:

$$\sum_{s \in L_j^+} v_s + g_j = \sum_{s \in L_j^-} V_s \tag{10.22}$$

$$v_s = \frac{X_s f_s V_{jk}}{\sum\limits_{s \in S_{ij}} f_s X_s} = X_s f_s w_{jk} \tag{10.23}$$

The first term of the objective function (10.21) represents the in-vehicle travel time while the second is the total waiting time. This objective function is linear in the variables v_s and w_{jk} and the main problem seems to be generated by the non-linear constraints (10.23). Spiess (1983) has shown that these constraints can be relaxed as follows:

$$v_s \leq f_s w_{jk} \tag{10.24}$$

We can further introduce constraints (10.23) into the objective function:

$$\text{Minimise} \sum_{jk} \frac{V_{jk}\{\sum\limits_s t_s X_s f_s + 1\}}{\sum\limits_{s \in S_{ij}} f_s X_s} \tag{10.25}$$

subject to (10.22). This is a $(0, 1)$ hyperbolic programming problem.

Two different approaches can be followed here. The one proposed by Spiess and Florian (1989) is based on the linear programming version of this problem, whilst that proposed by De Cea and Fernández (1989) uses the hyperbolic programming (non-linear) formulation. If there are no congestion or capacity problems, the tasks above can be simplified as the set of optimal strategies will not depend on the actual flows. The Florian–Spiess algorithm has been implemented in EMME/2 (Babin *et al*. 1982) and the De Cea–Fernández algorithm in Santiago. Tests show that the De Cea–Fernández approach is about 2.5 times faster than the Florian–Spiess method and nearly 50 times faster than the best conventional approach. This improvement in performance, which is crucial to model realistic size problems, is achieved at the cost of additional memory requirements.

10.6.4 Assignment of Transit Trips

Once the best set of line segments to join origin and destination have been identified, one needs to consider the assignment of trips to them. Most programs seek to obtain a reasonable and realistic spread of trips among feasible routes. Conventional approaches, not dealing with the common lines problem explicitly, adopted a number of measures to generate this wider spread of trips. For example: to distinguish explicitly the different access points (bus stops, stations) for each zone and to build trees from each of them (and not just from the centroids) to all destinations. In this way several alternative routes are identified, one via each different access point. Passengers can then be assigned to these routes using a multinomial logit function of the costs of joining origin and destination via each path.

Spiess and Florian (1989) perform the assignment stage following the identified optimal strategies. This is achieved by assigning to each link the proportion of the volume accumulated to the upstream node that corresponds to the frequency served by the link. De Cea and Fernández (1989) follow a similar approach but in two stages:

1. First, once the set of common lines for all (i, j) pairs have been identified a new network is built on the basis of *nodes* and *route sections*. Note that route sections contain only the lines that minimise the total expected travel time for the section; They have an associated travel time (t_r) and a frequency (f_r) corresponding to the sum of the attractive frequencies (those in the common lines). With these two elements it is possible to obtain a composite cost of travelling along this route section and therefore an efficient private-transport tree-building algorithm can be used to find the best paths. Loading onto these trees results in a set of *route section flows* v_r.

2. Second, we can decompose the route section flows into their *line section* components:

$$v_s = \frac{f_s v_r}{f_r} \tag{10.26}$$

The treatment so far has not discussed the problems associated with special fare systems. If the fare system is proportional to the distance travelled, this is not a major problem as it is normally possible to convert it to time units and add them to the travel time on each link. However, this type of fare structure is hardly common. A flat-fare system could also be accommodated but the treatment of more complex schemes (from a modelling point of view) may pose additional problems for algorithm design.

In most practical cases it will not be possible to model the whole complexity of fare systems and some approximate shortcuts will have to be taken in accordance with the most common type of ticket used. For example, in the case of a zonal fare system assignment may be performed on the basis of time alone and the fare cost added at the end. This may still ignore the importance of special pass holders but is probably good enough for places like London.

Finally, we must stress that public-transport assignment suffers, in general, from similar weaknesses to those identified for private networks. Furthermore, it is fair to say that congested assignment is less well developed for transit networks. There are two effects in play here: first, the limited capacity of the units (buses, trains) may prevent some travellers from implementing their optimal strategies, thus increasing their travel times; second, there is interaction between public transport and private cars sharing the same road network—increased traffic on one mode will affect travel times on the other as well. Current methods to deal with this problem are rather approximate; we will consider some advances in this area in the next chapter.

10.7 PRACTICAL CONSIDERATIONS

The assignment sub-model is critical in the implementation of the whole of the transportation modelling package. However, in contrast with the other three sub-models there is no standard calibration procedure to make sure the assignment stage reproduces observations as closely as possible. The most likely candidate for external validation of the model is the use of traffic or cordon counts. The following procedure seems applicable to all kind of assignment packages, including public-transport and equilibrium methods as discussed in the next chapter.

Check and Double-check the Network This is the most important source of error in traffic assignment. There are numerous potential errors in coding a network: the omission of links and nodes previously thought irrelevant, miscoding of distances, use of wrong directions, missing turning-movement penalties, specification of incorrect capacities and time-flow curves, etc. Good software packages will flag many of these errors on input; the use of graphic displays of the network and even better, graphic editing of networks is very important. It is easy to underestimate the time taken to input and check a network for a particular study. Any facility likely to speed up and increase the accuracy of the process is worth many professional days.

An additional method for checking a network is the loading of a unit trip matrix (i.e. with a single trip per cell) and then checking modelled flows. This will facilitate the identification of unused links (perhaps because they were coded with too slow speeds, or too long distances) and also heavily used ones; these serve as pointers for coding errors. The printing, or even better plotting, of minimum path trees is also a useful aid for network checking. Odd shortest routes and unreachable nodes will also help to identify sources of problems.

Improve the connection of centroids to the network if some routes look too strange. Keep in mind, however, that under congested conditions other routes will become attractive and be used. In the case of detailed (microsimulation) assignment models there will be additional sources of problems as more local data are needed. The same is true of public-transport assignment where the connection to bus stops or stations is critical for good route choice representations; the same is true of interchange facilities, frequencies and speeds. The basic rule is: before going to the next step in model fitting make sure all the observable (measurable) data are correctly represented in the network. Check connectivity first, then link attributes and then detailed data like saturation flows, signal timings, and so on.

Fit the Generalised Cost Function Assign weights to time, distance and any other variables included in it (link status, scenic quality, etc.). As a fitting criterion one could use a statistic of the chi-squared form:

$$\sum_a \{(\text{observed flow}_a - \text{modelled flow}_a)^2/\text{observed flow}_a\}$$

The lower this statistic, the better the fit. This can be applied to cordon counts or to groups of traffic counts on parts of the network thought to be most critical, say primary and secondary roads. The value of the statistic for the whole of the network also provides an indication of overall fit.

Usually a good starting point is to assume that time alone explains route choice: use this assumption, run a complete assignment and then calculate the statistics above. Then begin increasing the weight attached to distance (or other factors) and recalculate the statistics so that the choice of parameters that produces the best fit can be made. One must resist the temptation of improving the fit at one step by trivial alteration of link speeds or turning penalties at this stage as this reduces the value of the model for forecasting purposes. True errors discovered at this stage must, of course, be corrected; the model should then be re-run for other generalised cost coefficients as well.

Note that the statistic proposed above gives greater weight to a given absolute difference at low flow levels that at high ones. If this is undesirable, collect it for different flow ranges. The percentage of over- and under-estimations of flows can give some indication of bias which if present should be investigated more thoroughly. Note too, that if the link capacities were well identified and coded and there is considerable congestion then equilibrium assignment will tend to produce a good fit with observed flows, even to the extent of masking a few errors in other sub-models.

There may be evidence suggesting that different weights should be applied to different user classes, for example, that heavy lorries are more sensitive to distance and gradient than cars. In that case, the classes should be assigned separately onto the network using their best coefficients in each case.

In the case of public-transport assignment the relative weights of walking, waiting and in-vehicle time are part of this calibration process. Interchange penalties play a similar role and provide an additional element for making the model more realistic. Passenger counts at interchanges and stops should be considered separately for the calibration of these weights. An approach similar to that of Suh *et al*. (1990) may well prove advantageous in fitting generalised cost functions once all other errors have been reduced to a minimum.

Fine-tune the Assignment Model This involves finding the best dispersion parameters for stochastic assignment models. Particular care should be exercised at this stage, as depending on the implementation these parameters may have different interpretation and even dimensions. The documentation of the programs should be examined in detail to guide us in this task.

Detailed urban assignment models like those described in the next chapter offer additional opportunities for fine-tuning. These make them powerful but may also inadvertently hide more fundamental errors in coding. Examples of this type are the fine-tuning of gap acceptance parameters at some junctions, the representation of opposed turning movements at traffic signals, and so on. Particular care should be taken to make sure these modifications correspond to actual traffic engineering conditions on the ground and not to fudge factors simply to improve the fit of the model.

It must be recognised that no assignment model will ever reproduce the observations exactly. There will be always variability in the traffic counts themselves, errors in the trip matrices used and a proportion of the actual route choice behaviour which will remain unexplained. What matters, however, is that the resulting costs are as accurate as possible and that the model rests on a sound basis to compare alternative tactical or strategic schemes as required.

EXERCISES

10.1 The road network represented in Figure 10.10 links two residential areas A and B with two major shopping centres L and M. Travel times between nodes are depicted in minutes and all links are two-way. Assume first that the costs on these links do not depend on traffic levels.

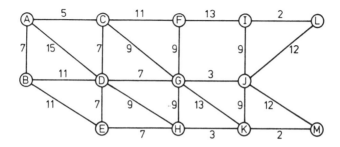

Figure 10.10 Simple network for Exercise 10.1

(a) Use a systematic procedure to find the quickest routes between origins A and B and destinations L and M; calculate the corresponding travel times.

(b) During a Saturday morning peak hour the numbers of vehicle movements from A and B to L and M are as follows:

$$A–L = 600 \qquad\qquad A–M = 400$$
$$B–L = 300 \qquad\qquad B–M = 400$$

Estimate the traffic flow on each link during this period.

(c) Consider now that travel time on each link increases by 0.02 of a minute for each vehicle/hour of flow. Use an incremental loading technique to obtain a capacity-restrained set of flows. Calculate final travel times for each O–D pair.

(d) Use an iterative loading procedure to obtain flows and costs under the conditions (c) above.

10.2 A study area contains two residential zones A and B and three workplace zones J, K and L. The zones are connected by a road network as shown in Figure 10.11, which also depicts travel costs in either direction; these are independent of the traffic flows.

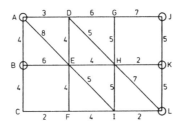

Figure 10.11 Simple network for Exercise 10.2

(a) Use a systematic procedure to find the cheapest routes from nodes A and B to destinations J, K and L and obtain the matrix of travel costs **C**.

(b) The total number of trips originating and terminating in each zone during the morning peak are given by:

Origin	Trips	Destination	Trips
A	1000	J	700
B	2000	K	1000
		L	1300

Run an origin-constrained gravity model in which the deterrence function is proportional to $\exp(-0.1 C_{ij})$ and obtain a trip matrix. Use this matrix to calculate flows on all the links of the network.

(c) Run a doubly constrained gravity model with the same type of deterrence function and obtain a new trip matrix and link flows. Compare your results of (b) and (c).

10.3 Consider the simple network in Figure 10.12 where there are 100 vehicles per hour travelling from A to X and 500 from B to X. The travel time versus flow

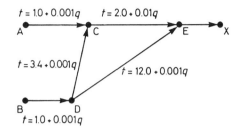

Figure 10.12   Simple network for Exercise 10.3

relationships are depicted in the figure in minutes and the flow q in vehicles per hour.

(a) Use an incremental loading technique with fractions 40, 30, 20 and 10% of the total demand to obtain an approximation to equilibrium assignment.

(b) Use an iterative loading procedure to achieve the same objective. How many iterations do you need to achieve a good degree of convergence?

11 Equilibrium Between Supply and Demand

11.1 INTRODUCTION

In Chapter 10 we presented assignment techniques for both private vehicles and public transport. We identified two main reasons for the spread of routes between each O–D pair that results in practice. The first one is the different perceptions of drivers about travel and link costs. The second reason resides in congestion effects, and we used Wardrop's principles as a general framework to discuss this issue. Wardrop's first principle states that under congested conditions drivers will choose routes until no one can reduce their costs by switching to another path; if all drivers perceive costs in the same way, this produces equilibrium conditions where all the routes used between two points have the same and minimum cost and all those not used have equal or greater cost.

Congested assignment techniques as discussed in the previous chapter try to approximate to this type of equilibrium. We saw that these heuristic methods often failed to achieve true Wardrop's equilibrium; therefore the problem deserves a better treatment. In section 11.2 we will cast equilibrium assignment in a more rigorous mathematical programming framework. This section is restricted to problems where the delay on a link depends only on flows on the link itself; however, extensions to stochastic user equilibrium and to social equilibrium will also be discussed there. Section 11.3 extends equilibrium assignment to problems where the delay on a link depends on the flow on the link itself and on other flows. This more general formulation is more appropriate to urban areas where the delay at, say a roundabout approach depends on circulating flows in the junction too. Section 11.4 extends the treatment of equilibrium to mode choice and distribution modelling; the objective here is to make sure that the travel times implied in the costs used to run these models are consistent with those generated during assignment. The naive iteration of the last three sub-models is known not to lead naturally to equilibrium conditions as it is somewhat akin to hard speed-change methods for congested assignment. Improved methods and practical considerations are included in this section.

11.2 EQUILIBRIUM

In this section methods specifically designed to achieve traffic assignment solutions satisfying Wardrop's first principle are discussed. We shall follow a combination of intuitive and analytical arguments but we shall not pursue the latter beyond what is necessary to understand and use equilibrium assignment techniques; readers interested in the more theoretical aspects of equilibrium assignment are directed to the excellent book by Sheffi (1985).

11.2.1 A Mathematical Programming Approach

Consider first some of the properties of Wardrop's selfish equilibrium, in particular that all routes used (for an O–D pair) should have the same (minimum) travel cost, and that all unused routes should have greater (or at most equal) costs. This can be written as:

$$c_{ijr} \begin{cases} = c_{ij}^* & T_{ijr}^* > 0 \\ \geqslant c_{ij}^* & T_{ijr}^* = 0 \end{cases}$$

where $\{T_{ijr}^*\}$ is a set of path flows which satisfies Wardrop's first principle and all the costs have been calculated after the T_{ijr}^* have been loaded. In this case the flows result from:

$$V_a = \sum_{ijr} T_{ijr} \delta_{ijr}^a \tag{11.1}$$

and the cost along a path can be calculated as:

$$C_{ijr} = \sum_{a} \delta_{ijr}^a c_a(V_a^*) \tag{11.2}$$

Although Wardrop presented his principles in 1952 it was not until four years later that Beckman *et al.* (1956) proposed a rigorous framework to express them as a mathematical program; it took several more years before suitable algorithms for practical implementations were proposed and tested.

The mathematical programming approach expresses the problem of generating a Wardrop assignment as one of minimising an objective function subject to constraints representing properties of the flows. The problem can be written as:

$$\text{Minimise } Z\{T_{ijr}\} = \sum_{a} \int_{0}^{V_a} C_a(v) \, dv \tag{11.3}$$

subject to

$$\sum_{r} T_{ijr} = T_{ij} \tag{11.4}$$

and

$$T_{ijr} \geq 0 \tag{11.5}$$

The objective function corresponds to the sum of the areas under the cost–flow curves for all links in the network. Why this is a sensible objective to minimise in order to obtain Wardrop's equilibrium, is something we will attempt to show below; but first we must consider the general properties of this mathematical programme.

The two constraints (11.4) and (11.5) have been introduced to make sure we work only on the space of solutions of interest, i.e. non-negative path flows T_{ijr} making up the trip matrix. The role of the second constraint (non-negative trips) is important but not essential as this level of discussion of the problem. The interested reader is referred to Sheffi's book or to some of the classic papers on the topic like Fernández and Friesz (1983) and Florian and Spiess (1982).

It can be shown that the objective function Z is convex as its first and second derivatives are non-negative:

$$\frac{\partial Z}{\partial T_{ijr}} = \frac{\partial}{\partial T_{ijr}} \sum_a \int_0^{V_a} C_a(v)\, dv$$

$$= \sum_a \frac{d}{dV_a}\left(\int_0^{V_a} C_a(v)\, dv \right) \frac{\partial V_a}{\partial T_{ijr}}$$

but from (11.1)

$$\frac{\partial V_a}{\partial T_{ijr}} = \delta_{ijr}^a$$

Now, as V_a only depends on T_{ijr} if the path goes through that link,

$$\frac{d}{dV_a} \int_0^{V_a} C_a(v)\, dv = C_a(V_a)$$

therefore,

$$\frac{\partial Z}{\partial T_{ijr}} = \sum_a C_a(V_a)\delta_{ijr}^a = c_{ijr} \tag{11.6}$$

and the second derivative of Z with respect to the path flows is:

$$\frac{\partial^2 Z}{\partial T_{ijr}^2} = \frac{\partial}{\partial T_{ijr}} \sum_a C_a(V_a)\delta_{ijr}^a$$

$$= \sum_a \frac{dC_a(V_a)}{dV_a} \frac{\partial V_a}{\partial T_{ijr}} \delta_{ijr}^a$$

$$= \sum_a \frac{dC_a(V_a)}{dV_a} \delta_{ijr}^a \delta_{ijr}^a \tag{11.7}$$

This expression is greater than or equal to zero only if the derivative of the cost–flow relationship is positive or zero. This is a general requirement for convergence of Wardrop's equilibrium to a unique solution. The meaning of this condition is that the cost–flow curve should not have sections where costs decrease when flows increase.

As the problem identified in (11.3)–(11.5) is a constrained optimisation problem, its solution may be found using a Lagrangian method. The Lagrangian can be written as:

$$L(\{T_{ijr}, \phi_{ij}\}) = Z(\{T_{ijr}\}) + \sum_{ij} \phi_{ij}[T_{ij} - T_{ijr}] \tag{11.8}$$

where the ϕ_{ij} are the Lagrange multipliers corresponding to constraints (11.4).

Taking the first derivative of (11.8) with respect to ϕ_{ij} one obtains, of course, the corresponding constraints. Taking the derivative with respect to T_{ijr} and equating it to zero (for optimisation), one has:

$$\frac{\partial L}{\partial T_{ijr}} = \frac{\partial Z}{\partial T_{ijr}} - \phi_{ij} = c_{ijr} - \phi_{ij}$$

Here we have two possibilities with respect to the value of T_{ijr}^* at the optimum. If $T_{ijr}^* = 0$ then

$$\frac{\partial L}{\partial T_{ijr}} \geq 0 \qquad \text{as the function is convex}$$

If $T_{ijr}^* \geq 0$ then

$$\frac{\partial L}{\partial T_{ijr}} = 0$$

This can be translated into the following conditions at the optimum:

$$\phi_{ij}^* \leq c_{ijr} \text{ for all } ijr \text{ where } T_{ijr}^* = 0$$
$$\phi_{ij}^* = c_{ijr} \text{ for all } ijr \text{ where } T_{ijr}^* > 0$$

In other words, the ϕ_{ij}^* must be equal to the costs along the routes with positive T_{ijr} and must be less than (or equal) to the costs along the other routes (i.e. where $T_{ijr} = 0$). Therefore, ϕ_{ij}^* is equal to the minimum cost of travelling from i to j: $\phi_{ij}^* = c_{ij}^*$.

In this way, the set of T_{ijr}^* which minimises (11.7) has the following properties:

$$c_{ijr} \geq c_{ij}^* \text{ for all } T_{ijr}^* = 0$$
$$c_{ijr} = c_{ij}^* \text{ for all } T_{ijr}^* > 0$$

Therefore, the solution satisfies Wardrop's first principle.

Example 11.1: Consider again the town-centre/bypass problem of Example 10.4. Figure 11.1 shows the cost–flow relationships and the shaded area is the objective function that we want to minimise. Of course one way to minimise this area is to have no flow $V_b = V_t = 0$, but this solution is not only trivial but of little interest. What we want is the solution that satisfies the total demand (2000 vehicles), and this is shown in Figure 11.2, where the two cost–flow functions are now displayed with the X-axis running in opposite directions and separated by the total flow that must be split between the two routes.

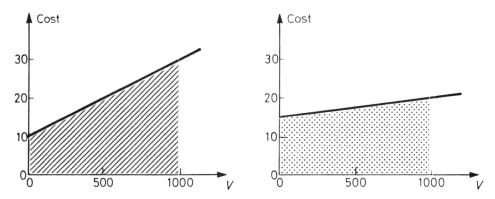

Figure 11.1   Two cost-flow relationships for bypass-town centre problem

It can easily be seen in Figure 11.2a that the sum of areas under the cost–flow curves is minimised for $C_b = C_t$; any departure from this point will simply add a new section to the area, as illustrated in Figure 11.2b. As can be seen, the equilibrium solution involves a flow via the two centre of 600 vehicles and 1400 via the bypass. It is worth noting that the cost via each route is 22 minutes and the total expenditure in the network is then 44 000 vehicle-minutes.

In this treatment of equilibrium assignment we have omitted a number of issues; for example, that of uniqueness of the solution. It can be shown that only the link costs c_a^*, inter-zonal costs c_{ij}^* and link flows V_a^* are unique in the optimum. The path flows T_{ijr}^*, however, are in general not unique at all. What this means is that there may be several combinations of paths and trips using them which result in the same link flows and costs; as all used routes (for an O–D pair) have the same minimum cost, the total inter-zonal costs are the same. This can be easily seen if one thinks of several external zones of origin feeding trips into junction A and then exiting to different destinations at junction B in Figure 10.2; although these trips can be distributed in many ways between town-centre and bypass routes under equilibrium conditions, the link flows and costs will remain the same.

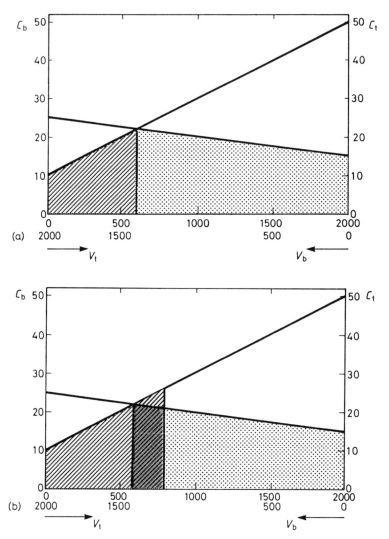

Figure 11.2   Equilibrium in simple network

11.2.2 Solution Methods

We have described a mathematical programme and shown its relevance in solving
the traffic assignment equilibrium problem. The mathematical programme is non-
linear and it can be solved by a number of methods. Although understanding the
theory of equilibrium assignment requires some mathematical background, the
actual application of the principles and solution algorithms is much less demanding.
The most commonly used algorithm is due to Frank and Wolfe. This algorithm can
be seen as a major improvements on the standard iterative method discussed in
section 10.5.4.

11.2.2.1 The Frank–Wolfe Algorithm

1. Select a suitable initial set of current link costs, usually free-flow travel times $C_a(0)$. Initialise all flows $V_a^0 = 0$; make $n = 0$.
2. Build the set of minimum cost trees with the current costs; make $n = n + 1$.
3. Load the whole of the matrix **T** all-or-nothing of these trees, obtaining a set of auxiliary flows F_a.
4. Calculate the current flows as:

$$V_a^n = (1 - \phi)V_a^{n-1} + \phi F_a$$

   choosing ϕ such that the value of the objective function Z is minimised.
5. Calculate a new set of current link costs based on the flows V_a^n; if the flows (or current link costs) have not changed significantly in two consecutive iterations, stop; otherwise proceed to step 2.

The main improvement over the iterative method is in step 4, where ϕ is calculated using the mathematical programming formulation instead of a fixed rule. This is enough to guarantee efficient convergence to Wardrop's equilibrium.

The Frank–Wolfe algorithm can be visualised as a descent approach to the problem of minimising the objective function. The problem is similar to the establishment of the rules to be followed to find the lowest point of an enclosed valley in thick fog (or more realistically perhaps, to find the peak of a mountain in thick fog, but then one has to use *up* instead of *down* in the rules below). A suitable set of rules for the valley problem would be:

1. Choose what looks like a good downhill direction; in thick fog this will depend essentially on local topography.
2. Walk in that direction until you start to go uphill again.
3. Stop at that point and choose another good downhill direction and proceed to step 2, unless you have found a point with no downhill directions, i.e. the bottom of the valley.

This is essentially what the Frank–Wolfe algorithm does, albeit in a space with may more dimensions. At each step in the iterations we have a current feasible solution (a location in the valley) and the algorithm uses the latest all-or-nothing assignment to provide a descent direction. The use of the latest all-or-nothing assignment to this end can be seen as a local approximation to minimising the objective function Z. Given that the current feasible solution is specified by the path flows $\{T_{ijr}\}$, Frank–Wolfe seeks a second attractive feasible direction $\{W_{ijr}\}$ using a linear (Taylor series expansion) approximation to Z:

$$Z'(\{W_{ijr}\}) = Z(\{T_{ijr}\}) + \sum_{ijr} \frac{\partial Z}{\partial T_{ijr}}(W_{ijr} - T_{ijr}) \tag{11.9}$$

$$= Z(\{T_{ijr}\}) + \sum_{ijr} C_{ijr} W_{ijr} - \sum_{ijr} C_{ijr} T_{ijr}$$

Here the only term which is not fixed by the feasible solution $\{T_{ijr}\}$ is $C_{ijr}W_{ijr}$; so if we wish to minimise a local approximation to Z we must choose routes W_{ijr} such that the corresponding multipliers C_{ijr} are minimised. A way to do this is to choose routes which are currently and locally minimum cost, i.e. all-or-nothing assignment on trees from current costs.

In general terms the Frank–Wolfe algorithm tends to converge rapidly over early iterations but less so as it starts to approach the optimum. This is a well-known problem and a number of improvements have been suggested to speed up convergence; see for example the work of Arezki (1986).

11.2.3 Social Equilibrium

Most of what has been discussed so far applies to Wardrop's first principle or *user equilibrium* (UE) problems. Wardrop's second principle specifies that drivers should be persuaded to choose routes in such a way that total (or average) costs are minimised. This is the *social optimum* solution and is a prescription for design rather than a model of driver's behaviour.

It is easy to see that Wardrop's second principle can be embodied in a mathematical programme of the form:

$$\text{Minimise } S\{T_{ijr}\} = \sum_a V_a c_a(v) \tag{11.10}$$

subject to (11.4) and (11.5).

This objective function can also be expressed in the following form:

$$\text{Minimise } S\{T_{ijr}\} = \sum_a \int_o^{V_a} Cm_a(v)\,dv \tag{11.11}$$

where Cm_a is the *marginal cost* of travelling along link a.

This problem can be solved with a simple adaptation to the Frank–Wolfe algorithm which consists in replacing the objective function used in the estimation of the parameter ϕ in step 4 by (11.10) or (11.11). It is easy to see that the solution to this problem makes all the marginal costs of all the routes used between two points to be equal and minimum.

The solutions to the two problems do not coincide; in other words, the user equilibrium solution generates higher total costs than the social equilibrium solution. The difference lies in the external effects due to congestion. Users perceive only their own personal costs and do not discern the additional delay incurred by other drivers due to extra vehicle on the road. One can envisage electronic road pricing as a possible method to make drivers perceive marginal rather than average costs.

Example 11.2: We take again our town-centre/bypass problem but now seek the flow pattern that minimises total expenditure (or what is equivalent in the case of a fixed trip matrix like this one, minimise average travel costs). The total expenditures are:

$$E_b = V_b(15 + 0.005V_b) \quad \text{via the bypass, and}$$

$$E_t = V_t(10 + 0.02V_t) \quad \text{via the town centre}$$

The respective marginal costs are

$$\frac{\partial E_b}{\partial V_b} = 15 + 0.01V_b$$

$$\frac{\partial E_t}{\partial V_t} = 10 + 0.04V_t$$

Equating the two and taking advantage of the fact that $V_b + V_t = 2000$, one can solve and find that for social equilibrium conditions:

	Town centre	Bypass	Total
Flow	500	1 500	2 000
Marginal cost	30	30	
Average cost	20	22.5	
Expenditure	10 000	33 750	43 750

Note that the total network expenditure is now 250 vehicle-minutes less than the user equilibrium solution found in Example 11.1. Of course, one cannot expect drivers to choose the bypass in these numbers as at least some could reduce their travel costs by choosing the town-centre route. In order to achieve this social optimum one would need to increase user costs by 2.5 minutes via the town-centre, for example by charging the equivalent as a town-centre toll. This would represent simply a transfer from private to social consumption resulting in a saving in the use of resources (time, fuel).

11.2.4 Stochastic Equilibrium Assignment

We have discussed pure stochastic and pure user-optimised equilibrium traffic assignment models. In the first case a spread of routes between two points is produced because of variability in the perceived routes costs, and in the second because of capacity-restraint effects. One would expect that in reality both types of effects should play a role in route choice. Models which try to include both effects are called stochastic user equilibrium (SUE) models and they seek an equilibrium condition where:

> Each user chooses the route with the minimum 'perceived' travel cost; in other words, under SUE no user has a route with lower 'perceived' costs and therefore all stay with their current routes.

The difference between stochastic and Wardrop's User equilibrium is that in SUE models each driver is meant to define 'travel costs' individually instead of using a single definition of costs applicable to all drivers.

In theory, models incorporating stochastic and equilibrium properties look particularly attractive; there are, however, operational and practical difficulties for applying them. From a practical point of view, the most important of these difficulties lies in the convergence properties of these algorithms. To examine this problem, let us define convergence here in the following way: an assignment algorithm is said to be convergent if:

- one starts with a particular set of link costs C_a, for example free-flow costs in the first iteration but calculated costs as a function of flows in subequent ones; and
- one assigns a matrix using specific rules, say Dial's, and produces new link flows $\{V_a\}$, and then one finds that:

$$C_a = C_a(V_a)$$

In other words, the costs resulting from the new flows are practically the same as those used to find routes and assign traffic. If an algorithm is not convergent the solution (flows and costs) will depend on when the iterative process was stopped, i.e. an arbitrary decision. For example, the next planner dealing with exactly the same problem but specifying a different number of iterations would find different costs; this is obviously not a desirable property for the assessment of transport projects.

It can be shown that under specific circumstances it is possible to formulate convergent SUE algorithms (Sheffi 1985). In fact, a practical algorithm to perform SUE assignment is just an extension of the iterative loading methods (MSA algorithm) described in section 10.5.4. Such as algorithm can be described as follows:

1. Set current costs $C_a = C_a(0)$, i.e. free-flow travel costs, initialise $V_a = 0$ for all a, make $n = 0$.
2. Make $n = n + 1$; build a set of minimum cost trees with the current costs.
3. Assign the trip matrix to the network using the current trees and a suitable stochastic method, e.g. Burrell's; obtain a set of auxiliary flows F_a.
4. Calculate current flows as:

$$V_a^n = (1 - \phi)V_a^{n-1} + \phi F_a$$

with $\phi = 1/n$.
5. Calculate a new set of current link costs based on the flows V_a^n; if the flows (or current link costs) have not changed significantly in two consecutive iterations, stop; otherwise proceed to step 2.

This algorithm will always tend to produce small changes in flows and costs as ϕ is small for large n. However, it is important to prove that it converges to the right SUE solution.

Sheffi (1985) has shown that this algorithm converges to a SUE solution in the long run, that is, for a large number of iterations, perhaps 50 or more. The convergence of this algorithm is not monotonic because the search direction is only a descent direction on average. The speed of convergence depends on the level of network congestion and on the dispersion parameter.

The convergence of the MSA algorithm for SUE problems is rather slow for congested networks. Sheffi (1985) has also shown that for very congested networks UE provides a good approximation to SUE and is faster in convergence. This suggests that the use of SUE would only be advantageous in low to medium congested assignment problems.

11.3 EXTENSIONS TO EQUILIBRIUM ASSIGNMENT

11.3.1 Limitations of Classic Methods

In previous sections we have described the most important classic methods for traffic assignment. Before considering more detailed and to some extent advanced methods, it is worthwhile reviewing what are seen as the main limitations of these approaches. These deficiencies may come from different sources.

11.3.1.1 Limitations in the Node-link Model of the Road Network

These include the fact that not all real road links are considered in the network (incomplete networks), 'end effects' due to the aggregation of trip ends into zones represented by single centroids, banned and penalised turning movements not specified in the network, and the fact that intra-zonal trips are not fully treated.

The main problem with incomplete networks arises in heavily congested areas where some of the medium- and long-distance trips will use minor roads as 'rat runs'; a new road scheme may relieve congestion and attract some of these rat-run trips, which will seem to be 'generated journeys' when they are not. Even when great care is taken in connecting the network to zone centroids, end effects are inevitable. These will make estimated link volumes less reliable in the vicinity of centroid connectors, probably overestimating the flows.

It is possible to expand simple nodes to represent all turning movements at a junction and then penalise or remove those links representing banned manoeuvres. An example of a fully expanded junction is given in Figure 11.3; any particularly

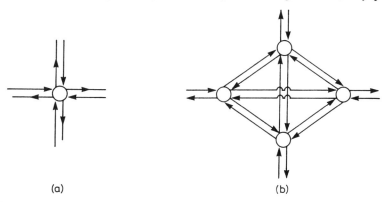

(a)   (b)

Figure 11.3 Representation of a junction as a simple node (a) and expanded showing all turning movements (b)

difficult manoeuvre, e.g. an opposed turn, can then be penalised by associating a longer delay to it. Good software provides efficient ways of automatically expanding junction representations and banning or penalising movements; alternatively, this must be done by hand in the network-building stage itself. In either case, it is likely that some turning movements will not be properly treated.

The treatment of intra-zonal movements is also a source of problems: some of them could make use of main links in the road network but they will not appear in the network model. It is difficult to devise a good method to account for them in assignment.

All of these problems are more difficult to handle when the zones are large and the network representation sparse. As usual, greater resolution in network and zonal definition will increase realism but also the costs of data collection and processing.

11.3.1.2 Errors in Defining Average Perceived Costs

We do not have enough evidence about how these are likely to change with time, journey purpose, length of journey, income, predictability and the environment. Moreover, when we wish to forecast components of cost, for example fuel consumption, we rely on simplifying assumptions which may give rise to additional errors.

11.3.1.3 Not all Trip Makers Perceive Costs in the Same Way

Our stochastic methods are an approximation to this phenomenon but even they must limit the number of randomisations for reasons of economy. Another possibility is to consider several different user classes, each with its own set of perceived costs.

11.3.1.4 The Assumption of Perfect Information about Costs in all Parts of the Network

Although this is common to all models it is essentially overoptimistic, at least until the widespread use of road transport informatics makes more realistic modelling a possibility. Drivers have only partial information about traffic conditions on the same route last time they used it and on problems in other parts of the network depending on their own experience, disposition to explore new routes and the use of traffic information services. Moreover, there is evidence that many drivers are heavily influenced by road signs in their choice of route and that sometimes signed routes are not the cheapest (Wootton *et al.* 1981). Current methods ignore these effects. The future influence of variable message signs and more advanced route guidance technology over part of the vehicle fleet, is likely to place new requirements for traffic assignment methods (see several articles in this field in Papageorgiou 1991).

11.3.1.5 Day-to-day Variations in Demand

These probably prevent true equilibrium ever being reached in practice. In that sense Wardrop's equilibrium represents 'average' behaviour if all travellers think alike and have perfect information. Its solution, however, has enough desirable properties of stability and interpretation to warrant its use in practice; however, it is still only an approximation to the traffic conditions on any one day.

In the same vein, there are time variations in demand and flow within each day. This makes 24-hour models very poor in terms of traffic assignment, and therefore travel times and costs. The use of peak and off-peak periods for modelling and assignment is essential in congested urban areas but even then we know that the build-up of congestion produces important changes in travel time in very short timeframes. Moreover, a 10-minute delay in departure for the same journey may produce a much greater delay on arrival at the destination because of increased congestion in the network. The costs on links change dynamically in response to traffic: some drivers understand this well and plan their journeys accordingly; others lack the necessary experience. In reality, the route choice problem has strong time-dependent elements but practical dynamic assignment techniques are as yet in their first steps.

11.3.1.6 Imperfect Estimation of Changes in Travel Time with Changes in the Estimated Flow on Links

This is partly due to the nature of the cost–flow relationships used. As stated in section 10.1.3, it is normally assumed that the travel time on a link depends only on the flow on the link itself. At least in urban areas, the delay on a link depends in general on flow on other links too, for example at a priority junction, thus creating interaction effects. This assumption will be discussed later as it requires better delay models than those assumed in conventional cost–flow relationships.

11.3.1.7 Input Errors

The accuracy of an assignment model depends also on the accuracy of other elements in the transport model, in particular that of the trip matrix to be loaded. Errors in the conversions from passengers to vehicle trip matrices also limit the accuracy of traffic assignment.

11.3.2 Junction Interaction Methods

As discussed above, some of the simplifying assumptions in classic traffic assignment models make them not realistic enough for congested urban areas. If one looks at the route choice and assignment problems in greater detail one should search for better delay models and a better treatment of dynamic problems. In addition there is a need to consider the interaction between traffic control and route choice, and the need to treat different vehicle classes separately. We shall discuss these issues in turn.

11.3.2.1 Improved Delay Models

So far we have considered traffic as a continuous variable operating under steady-state conditions. In reality, traffic is made up of discrete entities (vehicles) which in urban areas form queues at junctions and bottlenecks. If a particular assignment model puts more traffic on a junction than its capacity, it is very likely that the flows downstream will be overestimated; this happens because the junction will actually put an effective upper limit, not recognised by the model, and the modelled flows downstream will be greater than the actual flows. Therefore, potential routes using these links may well be ignored by the model. Double counting of delay and missing of potential routes are a perverse effect of this simplistic treatment of traffic delay.

Two types of improvement are needed here: first, to consider the physical nature of queues at junctions and their effects in limiting traffic downstream; second, the need to model the time-dependent nature of queues at junctions as demand builds up and decays before, during and after the peak period. The second problem can be treated using time-dependent queueing models as proposed by Kimber and Hollis (1979). These approaches model the way in which queues and delay change over time, as traffic demand evolves, and even allow for the presence of queues at the start of a time period of interest.

The first problem requires a physical model of queues and this can be undertaken through a simple conversion of vehicles queued into queue length or, in more detailed models, through the simulation of the actual queues. A critical issue is the ability of these models to represent the situation where a queue begins to block back an upstream junction and the additional delays this generates to other streams.

11.3.2.2 The Dynamic Nature of Queue Building and its Effect on Route Choice

This is a difficult problem which is treated in models like SATURN (Hall *et al.* 1980) by dividing the period of interest into shorter time intervals, typically 10 or 15 minutes long. Each time interval is then treated as a steady-state assignment problem. This captures some of the effects of the build-up of congestion but still assumes that all vehicles in the same time interval are faced with the same set of costs. Other models assign vehicles individually or in small groups (or packets) and release them sequentially throughout each time interval. This approach is followed by CONTRAM (Leonard and Gower 1982) thus providing a better representation of the delays facing vehicles at each stage in their passage through the network. Although SATURN does not treat vehicles individually but as part of platoons modelled at sub-minute time slices (1 to 5 seconds long), both models are classed as microsimulations because of the level of detail involved. Both these models still assume perfect information.

11.3.2.3 The Interaction Between Traffic Control and Delay

This is difficult to treat in detail. Most large urban areas are under area traffic control (ATC) systems, that is computer control of the traffic signals to reduce delay and, in some cases, create attractive 'green waves'. It is known that such systems are designed to cope well with existing traffic patterns and that travel time savings of between 10 and 20% can be achieved in comparison with non-coordinated systems. The problem is that the traffic flow patterns (flows on links) depend on the set of best routes available and that these depend, in turn, on signal timings at each junction. However, any model attempting to combine assignment and traffic control may run into a number of problems; see for example Allsop and Charlesworth (1977) and M. J. Smith (1979, 1981).

One possible solution is to run an assignment problem with fixed signal settings, obtain a future set of link flows and then run a program like TRANSYT (Robertson 1969) to optimise the setting for these new flows. The process should be repeated with the new settings, obtaining in turn new flows, with the hope that these iterations will converge to a stable and self-consistent solution. The problem is that the solution depends considerably on the starting point; if a corridor is heavily used in the first iteration, TRANSYT will produce signal timings to reduce delay there, thus encouraging more trips to prefer it in the next iteration. This also tends to favour all-or-nothing type of solutions to the traffic control/assignment problem.

11.3.2.4 Multiple Classes of Vehicles

Cars, buses, lorries and two-wheelers probably need a different treatment in detailed assignment models. The reasons for this is that they may have to be assessed separately: their environmental impacts are different, they may be subject to specific traffic management schemes (e.g. lorry bans) and they may use different criteria for route choice (minimise monetary costs rather than time).

Classic traffic assignment models consider a single trip matrix even if this may be represented by passenger car units (PCUs) through appropriate combination of vehicle matrices and their PCU values. Several models now allow for three or more vehicle classes to be treated separately, as well as to permit selective use of links, turning movements and costs; see for example Van Vliet *et al.* (1987). A special case of this are vehicles that are forced to follow a particular route, mainly buses. These are usually pre-loaded onto the network so that their impact on other traffic is considered for assignment purposes.

11.3.3 Microsimulation and Equilibrium

Two of the most popular models for detailed design of urban traffice management schemes have already been mentioned: CONTRAM and SATURN. Another model of the same family is NETSIM (Lieberman 1981), used mainly in the USA. These microsimulation models make greater requirements in terms of data collection, calibration and computer resources than the classic models discussed before. However, the data needed would be required for any traffic management scheme design anyway and there are now microcomputer versions of these programs. Demand for computer resources is likely to have a smaller role in determining the choice of model than training, calibration, validation and interpretation requirements.

Microsimulation models are not the only way of tackling the special needs of traffic assignment in urban areas. Simpler models, based on more classic approaches, have been developed to cover issues like link–junction interaction and to some extent dynamic queues. Examples of these are JAM (Wootton Jeffreys and Partners 1980) and TRIPS (MVA Systematica 1982). It is outside the scope of this book to discuss these commercial packages in greater detail.

We will use the example of SATURN to explore the way in which these models may tackle Wardrop's equilibrium. SATURN uses, in this respect, two models in order to achieve realistic assignments (Van Vliet 1982). The first one is a simulation model based on the use of cyclic flow profiles to represent the movement of platoons of vehicles over a network. This model takes good account of the interaction of different flows at roundabouts, signal-controlled and priority junctions. It needs information about the flow on each link of the network to estimate capacity, queues and delays. Therefore, an assignment model is required to load a trip matrix onto the network and obtain an estimate of these flows. This is achieved through a separate assignment model which can perform Wardrop's and stochastic user equilibrium assignment. The link between the two is through link volumes (from assignment to simulation) and through speed–flow relationships (from simulation to assignment), as depicted in Figure 11.4.

The simulation model is used, therefore, to generate suitable cost–flow relationships for the assignment problem. The cost–flow relationships are produced for each link in terms of the flow on the link itself, and take the form of a polynomial:

$$C(V_a) = a_0 + a_1 V_a^n \tag{11.12}$$

However, these relationships are calculated from the current simulation model so that they take into account the interaction and constraints generated by the flows on

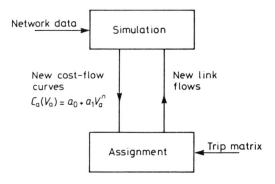

Figure 11.4   The simulation-assignment cycle in SATURN

the other links in the network. In fact, several iterations of the simulation–assignment cycle must be performed before the whole process converges to a self-consistent set of flows and costs.

In theoretical terms, what the SATURN approach attempts to do is to *diagonalise* the cost–flow relationships. In a congested and well connected network, like those existing in urban areas, the cost on a link depends not just on the flow on that link but on all other flows in the network (albeit especially on those joining the same junction). This can be written as:

$$C_a = C_a(V_1, V_2, \ldots, V_a, \ldots, V_n)$$

If we fix all flows but that on link a and we vary V_a between, say zero and the capacity at a, then we can 'calibrate' a cost–flow relationship that, in this iteration, depends only on V_a. We can then perform a conventional Wardrop equilibrium assignment using, for example, the Frank–Wolfe algorithm, obtain a new set of flows on all links and run the simulation program again. The strict condition for the convergence of this type of scheme requires that the delay on a link depends mainly on the flow on the link itself and more weakly on flows on the other links (Sheffi 1985). In practice, however, this condition is not satisfied as delays at, for example, priority junctions and roundabouts depend primarily on the flows on the links having priority (circulating and main-road flows respectively).

Example 11.3: Consider the simple network in Figure 11.5 corresponding to two routes from an origin to a destination merging into one. The total flow is 100 vehicles from A to Z.

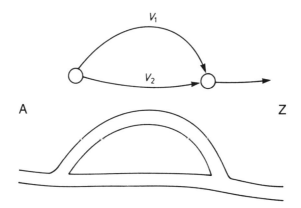

Figure 11.5 A simple network with a merge or give-way junction

Consider first the case in which both streams perform a merge operation; therefore delays on each depend also on flow on the other link. Assume now that the cost–flow relationships are as follows:

$$C_1(V_1, V_2) = 8 + 0.3V_1 + 0.2V_2$$
$$C_2(V_2, V_1) = 13 + 0.4V_2 + 0.2V_1$$

This can be solved to find a single equilibrium point at $V_1 = 83.5$ and $V_2 = 16.5$ with a minimum cost of 36.35. However, it is illustrative to show a range for values for V_1 and the corresponding link and total expenditure:

V_1	C_1	C_2	Expenditure
0	28.0	53.0	5300
10	29.0	51.0	4880
20	30.0	49.0	4520
30	31.0	47.0	4220
40	32.0	45.0	3980
50	33.0	43.0	3800
60	34.0	41.0	3680
70	35.0	39.0	3620
80	36.0	37.0	3620
83	36.3	36.4	3632
84	36.4	36.2	3636
90	37.0	35.0	3680
100	38.0	33.0	3800

As can be seen, the solution is a unique, stable equilibrium point. If some flow switches to link 2 then that link has increased delay and therefore drivers will come back to the original route. The same is true if more traffic switches to link 1. The fact that the total expenditure is minimal at another point, approximately $V_1 = 75$, is another example of the difference between social and selfish user equilibrium.

Consider now a slightly different problem with the same type of network. Now the junction is of a give-way type for link 1; link 2 has right of way and therefore its travel time does not depend on flow on link 1. The new relationships are now:

$$C_1(V_1, V_2) = 8 + 0.1V_1 + 0.2V_2$$
$$C_2(V_2, V_1) = 20 + 0.05V_2$$

The same type of table can be used to illustrate possible solutions to this assignment problem as shown on the following page.

In this case, the solution $V_1 = 60$ and $V_2 = 40$ is not stable. A switch to link 1 will decrease costs on that link faster than on link 2, therefore precipitating the solution $V_1 = 100$ and $V_2 = 0$. However, a switch in the other direction, that is to link 2, has the opposite effect, increases costs on link 2 slower than on link 1 therefore leading to another solution: $V_1 = 0$ and $V_2 = 100$. These two extreme solutions are stable albeit not with equal costs by each route; however, these two are UE solutions as the costs of the paths not used are greater than the costs on the paths used. Any departure from these extreme points will result in new cost pulling the solution back to the starting point. Note that the equations chosen are simple but not unreasonable. Observe too, that the equation for the non-priority flow shows that delay depends mainly on flow on the priority link, therefore violating the requirement for a unique solution.

V_1	C_1	C_2	Expenditure
0	28	25.0	2500
10	27	24.5	2475
20	26	24.0	2440
30	25	23.5	2395
40	24	23.0	2340
50	23	22.5	2275
60	22	22.0	2200
70	21	21.5	2115
80	20	21.0	2020
90	19	20.5	1915
100	18	20.0	1800

The fact that the solution $V_1 = 100$ is preferable because of lower overall expenditure is only relevant in terms of network design. For example, we may wish to direct drivers to choose link 1 and ignore link 2. Without this advice drivers may find either of the two extremes, or even one on a particular occasion and the other the following day. Reality may be non-convergent to a stable equilibrium solution; good assignment models may fail to converge simply because they represent well this feature of reality.

SATURN and models like it therefore, can only be said to provide a reasonable practical approximation to the ideal of Wardrop's equilibrium in congested urban areas. They normally offer practical indicators to estimate how close to a possible equilibrium the iterative process has been able to reach at any one stage. Microsimulation models do represent, however, the state of the art in detailed traffic assignment for the design of traffic management and other schemes in urban areas.

11.4 TRANSPORT SYSTEM EQUILIBRIUM

11.4.1 Introduction

The type of equilibrium problems we have discussed so far concern just a single mode in a network. Wardrop's first principle models this type of behaviour and a suitable algorithm permits the identification of the routes and flows that will generate consistent costs for all users. As stated before, a similar principle applies to congestion or capacity problems in public transport networks.

The problem becomes more complex when one considers interactions between two or more modes. These may take the following forms:

- Congestion generated by cars will affect bus travel times in certain routes and therefore change assignment strategies for public transport users; congestion generated by buses (and street-running LRT systems) and bus stop operations will affect capacities and speed for cars, and therefore their route choices;
- Interaction due to park-and-ride and kiss-and-ride operations for buses and for segregated track systems. The attractiveness of these mixed-mode operations will

depend on road congestion, service frequency and fares (mode and parking) and all of these are, in general, mutually related.

Pragmatic approaches to this problem are usually of the hard or soft speed-change nature discussed in section 10.5: assume bus times and flows fixed and known, assign cars to network to equilibrium, assign passengers to transit network, obtain new speeds and travel times and fix them, re-assign, obtain new speeds, re-assign, etc. Of course, if one is not prepared to change the bus frequencies in accordance to demand, the problem may converge soon at this level.

In the case of mixed-mode users the problem is more difficult because they may decide to change their park-and-ride station as a result of congestion in the road network and therefore change the same levels of congestion when they do so. Even if mixed-mode movements are few at present, not including them in the equilibrium procedure may cause severe problems for design-year forecasts in heavily congested networks (see ESTRAUS 1989).

In all the cases above we have kept the assumption of a fixed trip matrix (inelastic demand) for each mode. However, what we have seen in earlier chapters must lead us to treat the assumption of a fixed matrix with caution, at least when we are considering major changes to the transport network or longer timescales. We must accept in some cases that demand is elastic, in particular to travel aned route costs. This leads us to consider the influence of congestion and delay on mode and destination choice.

At this higher level, travel times by different modes between points are used to model mode choice. If the levels of congestion and delay change as a result of new assignments, then the models should be run again to obtain improved modal splits. The same is true of trip distribution models which use time as an important explanatory variable in their deterrence functions.

What we have now is a nested set of models and we need to make sure that the travel costs used by all of them are consistent. An iterative strategy of the type *run all the models first, obtain new travel times, repeat until convergence* has all the makings of a non-convergent approach. Oscillations are likely to be a more or less permanent feature of this type of technique unless we pay considerably more attention to the development and use of sound algorithms.

One of the best reviews of the state of the art is still Fernández and Friesz (1983). Rather than summarise this paper, we are taking a couple of examples, consistent with the scope of this book, to show how system equilibrium modelling can be developed.

11.4.2 Combined Mode Choice and Assignment

A reasonable way of tackling the above problem is to collapse as many sub-models as possible into one, in particular if one can include assignment in the same process. What may be important, however, is not to compromise too much the realism of the modelling process for the sake of expedience in equilibrium, particularly in short-term tactical decision making.

Consider first the problem in general terms where a typical demand curve may be inverted to give travel costs as a function of number of trips $C_{ij} = g_{ij}(T_{ij})$, as shown in Figure 11.6.

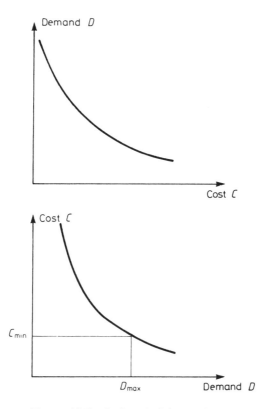

Figure 11.6   An inverted demand curve

Of course travel costs between i and j would never be zero. Therefore it is reasonable to assume that there will be a minimum cost C_{ij}^{min} and associated with it a maximum demand T_{ij}^{m}. It is interesting to try to construct a model that combines demand and assignment.

Consider first the following objective function:

$$\text{Minimise } Z = \sum_a \int_0^{V_a} c_a(v)\,\mathrm{d}v - \sum_{ij} \int_0^{T_{ij}} g_{ij}(t)\,\mathrm{d}t \tag{11.13}$$

subject to

$$T_{ijr} \geqslant 0$$

$$T_{ij} = \sum_r T_{ijr} \tag{11.14}$$

$$V_a = \sum_{ijr} T_{ijr}\,\delta_{ijr}^a \tag{11.15}$$

The derivative of Z with respect to T_{ijr} is:

$$\frac{\partial Z}{\partial T_{ijr}} = C_{ijr} - g_{ij}$$

We can now consider the behaviour of Z at T^*_{ijr} directly:

$$\text{If } T^*_{ijr} = 0 \quad \text{then} \quad \frac{\partial Z}{\partial T_{ijr}} \geqslant 0 \quad \text{and} \quad c_{ijr} \geqslant g_{ij} \tag{11.16a}$$

$$\text{If } T^*_{ijr} > 0 \quad \text{then} \quad \frac{\partial Z}{\partial T_{ijr}} = 0 \quad \text{and} \quad c_{ijr} = g_{ij} \tag{11.16b}$$

Therefore, if a particular path is used, then the path cost specifies a value for the demand curve, so we must have:

$$g_{ij}(T_{ij}) = c^*_{ij}$$

The inverted demand function could be of a very general form. However, it may not be possible to obtain neat analytical functions in some cases.

Consider now the simpler problem of combining only mode choice and assignment. In this case we can reformulate the problem of assignment with elastic demand as an *extended* assignment problem with *inelastic* demand.

If T_{ij} are now the total number of trips by all modes (public transport and car) between i and j and T^c_{ij} are those made by car, the remainder $T^b_{ij} = T_{ij} - T^c_{ij}$ will be made by public transport. We could create some new, notional, links in an extended network to carry those trips not made by car. We could simplify these links and have only one aggregate link between each O–D pair to represent these journeys. This is illustrated in Figure 11.7.

The public-transport trips can also be written as:

$$T^b_{ij} = T_{ij} - \sum_r T_{ijr} \tag{11.17}$$

and we may now rewrite our optimisation problem as:

$$\text{Minimise } Z = \sum_a \int_0^{V_a} c_a(v)\, dv - \sum_{ij} \int_0^{T_{ij}} g_{ij}(t)\, dt + \sum_{ij} \int_{T^c_{ij}}^{T_{ij}} g_{ij}(t)\, dt$$

subject to the same constraints.

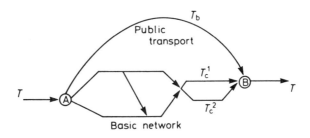

Figure 11.7   An extended network

The first integral is the conventional equilibrium assignment objective function. The second one is, in this context, an integral between two constants, 0 and the total number of trips, and therefore it can be ignored in the optimisation process.

We can now set up a *pseudo cost–flow* curve for the car trips through a change of origin, replacing the range T_{ij}^c to T_{ij} by 0 to T_{ij}^b and making:

$$g'_{ij}(t) = g_{ij}(T_{ij} - t)$$

so that the objective function can now be rewritten as:

$$\text{Minimise } Z = \sum_a \int_0^{V_a} c_a(v)\,\mathrm{d}v - \sum_{ij} \int_0^{T_{ij}^b} g'_{ij}(t)\,\mathrm{d}t \tag{11.18}$$

This is now a composite objective function but both terms can be interpreted as areas under the cost–flow curves; in the second case, a pseudo cost–flow curve. Therefore the problem can now be solved using a standard equilibrium assignment algorithm for example Frank–Wolfe. However, it is not necessary to set up the notional network explicitly; it is only the corresponding cost–flow curves that are needed:

1. Set up pseudo cost–flow curves for each (i, j) pair following an existing modal-split model, for example, a binomial logit model.
2. Initialise $n = 0$; assign all T_{ij} to free-flow minimum-cost routes and express the resulting flows as V_a on the real links and as $T_{ij}(n)$ on the notional links.
3. Make $g'_{ij}(n) = g'_{ij}[T_{ij}(n)]$, i.e. the current pseudo link-cost as implied by the total trips minus the number of car trips from i to j.
4. Define a matrix $W_{ij}(n)$ to be assigned to the real network such that:

$$W_{ij}(n) = \begin{cases} T_{ij} & \text{if } g'_{ij}(n) \geqslant c_{ij}^* \\ 0 & \text{otherwise} \end{cases}$$

5. Estimate new minimum-cost routes, assign $W_{ij}(n)$ to them and obtain new real link flows $F_a(n)$.
6. Combine the trips between i and j and the real flows in the following way:

$$T_{ij}(n + 1) = (1 - \phi)T_{ij}(n) + \phi W_{ij}(n)$$
$$V_a(n + 1) = (1 - \phi)V_a(n) + \phi F_a(n)$$

where ϕ is chosen so as to minimise the objective function (11.18).
7. Make $n = n + 1$ and return to step 2 unless some convergence criterion has been met.

In steps 3 and 4 we are calculating the pseudo link-costs and comparing them with the current minimum costs via the real network. We then assign T_{ij} (total trips) to the notional links if $g'_{ij} < C_{ij}^*$; otherwise we assign them to the real network. This is, again in the Frank–Wolfe algorithm, an auxiliary all-or-nothing assignment to the extended network. The implementation of this algorithm requires only small modifications to the code of the Frank–Wolfe equilibrium assignment method.

11.4.3 Mode, Destination and Route Choice Equilibrium Methods

11.4.3.1 Combined Distribution and Assignment

Such a model results from the following mathematical programme:

$$\text{Minimise } Z = \sum_a \int_0^{V_a} c_a(v)\,dv - \frac{1}{\beta}\sum_{ij} T_{ij}(\log T_{ij} - 1) \tag{11.19}$$

subject to:

$$T_{ijr} \geq 0$$

$$\sum_j T_{ij} - O_i = 0 \tag{11.20}$$

$$\sum_i T_{ij} - D_j = 0 \tag{11.21}$$

It can be shown that equation (11.19) is a convex function and that using a Lagrangian technique the solution is found to be:

$$T_{ij}^* = a_i b_j \exp\left(-\beta c_{ij}^*\right)$$

where equilibrium assignment is used for the estimation of c_{ij}^*. Again it may be possible to adapt the Frank–Wolfe algorithm to solve this problem. However, Evans (1976) has devised a more efficient algorithm involving iterative solutions of the gravity model and equilibrium assignment problems with inelastic demand.

11.4.3.2 Combined Distribution, Mode Split and Assignment

A next step would be to combine distribution, modal split and assignment in a single optimisation framework, and solve the three problems simultaneously. This approach has been followed by several authors, among them Florian and Nguyen (1978), who take advantage of the combined distribution modal-split model discussed in Chapter 6 to cast the problem in the following form:

$$\text{Min } Z = \tau \sum_{ij} T_{ij}^c \log T_{ij}^c + \sum_{ij} T_{ij}^b(\tau \log T_{ij}^b + u_{ij}^b) + \sum_a \int_0^{V_a} c_a(v)\,dv \tag{11.22}$$

subject to:

$$\sum_j (T_{ij}^c + T_{ij}^b) = O_i \tag{11.23a}$$

$$\sum_i (T_{ij}^c + T_{ij}^b) = D_j \tag{11.23b}$$

$$\sum_r T_{ijr}^c = T_{ij}^c \tag{11.24}$$

$$V_a = \sum_{ijr} \delta_{ijr}^a T_{ijr}^c + V_a^b \tag{11.25}$$

plus non-negativity constraints on T_{ij}^c, T_{ij}^b and T_{ijr}^c where the superscripts c and b indicate car and public transport respectively, and:

u_{ij}^b is the travel time by public transport, assumed independent of traffic volumes;

V_a^b is the public transport contribution to flow on link a: it may well be zero for modes with segregated track.

Florian and Nguyen (1978) have shown that with the usual assumptions for $C_a(V_a)$ this objective function is the sum of convex functions, is therefore convex and the problem has, in general, a unique solution of the form:

$$T_{ij}^c = a_i b_j \exp(-\beta c_{ij}^{*c}) \tag{11.26a}$$
$$T_{ij}^b = a_i b_j \exp(-\beta u_{ij}^b) \tag{11.26b}$$

and mode split becomes a binomial logit model with scale parameter β. The cost c_{ij}^{*c} is the user equilibrium assignment cost of travelling by car between i and j.

These authors also developed an adaptation of the Frank–Wolfe algorithm to solve this problem. Their main innovation was the method to find the search direction using a linear programming approach which in this case translates into the simpler Hitchcock algorithm for the *transportation problem*. This poses little computational requirements for this size of problem. The whole algorithm may be summarised as follows:

1. Obtain an initial feasible solution for T_{ij}^c, T_{ij}^b and V_a^c.
2. For each arc a compute the current cost $C_a = C_a(V_a)$.
3. For each O–D pair determine the shortest paths Γ_{ij} and let c_{ij}^{*c} be the cost of travel along that path.
4. Compute:

$$C_{ij}^c = C_{ij}^{*c} + \tau + \tau \log T_{ij}^c$$
$$C_{ij}^b = C_{ij}^{*b} + \tau + \tau \log T_{ij}^b$$

5. If $c_{ij}^c > c_{ij}^b$ then set $y_{ij}^c = 0$; otherwise set $y_{ij}^b = 0$.
6. Solve Hitchcock's transportation problem to obtain new auxiliary demands y^c and y^b.
7. Initialise $w_a = 0$ for all arcs a; for all (i, j) set $w_a = w_a + y_{ij}^c$ for a in the path Γ_{ij}.
8. Obtain the optimal step length ϕ which minimises the objective function (or a linearised version of it) and revise the demands and flows as follows:

$$T_{ij}^c = T_{ij}^c + \phi(y_{ij}^c - T_{ij}^c)$$
$$T_{ij}^b = T_{ij}^b + \phi(y_{ij}^b - T_{ij}^b)$$
$$V_a = V_a + \phi(w_a - V_a)$$

9. If a suitable convergence has been reached, stop; otherwise continue to step 2.

In this problem, the only parameter for calibration is τ; Florian and Nguyen (1978) suggest the use of the weighted mean trip cost as a means of calibrating this parameter. They also acknowledge that in some cases it may be advantageous to use two different parameters for calibration, one for public transport and one for cars; the same solution algorithm applies in this case.

11.4.3.3 Combined Trip Generation, Distribution, Mode Choice and Assignment

Safwat and Magnanti (1988) have gone one step further with the development of a single mathematical program called STEM (Simultaneous Transportation Equilibrium Model). They also provide functional forms of greater flexibility than those provided by the normal entropy-maximising formalism. However, their demand functions are still less general than those used in sequential transportation models. They use a flexible measure of disutility of travel U_{ij} defined as:

$$U_{ij} = -\beta u_{ij} + A_j \qquad (11.27)$$

where β is a parameter for calibration, u_{ij} is the perceived cost of travel between i and j, and A_j is the composite effect of several socioeconomic variables influencing trip attraction, for example a weighted sum of employment, shopping area, etc.

Safwat and Magnanti (1988) link trip generation to a systematic measure of accessibility specified as follows:

$$S_i = \max\left\{0, \log \sum_j \exp\left(-\beta u_{ij} + A_j\right)\right\} \qquad (11.28)$$

Trip distribution is given by a logit model:

$$T_{ij} = \frac{\exp\left(-\beta u_{ij} + A_i\right)}{\sum_j \exp\left(-\beta u_{ij} + A_j\right)} G_i \qquad (11.29)$$

This form, they argue, is more general than the conventional gravity model as it includes socioeconomic variables in a flexible utility function. It is easy to see that if one lets $A_j = \log D_j$, as we did in Chapter 5, then the model becomes a singly constrained gravity model with exponential deterrence functions. A doubly constrained gravity model can be approximated with a very small but positive value for β. Mode choice in this model is analogous to route choice. Each individual user selects the mode that turns out to be more attractive. Route choice and mode choice then become embedded into an equilibrium assignment framework.

The solution method proposed by Safwat and Magnanti (1988) is again an adaptation of the Frank–Wolfe algorithm; the reader is referred to their paper for the full details of this method. The STEM model has been applied to a small number of studies, including Egypt and Texas. These applications have shown that the requirements of computer resources imposed by the model are not excessive.

11.4.4 Practical Considerations

Modelling transport systems involves striking a balance between behavioural richness, data-collection costs, computational tractability and forecasting ability. The type of equilibrium models outlined above attempt to achieve computational efficiency without compromising too much the behavioural aspects of demand estimation. However, these global equilibrium models impose some limitations on the type of demand functions that can be used. These limitations are, in effect, analogous to those imposed to supply cost–flow functions in order to guarantee good properties of uniqueness and convergence to equilibrium assignment. In particular, it should be possible to invert the demand functions; they should be symmetric and negative definite.

Safwat and Magnanti (1988) have explored how much successful flexibility can be preserved within those constraints, and their model approaches the behavioural richness attained by most aggregate transport models. However, one would like to exploit the full behavioural strength of disaggregate demand modelling techniques and this calls for less concise and more demanding approaches to transport system equilibrium.

Experiments have been carried out in implementing more ambitious system equilibrium frameworks, one of them as part of the Santiago Transportation Study (ESTRAUS 1989). However, their discussion is somewhat beyond the scope of this book. Nevertheless, two key questions can be partly answered at this stage in the state of the art:

- Does it really matter whether or not system equilibrium is attained?
- Does it require considerable computer resources?

The answer to the first question seems to be Yes. If system equilibrium is not sought, the final 'solution' to a transport model can depend heavily on the initial costs assumed in running each of the models; so much so, that even alternative strategic schemes may be preferred, (see, for instance, Arezki *et al.*, 1991). This is a point that should influence the balance to be struck in modelling transport problems. Modern software, like the combination of SATURN and SATCHMO (Willumsen *et al.*, 1993), now provide the facilities to seek equilibrium solutions involving route, mode, destination and time-of-departure choice.

The answer to the second question seems to be Yes too. In the case of the Santiago Study some 12 hours in a DEC 3000 with 96 MBytes RAM are needed to run the full system equilibrium model with 260 zones and some 11 modes. Depending on the range of choices allowed into the equilibration framework this type of approach multiplies cpu times by a factor between 4 and 40. Judgement is still required in the identification of the behavioural responses allowed in the equilibrium model in order to keep computer times at a reasonable level. On the positive side, the data requirements are not greater than for non-equilibrium modelling and one of the clearest trends in technology is to increase computer power at a fraction of the cost. This is a case where investment in modest programming costs will permit the benefits of increased computer power to be reaped in more consistent and sound transport modelling applications.

It is likely that a good deal of transport research during the last decade of this century will concentrate on how to preserve and improve the behavioural richness of transport models whilst at the same time ensuring good convergence and equilibrium properties for the whole package of models applied to a study.

Equilibrium in transport systems and markets is not an end in itself. There are good reasons to suspect that equilibrium does not happen in practice, not even at the simplest network level. The real systems are in a permanent state of change, with travellers experimenting new routes, modes and destinations. Families change residences, jobs, shopping and social patterns and lifestyles. However, the state of the art in dynamic modelling of these phenomena is still many years behind that of equilibrium modelling.

The 'bottom line' is that one needs to use models to provide advice on transport decisions and this requires comparing alternative ways of intervening in the transportation system. Consistency in the use of models to estimate the performance of these interventions is then of capital importance as we wish to compare 'like with like'. Casting the transport modelling effort into a general equilibrium framework seems a prerequisite for ensuring this consistency. It is not, of course, a sufficient condition: there will be cases where partial modelling of the system will be enough to discriminate a good scheme from one that is not so good. However, the state of the art of equilibrium modelling is such that one seldom has to sacrifice too much realism to achieve it. The state of the art is now permeating the state of practice and we are likely to see much improved modelling approaches in the near future.

EXERCISES

11.1 A 12-kilometre expressway connects two urban areas. The supply function for each of the three lanes per direction of the link may be approximated by

$$t = 20 + q/200$$

where t is the travel time in minutes and q the flow per lane in passenger car units (PCU) per hour. The road is normally used by cars and express (non-stop) buses only; the corresponding vehicle travel times are t_c and t_b. The bus service has a peak-hour frequency of one bus per minute. The demand function for car travel has been estimated to be:

$$V_c = 3480 - 60t_c$$

where V_c is the total car flow per hour and direction. In a similar way, the demand function for bus trips is thought to be:

$$V_b = 4200 - 75t_b$$

where V_b is the number of passengers per hour and direction. You may assume that both t_c and t_b can be calculated from the above supply functions and that a bus is equivalent to 2 PCUs.

(a) What is the initial equilibrium state? If a bus has 60 seats, what is their load factor (occupancy divided by capacity)?

(b) One of the lanes is now taken for exclusive use by buses. What is the new equilibrium state and the new load factor for buses?

(c) Discuss the assumptions implicit in the demand functions used above.

11.2 Two cities 60 kilometres apart are connected by a two-way road over which cars operate throughout the day. The peak-hour demand for travel by car between the two cities is thought to be well described by the following function:

$$q = 6000 - 1500t$$

where q is the demand in vehicles per hour and t the travel time in hours. The travel times versus flow relationship for the road is:

$$t = 0.90 \exp(0.0003q)$$

(a) Estimate how many vehicular and person trips per hour are made under equilibrium conditions if each car carries 1.5 passengers on average.

(b) A frequent (but slow) rail service is now implemented between the cities, where each train has a nominal capacity of 300 passengers. During the peak hour the rail company is prepared to run a train every 10 minutes with an estimated travel time of 90 minutes. If passengers are assumed to use the fastest mode available, is this a sensible level of service?

11.3 Consider the network and conditions described in Exercise 10.3.

(a) Express the objective function of the mathematical programme corresponding to Wardrop's selfish equilibrium in terms of the flows and travel time–flow relationships in the figure.

(b) Calculate the equilibrium flows on each link and the travel time for each group of travellers. Calculate the value of the objective function above under equilibrium conditions and the total expenditure in travel time in the system.

(c) Local traffic engineers have decided to install speed restrictions on link C–D so that the new travel time versus flow function is:

$$t = 5.2 + 0.001q$$

Calculate the new equilibrium conditions in terms of flows and travel times and show that under these conditions the total expenditure in travel time in the system is less than in (b).

11.4 The network in Figure 11.8 is loaded during the peak hour with 100 vehicles travelling from A to D. The equations in the network show the travel time on each link in minutes as a function of the flow q on the link in vehicles per hour. All links are unidirectional as shown.

(a) Identify the minimum-cost routes used, their flows and their corresponding equilibrium costs. What is the total expenditure in travel time in the network?

(b) Assume that link CB is pedestrianised and therefore unavailable to vehicular traffic. Identify the new equilibrium flows, costs and total expenditure in travel time in the network.

(c) Discuss your results.

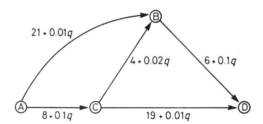

Figure 11.8   Simple network for Exercise 11.4

12 Simplified Transport Demand Models

12.1 INTRODUCTION

For many years the main emphasis in transport modelling has been to enrich their behavioural content and improve data-collection methods as a means to enhance their accuracy, realism and reduce costs. A parallel line of research has sought to improve transport modelling by emphasising the use of readily available data and the communicability of simpler model features and results. This stream of research has had an important impact in practice as it offers not only reduced costs but also simplified data-collection and processing requirements. The interest in simplified modelling techniques is unlikely to diminish in the future (see a recent compilation in Ortúzar 1992). Consultants and local authority modellers are often asked to study transport proposals in almost ridiculously short time spans: the development of better and sounder simplified methods to achieve this will always be welcome.

The idea of using simpler and quick response models is not new. The practice of not using any formal model for transport project assessment is much more prevalent than what official documents and technical literature would lead one to believe. Of course, the idea of not using any formal model simply means that decision makers are using their own, mental models, to make decisions. These may be quite powerful and certainly more sensitive to political and social variables than any formal mathematical effort.

Mental models are formed and refined through observation, analogies, discussions, experimentation and mistakes. Mental models are indeed essential to make use of formal ones, interpret their results and add considerations normally outside their scope. For this end, the limited numerical processing ability of mental models is not a major limitation. However, mental models have two major weaknesses: sometimes they fail completely, for example to consider the explosive implications of exponential growth or the interconnections between seemingly unrelated decisions on taxation and mode choice; the second disadvantage of mental models is that they cannot normally be 'opened up' to discuss them and qualify the recommendations resulting from their use. They are, therefore, more difficult to transfer to other users.

There is a whole range of modelling approaches in between the extremes of using no formal models at all and employing the most advanced and complex simulation techniques. One of the ways of looking at these is to consider the manner in which they represent space and hence distance, the key element in transport. Some models ignore space completely. These are usually of the kind concentrating on the financial implications of subsidies, taxation, and so on. They may be simple elasticity models, sometimes used to discuss fare increases or changes to petrol prices and car taxes. In other cases they may include more complex interactions, for example, between road, petrol and car taxes, and car ownership and use.

Some authors have advocated the use of structural modelling techniques; see for example the interesting work of Roberts (1975) in respect of fuel consumption. In this case a directed graph is often used to connect elements in the transport system, for example, the number of cars, fuel tax, improved fuel consumption, pollution emissions and costs. Weights could be attached to these linkages to represent the relative strength of each relationship.

If weights are replaced with formal equations, calibrated from actual observations, one ends up with a non-spatial interaction model. Khan and Willumsen (1986) developed a model of this kind to enhance the study of car ownership in less developed countries; the philosophy behind their model was that in developing countries car ownership should not just be forecast but examined together with its implications for resource allocation to roads and fuel consumption. The model included, in addition to the variables above, functions representing fuel consumption and the need for additional expenditure on road maintenance and new construction. Some of these, in particular construction and the importation of new cars, have severe implications for the balance of payment in these countries and should be explored before deciding on a policy relaxing restrictions to car ownership and use.

A better representation of space can be obtained with idealised models of the type first proposed by Smeed (1968) and also used by Wardrop (1968) to study, among other policy issues, the limits of car commuting in urban areas. As more people use cars for the journey to work, more space needs to be devoted to roads and parking until radical changes are needed to the nature of the urban area. These models have seldom been used for decision making but have served to illustrate important policy issues.

The next stage in space modelling involves simplifications to more conventional modelling approaches as addressed in this book. Sketch planning models have been developed specifically to provide quick response and limited data-collection requirements; they are discussed in section 12.2. Increasing the degree of realism, we then discuss the idea of using simplified modal-split models in section 12.3. Section 12.4 covers an important group of models which make use of readily available data, in particular traffic counts. The special characteristics of transport systems in corridors enable another type of simplification, as discussed in section 12.5. Finally, the interpretation of model output and the use of models would also benefit from special training techniques; gaming simulation has been put forward as assisting in this area and it is discussed in the last section of this chapter.

12.2 SKETCH PLANNING METHODS

Sketch planning models have been put forward as tools for long-range planning by many authors, as reported in OECD (1974) and Sosslau *et al.* (1978). They are modcls with a greater level of detail than the idealised network approaches mentioned in the previous section but much simpler than conventional suites of programs. This feature facilitates the analysis of broad transport and land-use strategies at a coarse level of resolution, without requiring large amounts of data or the rigid assumptions of ideal space models. Their practical implementation ranges from scaled-down conventional aggregate modelling suites of programs to *ad hoc* approaches developed from some simple ideas and assumptions.

Most sketch planning methods rely considerably on the transfer of parameters and relationships from one area or country to another. Only certain aspects of the models are made location dependent, usually network characteristics, population, income levels, and so on. Perhaps at one extreme of sketch planning models are those relying heavily on assumed regularities in human behaviour in the transport field. A typical example of this is the UMOT (Unified Mechanism of Travel) model proposed by Zahavi (1979). This model is based on the assumption that the following relationships are transferable over time and space (regions, countries):

- the average daily travel time per traveller, i.e. an assumption of constant travel time budgets;
- the average daily travel expenditure (money) as a function of income and car ownership, i.e. a money budget relationship;
- the average number of travellers per household as a function of household size and car ownership;
- the unit cost of owning and running a car;
- the speed-flow relationship by road type;
- the threshold of daily travel distance that justifies owning a car.

These relationships were developed by Zahavi following an extensive compilation of data bases from all over the world. UMOT only requires as location-specific input the following:

- the number of households and their sizes in the study area;
- the income distribution of households;
- the unit cost of travel by mode;
- the length of the road network in the study area.

An interesting feature of UMOT is that it produces the following results as output:

- car ownership per household by income group;
- aggregate modal choice for the whole city;
- average travel times and speeds;
- other performance indicators like total expenditure and travel times.

UMOT gained some support as a tool for testing broad policy options, for example on fiscal policy (taxation), on fuel and car ownership, pricing policy for public transport and even broad infrastructure investment programmes. However, the model has been tested by Downs and Emmerson (1983) and Willumsen and Radovanac (1988), among others, who found that, in general, it did not represent situations in other countries well, not even at the very high level of aggregation used. In fact, the transferability of relationships and budgets was not found to be consistent enough to warrant the use of UMOT, even after improvements to the models were implemented by the authors.

At another extreme, there are now microcomputer programs capable of running many of the classic four-stage models using borrowed parameters, in particular trip generation with standard trip rates, trip distribution using gravity models, binomial logit modal split and traffic assignment. These are usually restricted in terms of the number of zones, links and modes that can be used, but offer an approximate method for the analysis of transport problems with a very fast turnaround (see Sosslau *et al.* 1978).

Sketch planning techniques seem to offer advantages in terms of simplicity, fast response and low data requirements. However, very often they rely too heavily on the transfer of relationships and parameters from one context to another. This detracts from the analysis unless it is performed only as an initial coarse sketch to select possible solutions for more detailed consideration.

12.3 INCREMENTAL DEMAND MODELS

A number of approaches have been put forward to perform quick demand analysis of the impact of changes in fares, levels of service (LOS), or other attributes of a particular mode. The best known of these methods fall under the headings of incremental elasticity analysis and pivot-point modelling. In both cases, the aim is to estimate small changes in demand as a result of (small) changes in one (seldom more) of the LOS attributes, at a given point in time.

12.3.1 Incremental Elasticity Analysis

Consider an initial situation where the level of demand for a mode is T_0, its level of service S_0 (probably a vector including attributes like travel time, fare, waiting time, etc.). The elasticity of demand with respect to LOS (at a given level of demand and LOS) is given by:

$$E_s = \frac{S_0}{T_0} \frac{\partial T}{\partial S} \approx \frac{S_0}{T_0} \frac{T - T_0}{S - S_0} \tag{12.1}$$

therefore

$$T - T_0 = \frac{E_s T_0 (S - S_0)}{S_0} \tag{12.2}$$

The left-hand side of this equation is the change in demand for the mode to be achieved by a relative change in the level of service of size $(S - S_0)/S_0$. This type of calculation is often used during fare or frequency reviews for public-transport services.

This is, of course, an approximation which assumes that we have calculated E_s beforehand (perhaps from time series data), that this elasticity is constant (or that the demand function is linear—not very likely) and that everything else remains the same. This result is a reasonable approximation for small changes in the LOS variables.

Example 12.1: The fare/demand elasticity of public transport is often taken to be -0.30. If a public-transport system carries 200 000 passengers in the peak period at an average fare of 80 pence/trip:

- Estimate the fall in the demand if the average fare increases by 2.5%.
- Find out how sensitive is the result to the elasticity value.

In this case $T_0 = 200\,000$; $E_s = -0.30$, and $(S - S_0)/S_0 = 0.025$, so using (12.2) we get:

$$T - T_0 = -0.30 \times 200\,000 \times 0.025 = -1500 \text{ passengers}$$

If $E_s = -0.2$, the expected reduction in patronage would be 1000 passengers; if it is -0.4, it would then be 2000 passengers.

12.3.2 Pivot-point Modelling

This method has been developed to estimate future travel demand on the basis of knowledge of the current levels of demand and changes in the LOS variables for each alternative. In this case we require to know the demand function but not the specific values of the levels of service variables which are not to change; for example, that of parking charges in different parts of a city. The only data needed are the current market shares of each mode and the proposed changes in the LOS variables; then, an incremental form of the demand model is used to 'pivot' around the current situation.

The incremental form of the multinomial logit mode choice model is given by (Kumar 1980):

$$p'_k = \frac{p^0_k \exp(V_k - V^0_k)}{\sum_k p^0_k \exp(V_k - V^0_k)} \tag{12.3}$$

where p'_k is the new proportion of trips using mode k; p^0_k is the original proportion of trips by mode k; and $(V_k - V^0_k)$ is the change in the utility of using mode k, in our case generated by changes to the LOS attributes of mode k.

Incremental forms for more complex functions, such as the nested logit model, are also available (see Bates *et al.* 1987; Martínez 1987).

Example 12.2: Consider a transport system with three modes: car, bus and rail with proportions 0.4, 0.45 and 0.15. Assume that the utility function has the following linear form:

$$V_k = -0.10t_k - 0.20w_k - 0.05C_k/I + \delta_k$$

where t_k stands for in-vehicle travel time, w_k for waiting time and C_k/I, cost divided by income; δ_k is a modal penalty.

Assume also that we are only interested in changes in frequency that would reduce expected waiting time by rail from 10 minutes to 7.5 minutes and increase that of bus from 3 to 4 minutes; therefore we would have for rail:

$$V_r - V_r^0 = -0.2(7.5 - 10) = 0.5$$

and for bus:

$$V_b - V_b^0 = -0.2(4 - 3) = -0.2$$

The change in modal share would then be:

$$p_r' = \{0.15 \exp{(0.5)}\}/\{0.15 \exp{(0.5)} + 0.45 \exp{(-0.2)} + 0.4\}$$

the reader can verify that this produces:

$$p_r' = 0.24 \quad \text{and} \quad p_b' = 0.36$$

In the same vein, the singly constrained incremental gravity model can be written as:

$$T_{ij} = \frac{G_i T_{ij}^0 a_j \exp{(-\beta \Delta GC_{ij})}}{\sum_{ij} T_{ij}^0 a_j \exp{(-\beta \Delta GC_{ij})}} \tag{12.4}$$

where G_i is the total trips generated at zone i, ΔGC_{ij} the difference in generalised cost between the base and design years, and a_j growth factors reflecting changes in the destinations j.

Incremental forms for most travel choice models are not, in general, difficult to develop or implement. For example, Abraham *et al.* (1992) report on an incremental model for the whole of London handling both mode and doubly constrained gravity models for different person types and modes. This was implemented in EMME/2 taking advantage of its macro facilities. Similarly, SATCHMO has modules to implement incremental mode, distribution and other logit choice models (see Willumsen *et al.* 1993).

Incremental or pivot-point model formulations are helpful as we only need to account for changes in the generalised costs or utility functions, not their complete values. Therefore, if we are not introducing new modes modal penalties can be ignored as they cancel out in ΔGC. An additional advantage is that the model preserves the current (or base) matrices, therefore retaining any special associations detected in the data but never completely accounted for in a model; this is particularly valuable when dealing with destination choice where the gravity model has never performed sufficiently well, see section 5.8.7. The incremental gravity model will be expected to represent changes in the trip pattern resulting from changes in travel costs and generations and attractions.

12.4 MODEL ESTIMATION FROM TRAFFIC COUNTS

12.4.1 Introduction

Conventional methods for collecting origin-destination information from, for example, home or roadside interviews tend to be costly, labour intensive and time disruptive to the trip makers. The problem is even more acute in developing countries, where rapid changes in land use and population shorten the 'shelf-life' of data. The need for developing low-cost methods to estimate the present and future O–D matrices is apparent.

Traffic counts can be seen as the result of combining a trip matrix and a route choice pattern. As such, they provide direct information about the sum of all O–D pairs which use the counted links. Traffic counts are very attractive as a data source because they are non-disruptive to travellers, they are generally available, they are relatively inexpensive to collect, and their automatic collection is well advanced. The idea of estimating trip matrices or demand models from traffic counts deserves serious consideration and the last decade has seen the development of a number of approaches attempting just that.

Consider a study area which is divided into N zones inter-connected by a road network which consists of a series of links and nodes. The trip matrix for this study area consists of N^2 cells, or $(N^2 - N)$ cells if intra-zonal trips can be disregarded. The most important stage for the estimation of a transport demand model from traffic counts is to identify the paths followed by the trips from each origin to each destination. The variable p_{ij}^a is used to define the proportion of trips from zone i to zone j travelling through link a. Thus, the flow (V_a) in a particular link a is the summation of the contributions of all trips between zones to that link. Mathematically, it can be expressed as follows:

$$V_a = \sum_{ij} T_{ij} p_{ij}^a, \qquad 0 \leqslant p_{ij}^a \leqslant 1 \tag{12.5}$$

The variable p_{ij}^a can be obtained using various trip assignment techniques ranging from a simple all-or-nothing to a more complicated equilibrium assignment. Given all the p_{ij}^a and all the observed traffic counts $(\hat{V}_a)$, there will be N^2 unknown T_{ij}'s to be estimated from a set of L simultaneous linear equations (12.4), where L is the total number of traffic counts.

In principle, N^2 independent and consistent traffic counts are required in order to determine uniquely the trip matrix **T**. In practice, the number of observed traffic counts is much less than the number of unknowns T_{ij}'s. Therefore it is impossible to determine a unique solution to the matrix estimation problem. In general, there will be more than one trip matrix which, when loaded onto the network, will reproduce (satisfy) the traffic counts. There are two basic approaches to resolve this problem: structured and unstructured methods. In the structured case, the modeller restricts the feasible space for the estimated matrix by imposing a particular structure which is usually provided by an existing travel demand model, for example a gravity or direct-demand model. The unstructured approach relies on general principles like maximum likelihood or entropy maximisation to provide the minimum of additional information required to estimate a matrix. These two general approaches will be discussed below, but first we must consider the relationship between route choice and model estimation.

12.4.2 Route Choice and Matrix Estimation

Robillard (1975) classified assignment methods for trip matrix estimation from counts under two main groups: *proportional* and *non-proportional* assignment. Proportional assignment methods make the proportion of drivers choosing each route independent from flow levels. The most common example is all-or-nothing assignment and in this case p_{ij}^a is defined as:

$$p_{ij}^a \begin{cases} 1 & \text{if trips from origins } i \text{ to destinations } j \text{ use link } a \\ 0 & \text{otherwise} \end{cases}$$

Pure stochastic assignment methods such as Burrell's and Dial's also fall into this group but in these cases p_{ij}^a can also take intermediate values between 0 and 1.

Non-proportional assignment techniques take explicit account of congestion effects and therefore the proportion of travellers using each link does depend on link flows. Equilibrium and stochastic user equilibrium assignment methods are members of this group.

Non-proportional assignment techniques are thought to be more realistic for congested conditions. However, the advantage of proportional assignment methods is that they permit the separation of the route choice and matrix estimation problem; the proportion of trips using each link p_{ij}^a can be assumed to be independent of the trip matrix to be estimated. In contrast, non-proportional route choice requires the joint or iterative estimation of route choice and trip matrices so that both are consistent. In what follows, we shall assume that proportional assignment methods are a reasonable approximation to route choice; we shall discuss later the extensions needed to cover non-proportional methods.

12.4.3 Transport Model Estimation from Traffic Counts

The calibration of a gravity model was one of the first methods put forward for estimating trip matrices from traffic counts. The basic idea is to postulate a particular form of gravity model and examine what happens when it is assigned onto the network. For example, in the case of inter-urban travel the trip matrix could be:

$$T_{ij} = \frac{\alpha P_i P_j}{d_{ij}^2}$$

where P_j is the population of urban area j, d_{ij} is the distance between both areas and α is a constant for calibration, in this case the only one. If a matrix of this kind is assigned on the network we get:

$$V_a = \sum_{ij} \frac{p_{ij}^a \alpha P_i P_j}{(d_{ij})^2} = \alpha \sum_{ij} \frac{p_{ij}^a P_i P_j}{(d_{ij})^2} \tag{12.6}$$

Note that on the right-hand side of this equation the only unknown is α: the other variables are provided by external data or a good route choice model. One can generalise this model slightly and include other trip generation/attraction factors like employment, industrial production, shopping floorspace, and so on. If we denote the gravity part of this model by:

$$G_{ij} = \frac{O_i D_j}{d_{ij}^2}$$

and allow several journey purposes k (or commodities if one is dealing with freight movements), one can write:

$$V_a = \sum_k \sum_{ij} p_{ij}^a \alpha_k O_i^k D_j^k / (d_{ij})^2 = \sum_k \alpha_k \sum_{ij} p_{ij}^a G_{ij}^k \tag{12.7}$$

Here the α_k are parameters for calibration but the rest of the data are, once more, assumed to be available. It is relatively simple to see that the α_k may be estimated using least squares techniques. In this case we postulate that $V_a' = V_a + \epsilon_a$, where ϵ_a is an error term. A change of variable:

$$X_k = \sum_{ij} p_{ij}^a G_{ij}^k$$

permits writing:

$$V_a' = \alpha_0 + \sum_k \alpha_k X_k \tag{12.8}$$

where α_0 is the intercept and may be deemed to depict the part of the flow not represented by the gravity model, for example local or intra-zonal traffic. This type of approach was followed by the first researchers in this area, Low (1972) for urban areas and Holm *et al*. (1976) for planning inter-urban networks in Denmark.

Equation (12.7) has at least one obvious deficiency. If a particular O_i and a particular D_j are each doubled, then the number of trips between these zones would quadruple when it would obviously be more likely that it should double also. To improve on this the following more conventional model can be used:

$$T_{ij} = \sum_k [\alpha_k O_i^k D_j^k A_i^k B_j^k f_{ij}^k] \tag{12.9}$$

where:

α_k is a scaling parameter which enable us to use different units for T_{ij} and O_i^k, D_j^k.

A_i^k and B_j^k are the balancing factors expressed as:

$$A_i^k = \left[\sum_j (B_j^k D_j^k f_{ij}^k)\right]^{-1}$$

$$B_j^k = \left[\sum_i (A_i^k O_i^k f_{ij}^k)\right]^{-1}$$

f_{ij}^k is a deterrence function, for example $\exp(-\beta_k C_{ij})$

Estimating this more conventional model from traffic counts represents a greater effort as the parameters for calibration are now A_i^k, B_j^k, β_k and α_k. This calls for alternative calibration methods, for example non-linear regression as used by Högberg (1976) or Robillard (1975).

Tamin and Willumsen (1989) generalised this approach following suggestions from Wills (1986) to combine in a single model features of the gravity and the intervening opportunities (OP) model. Wills has proposed a flexible gravity-opportunity (GO) model for trip distribution in which standard forms of the gravity and opportunity models are obtained as special cases. The choice between gravity or opportunity approaches is decided empirically by allowing the estimation of parameters which control the global functional form of the trip distribution mechanism.

We can define a transformation δ_{dj}^i such that δ_{dj}^i equals 1 if destination j is the dth position in ascending order of distance away from i, and zero otherwise, then the ordered (opportunities) trip matrix can be obtained by the following transformation:

$$Z_{id} = \sum_j [\delta_{dj}^i T_{ij}] \tag{12.10}$$

While the ordering transformation δ_{dj}^i produces an ordered trip matrix, its inverse $(\delta_{dj}^i)^{-1}$ allows the observed trip matrix to be recovered by

$$T_{ij} = \sum_d [(\delta_{dj}^i)^{-1} Z_{id}] \tag{12.11}$$

It should be noted that this class of transformation is applicable to any variable based on the O–D matrix, notably the cost matrix and the proportionality factor, in addition to the trip matrix. We can also define a direct Box–Cox transformation such as (8.2) on a variable y as:

$$y^\tau = \begin{cases} (y^\tau - 1)/\tau & \tau \neq 0 \\ \log y & \tau = 0 \end{cases}$$

and in inverse Box-Cox transformation as

$$y^{(1/\tau)} = \begin{cases} (y\tau + 1)^{1/\tau} & \tau \neq 0 \\ \exp y & \tau = 0 \end{cases}$$

These transformations may be combined into a new function which we introduce as a convex combination in μ,

$$y^{(\tau,\mu)} = \mu y^{(\tau)} + (1 - \mu)y^{(1/\tau)}, \qquad 0 \leqslant \mu \leqslant 1 \tag{12.12}$$

The proposed model can finally be written then as:

$$T_{ij} = \sum_k [\alpha_k O_i^k D_j^k A_i^k B_j^k f_{ij}^k] \tag{12.13}$$

where:

$$f_{ij}^k = \sum_d [(\delta_{dj}^i)^{-1} F_{id}^k] \tag{12.14}$$

$$F_{id}^k = \left(\sum_p^d U_{ip}^k\right)^{(\tau,\mu)} - \left(\sum_p^{d-1} U_{ip}^k\right)^{(\tau,\mu)} \tag{12.15}$$

$$U_{ip}^k = \exp\left[(1 - \tau)\gamma_m \log D_{pk}^i - \beta_m C_{ip}\right] \tag{12.16}$$

and

$$D_{dk}^i = \sum_j [\delta_{dj}^i D_j^k] \tag{12.17}$$

From this general form several special cases may be derived by setting τ and μ to particular values. Three extreme cases generating specific models are easily identified: the gravity (GR), the pure logarithmic-opportunity (LO) and the pure exponential-opportunity (EO) models.

Three estimation methods were implemented by Tamin and Willumsen (1989) to calibrate the general form from traffic counts, namely: non-linear least squares (NLLS), weighted non-linear least squares (WNLLS) and maximum likelihood (ML). The general model was tested for both freight transport in Bali, Indonesia (Tamin and Willumsen 1988) and passenger traffic in Ripon, UK (Tamin and Willumsen, 1989). In the case of road haulage, even if the traffic counts were not classified by lorry type it was possible to discriminate up to nine different commodity types, one of them empty trucks. In this case proxy data to the O_i^k and D_j^k are required, for example production levels of certain commodities. The parameter α_k then plays the double role of converting these proxies first to tonnes and then to lorries.

The main conclusions from this research were:

- The GO and OP model are more time consuming than the GR model since they require more complicated algebra and procedures which take longer to solve.
- Good fit at the traffic count level produced a general good fit at the trip matrix level as well.
- Although Burrell's stochastic assignment was also used to estimate the p_{ij}^a, it gave no better fit to the traffic counts than all-or-nothing assignment.

- Although the GO was the best model in terms of matching the observed traffic counts, it cannot be guaranteed that it will also produce the best-fit to an independently observed trip matrix. In fact, it was found that the model which gives the best fit at the trip matrix level is the GR gravity model with the NLLS method and Burrell assignment.

Holm *et al.* (1976) have extended the gravity model approach to include some features of equilibrium assignment. They make use of an iterative loading with $\phi = 1/n$ (see section 10.5.4) to obtain the proportion of trips using each link. However, this is only a heuristic approximation as under strict equilibrium conditions the proportions are not, in general, unique.

Of course other, perhaps direct-demand, models could also be used in this type of estimation method. One interesting advantage of this approach is that once a demand model is calibrated it may be used for forecasting purposes too, provided future values for parameters like O_i and D_j are available or estimable.

12.4.4 Matrix Estimation from Traffic Counts

Entropy-maximising and information-minimising techniques have been used as model-building tools in urban, regional and transport planning for many years, particularly after the work of Wilson (1970). For example, we discussed the derivation of the conventional gravity model from an entropy-maximising formalism in Chapter 5. In this context, the entropy-maximising formalism provides a naive, least-biased, trip matrix which is consistent with the information available represented as constraints to a maximisation (of an entropy function) problem. In the case of the gravity model the constraints represent trip-end and total cost information.

This idea was used by Willumsen (1978) to derive a model to estimate trip matrices from traffic counts. The problem can be written as:

$$\text{Maximise } S(T_{ij}) = -\sum_{ij}(T_{ij} \log T_{ij} - T_{ij}) \tag{12.18}$$

subject to:

$$\hat{V}_a - \sum_{ij} T_{ij} p_{ij}^a = 0 \tag{12.19}$$

for each counted link a, and:

$$T_{ij} \geq 0$$

Constraints (12.19) replace the trip-end and cost constraints of the gravity model derivation. The use of Lagrangian methods permits the formal solution to this problem to be found as:

$$T_{ij} = \exp \sum_a (-\tau_a p_{ij}^a) = \prod_a X_a^{p_{ij}^a} \tag{12.20}$$

where τ_a are the Lagrange multipliers corresponding the constraints (traffic counts) and,

$$X_a = \exp(-\tau_a)$$

The availability of an old matrix, or simply a matrix estimated (or cordoned off) from another study could be accommodated to some advantage. Let **t** be this prior matrix, sometimes called a 'reference trip matrix'; the new objective function becomes:

$$\text{Maximise } S_1(T_{ij}/t_{ij}) = -\sum_{ij}(T_{ij}\log T_{ij}/t_{ij} - T_{ij} + t_{ij}) \qquad (12.21)$$

subject to the same constraints (12.19) and non-negativity. This objective function is, of course, convex and the term t_{ij}, being a constant, is only there for convenience; it can actually be eliminated from the derivation of the model.

Using the same methodology and change of variables, the formal solution can be seen to be:

$$T_{ij} = t_{ij}\exp\sum_a(-\tau_a p_{ij}^a) = t_{ij}\prod_a X_a^{P_{ij}^a} \qquad (12.22)$$

Example 12.3: Consider the simple network depicted in Figure 12.1. This network has two origins (1 and 2) and two destinations (3 and 4). The flows on all links are also shown in this figure.

It can be seen that there are only six (integer) trip matrices that can reproduce the observed flows as shown below.

Matrix	First		Second		Third		Fourth		Fifth		Sixth	
i \ j	3	4	3	4	3	4	3	4	3	4	3	4
1	8	0	7	1	6	2	5	3	4	4	3	5
2	2	5	3	4	4	3	5	2	6	1	7	0
$S(T_{ij})$	−11.07		−7.46		−5.98		−5.78		−6.84		−9.96	
$S_1(T_{ij}/t_{ij})$	−5.79		−3.69		−3.70		−5.07		−7.22		−12.20	

The entropy-maximising formalism seeks to identify the most probable trip matrix consistent with the information available, in this case five traffic counts. Incidentally, the reader can verify that only three of these counts are independent (see section 12.4.5); therefore the problem is, indeed, underspecified.

The values of the objective function $S(T_{ij})$ are also shown in this table. According to this, the most probable trip matrix would be the fourth, {5, 3, 5, 2}, as it has maximum entropy value. If a prior matrix is available then a second objective

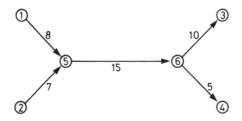

Figure 12.1   Simple network with traffic counts

function (12.21) should be used. Assume the prior matrix {3, 2, 1, 3} is available; the new values from the entropy function are also depicted above. The most probable trip matrix in these circumstances is the second one, {7, 1, 3, 4}. Of course, in more practical problems we cannot hope to calculate directly the entropy values of all possible matrices. Note, for instance, that reducing the number of counts increases the number of feasible trip matrices. More importantly, flows of the order of hundreds or thousands increase the number of possible (integer) trip matrices enormously. What is needed is an effective solution method not requiring matrix identification.

There are several possible methods to solve model (12.22). The most widely used one is the multi-proportional approach. This is, in essence, an extension of the bi-proportional and tri-proportional methods discussed in Chapter 5. In this case, instead of balancing the trip matrix trying to match trip-end totals (and cost-bin totals in the tri-proportional case), we undertake successive corrections to the prior trip matrix in order to reproduce the observed traffic counts. There is one correction factor X_a for each traffic count and its calculation involves the iterative estimation of these factors until the observed link flows are replicated to within an acceptable tolerance.

If no prior matrix is available, **t** can be taken as unity; in effect, an entropy-maximising formalism may be considered to generate as the most likely trip matrix, one that has the same number of trips in each cell, unless being prevented from achieving this by the constraints. In other words maximising entropy is equivalent to minimising the difference between a uniform target and the estimated matrix.

The detailed analysis of this maximum entropy matrix estimation (ME2) model and that of a related approach, based on information-minimising principles, is given by Van Zuylen and Willumsen (1980). Both models are practically equivalent and share most of their properties. The ME2 model will always reproduce the observations V'_a to within a given tolerance provided the constraints define a feasible space, i.e. equations (12.19) must have at least one solution in non-negative T_{ij}. An additional condition for the prior matrix **t** is discussed below.

It can be shown that minimising the negative of the objective function (12.21) is approximately equivalent to minimising:

$$S_2(T_{ij}/t_{ij}) = \frac{0.5(T_{ij} - t_{ij})^2}{T_{ij}} \tag{12.23}$$

This is an error-like measure of the difference between the values of t_{ij} and T_{ij}. In effect, the negative of $S_1(T_{ij}/t_{ij})$ is also a natural measure of the difference between these cell values: it is zero when $t_{ij} = T_{ij}$ and increasingly positive as the difference increases. In this sense, the estimated matrix is that closest to the prior matrix which when loaded onto the network can reproduce the traffic counts.

The model can accommodate other sources of data provided they can be incorporated as linear constraints. An example of this type may be information about the trip length distribution (TLD) thought to be realistic for the study area. This type of information can be translated into constraints equivalent to those of cost bins, as discussed in Chapter 5; for example:

$$\frac{1}{T}\sum_{ij} T_{ij}\delta_{ij}^k = P_k \tag{12.24}$$

where T is the total number of trips, P_k is the proportion of trips in cost (length) range (bin) k, δ_{ij}^k is 1 if trips between i and j have cost in range k, and zero otherwise.

Public-transport systems with a zonal or other variable fare system permit the introduction of constraints of this type to help estimate the corresponding trip matrices using passenger counts and ticket sales data (see de Cea and Cruz 1986).

Moreover, the mathematical program can also be written with a combination of equality and inequality constraints, thus enhancing the value of this type of approach. For example, the planner may know that the capacity of a link is Q_a but not have a traffic count for it; or that no more than D_j' vehicles can go to a particular destination because of parking capacity there. This type of information can be incorporated as inequality constraints, for example:

$$\sum_{ij} T_{ij}p_{ij}^a \leq Q_a \quad \text{for some links } a \tag{12.25}$$

$$\sum_i T_{ij} \leq D_j' \quad \text{for some destinations } j \tag{12.26}$$

The solution to this program is still a multiplicative model; Lamond and Stewart (1981) have shown how the multi-proportional algorithm can be extended to handle inequality constraints; therefore the same solution method may be used for this expanded model.

One of the features of the (extended) ME2 model is its multiplicative nature. This means that if a cell in the prior matrix is zero it will remain zero in the solution as well. This may be a source of problems if the cell in the prior matrix was zero by chance (i.e. because of the sampling rate adopted in the study) instead of representing an O–D pair with no trips at all. One pragmatic solution to this problem, for very sparse prior matrices, is to 'seed' the empty cells with a small value, for example 0.5 trips. The constraints, through the multi-proportional or other solution algorithm, will then ensure that some of these trips 'grow' to one or more full trips while others regain a zero value.

Example 12.4: Consider the same network as in Example 12.3 but assume now that we only have two traffic counts, on links 5–6 and 2–5 (15 and 7). The table below shows the multi-proportional algorithm as applied to this problem. The table shows first the full solution for the case of uniform (no) prior matrix, Case A.

		Traffic count	Modelled flow	Ratio	Trips per O–D pair			
					1–3	1–4	2–3	2–4
A	Prior Matrix				1.00	1.00	1.00	1.00
	Iteration	15	4.00	3.750	3.75	3.75	3.75	3.75
	1	7	7.50	0.933			3.50	3.50
	Iteration	15	14.50	1.034	3.88	3.88	3.62	3.62
	2	7	7.24	0.967			3.50	3.50
	Iteration	15	14.76	1.016	3.94	3.94	3.56	3.56
	3	7	7.11	0.984			3.50	3.50
	Iteration	15	14.89	1.008	3.97	3.97	3.53	3.53
	4	7	7.05	0.992			3.50	3.50
	Iteration	15	14.95	1.004	3.99	3.99	3.51	3.51
	5	7	7.03	0.996			3.50	3.50
B	Prior matrix				3.00	2.00	1.00	3.00
	Iteration	15	15.03	0.998	4.81	3.21	1.75	5.24
	5	7	6.98	1.002			1.75	5.25
C	Prior matrix				3.00	2.00	0.00	3.00
	Iteration	15	15.06	0.996	4.82	3.21	0.00	6.97
	6	7	6.97	1.004			0.00	7.00
D	Prior matrix				3.00	2.00	0.50	3.00
	Iteration	15	15.04	0.998	4.81	3.21	1.00	5.99
	6	7	6.98	1.002			1.00	6.00

As can be seen, it takes only five iterations to reach convergence within 5% tolerance. The solution {3.99, 3.99, 3.5, 3.5} does not coincide with the maximum-entropy solution in Example 12.3 because the number of traffic counts is not the same. Case B shows the problem with the prior matrix {3, 2, 1, 3}; again, it takes only five iterations to reach satisfactory convergence. The solution {4.81, 3.21, 1.75, 5.25} is indeed different, thus showing how the information contained in an outdated trip matrix can be used to advantage in matrix estimation; there is something of value in past information worth making use of.

Case C illustrates what happens when there is a zero entry in the trip matrix. There is still a solution but the zero is preserved in it. Finally, Case D shows the effect of 'seeding' the zero in the prior matrix with 0.5. The solution this time, {4.81, 3.21, 1.0, 6.0} affects only trips from the origin previously containing the zero.

Consider now the effect of increasing the number of counts to three by including link 2–5. The corresponding results are depicted in the table overleaf.

First, note that the number of iterations required has now increased. This seems to depend not so much on the actual number of counts used but on how close to removing all flexibility in the matrix these are. In this case three out of four degrees of freedom are removed by these counts. The solution in case A, {5.33, 2.68, 4.67, 2.35}, is the one that maximises $S(T_{ij})$ and if rounded to integers coincides with the solution in Example 12.3.

		Traffic count	Modelled flow	Ratio	Trips per O–D pair			
					1–3	1–4	2–3	2–4
A	Prior matrix				1.00	1.00	1.00	1.00
	Iteration	15	4.00	3.750	3.75	3.75	3.75	3.75
	1	7	7.50	0.933			3.50	3.50
		10	7.25	1.379	5.17		4.83	
	. . . . . .							
	Iteration	15	15.05	0.997	5.32	2.68	4.65	2.35
	10	7	7.00	1.000			4.65	2.35
		10	9.97	1.003	5.33		4.67	
B	Prior matrix				3.00	2.00	1.00	3.00
	Iteration	15	15.11	0.992	6.51	1.51	3.41	3.56
	14	7	6.97	1.004			3.42	3.58
		10	9.94	1.006	6.55		3.45	
C	Prior matrix				3.00	2.00	0.00	3.00
	Iteration	15	17.15	0.875	8.75	0.13	0.00	6.12
	20	7	6.12	1.143			0.00	7.00
		10	8.75	1.143	10.00		0.00	
D	Prior matrix				3.00	2.00	0.50	3.00
	Iteration	15	15.10	0.994	6.98	1.05	2.96	4.01
	19	7	6.97	1.004			2.97	4.03
		10	9.95	1.005	7.01		2.99	

The solution for case B, {6.55, 1.51, 3.45, 3.58}, has the same properties in respect of $S_1(T_{ij})$. Case C is interesting as it shows that in this opportunity with the inclusion of a zero in the prior the algorithm fails to converge, even after 20 itcrations. The reader may verify that forcing cell 2–3 to zero makes the problem unfeasible: there are seven trips out of node 2 but only five are permitted to reach their destination. Case D illustrates the effect of seeding the empty cell with 0.5 trips; the algorithm now converges to a reasonable solution.

12.4.5 Traffic Counts and Matrix Estimation

One can ask at this stage whether any set of counts is suitable for trip matrix estimation. For example, is it possible that certain combinations of counts make it impossible to estimate a matrix which satisfies them? These problems will be discussed under the headings of independence and inconsistency of traffic counts.

12.4.5.1 Independence

Not all traffic counts contain the same amount of 'information'. For example, in Figure 12.2 traffic link c is made up of the sum of traffic on links a and b. Counting traffic on link c is then redundant and only two counts there can be said to be independent.

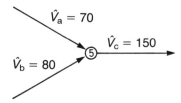

Figure 12.2   Dependent counts

Wherever a flow continuity equation of the type 'flows into' a node equals 'flows out of' the node can be written, its counts will be linearly dependent. In this case it will always be possible to describe one link flow as a linear combination of the rest. Note that a centroid connector attached to node 5 will remove the dependency in Figure 12.2.

12.4.5.2 Inconsistency

Counting errors and the fact that often traffic counts are obtained on different occasions (hours, days, weeks) are likely to lead to inconsistencies in the flows. In other words, the expected flow continuity relationships will not be met. If the count V_c in Figure 12.2 were to be 160 instead of 150, the corresponding equations would be inconsistent and no trip matrix could possibly reproduce these flows. One way of reducing this problem is to allow an error term in the equations or to remove the inconsistencies beforehand.

It is possible to identify two sources for inconsistencies in the link flows. The first one is simply the fact that errors in the counts may lead to situations in which the 'total flow into' a node does not equal the 'total flow out of' the same node, thus not meeting link flow *continuity* conditions. The second source is a mismatch between the assumed traffic assignment model and observed flows. For example, an assignment model may allocate no trips on a link having an observed (perhaps small) flow. In these conditions there will be no trip matrix capable of reproducing the observed link flows using that route choice model.

Example 12.5: It is useful to distinguish between these two types of inconsistency, first at *flow level* and then at *path flow level*. Assume we have observations on the flow of four links (identified by the pair of nodes delimiting them) and we would like to find non-negative trip matrices satisfying these and a route choice model as depicted in Figure 12.3.

Consider first the case where the count *x* has been found to be 8, thus making the

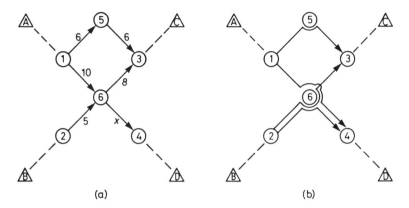

Figure 12.3   An example of path flow inconsistencies with counts: (a) network and flows, (b) assumed route choices

total flow into node 6 equal to 15, and the flow out of this node equal to 16. These counts are then inconsistent, perhaps because they were taken on different days or simply because of counting errors. We can remove this inconsistency by arbitrarily increasing the flows on links (1, 6) or (2, 6) by one, or by reducing the flows on links (6, 3) or (6, 4) by one. We can be more systematic and make the least adjustments necessary to preserve flow continuity conditions. For example, if what we want to minimise is the sum of the squares of the increments/reductions, then the optimum change is 0.25 on each link.

An alternative approach is to seek a maximum-likelihood solution to this problem, as put forward in Van Zuylen and Willumsen (1980). This assumes that link flows are Poisson distributed and that the observations available are samples on these distribution. Maximum likelihood is then used to generate a model for producing improved and consistent estimates of the flows. On the other hand, model calibration from traffic counts, as discussed in the previous section, makes an explicit allowance for errors in the observed link flows. These methods are not limited, therefore, by independence and consistency problems.

Consider now the case when the count x is 7. It can be seen that the link flow continuity conditions are now met. However, the assumed assignment depicted in Figure 12.3b is incompatible with the flows shown in Figure 12.3a. No feasible trip matrix can reproduce the count of 8 at link (6, 3) because the only path using it, B–C, is limited to a maximum of 5 by link (2, 6).

The set of linear equations corresponding to this example is given by:

link (1, 5)	$T_{AC} = 6$	(12.27)
link (5, 3)	$T_{AC} = 6$	(12.28)
link (1, 6)	$T_{AD} = 10$	(12.29)
link (2, 6)	$T_{BC} + T_{BD} = 5$	(12.30)
link (6, 3)	$T_{BC} = 8$	(12.31)
link (6, 4)	$T_{AD} + T_{BD} = 7$	(12.32)

Clearly equations (12.30) and (12.31) are incompatible with the non-negativity of T_{BC}. The same applies to equations (12.29) and (12.32), making it impossible to solve this set of equations. In simple problems like this, inconsistencies can be ascertained by inspection but in more complex networks they can only be identified by means of row and column operations on the linear equations. For large systems these operations are likely to be expensive in terms of computer requirements.

In this simplistic example it is not difficult to see that the problem originates in the assumed single route between A and C. If two paths were allowed, one via node 5 and the other via node 6, the inconsistency could be removed. Furthermore, the value of the resulting variable p_{AC}^6 cannot be arbitrarily chosen; in effect, a feasible solution requires

$$0.2 \leqslant p_{\mathrm{AC}}^6 \leqslant 0.5$$

The fact that the *path flow* continuity conditions are not met seems to reflect errors in assignment, whereas the *link flow* discontinuities are a reflection of errors in the traffic counts alone. It seems reasonable then to develop a technique for removing the link flow inconsistencies in the counts in order to ensure that the link flow continuity conditions are met. On the other hand, a reasonable approach to deal with the lack of consistency at the path flow level seems to be the adoption of a better route choice model. In general terms, consistency at the link flow level is a necessary but not sufficient condition for consistency at path flow level. Consistency at path flow level is, however, a sufficient condition for link flow consistency.

The interested reader may verify that there are only seven different (integer) trip matrices which can satisfy the observed flows in the example above.

12.4.6 Limitations of ME2

ME2, probably because of its simplicity, relative efficiency and ease of programming, has been widely implemented and used, particularly in the UK. The model has, however, some known limitations and it is worth exploring them before discussing opportunities to improve it.

One of the limitations arises when traffic has grown (or declined) markedly between the prior (or old) trip matrix and the present. As the model estimates the matrix closest to the prior which, when loaded on the network, reproduces the traffic counts, this may lead to distortions. In these cases it is probably better to consider the structure of the prior matrix, say through the proportion of total trips which appear in each cell, and not the absolute number of trips in each O–D pair. One would then try to find a matrix with the closest structure to that of the prior matrix which reproduces the traffic counts when loaded onto the network. This can be approximated by means of a general growth factor first, for example:

$$\tau = \frac{\displaystyle\sum_a \hat{V}_a}{\displaystyle\sum_a \sum_{ij} t_{ij} p_{ij}^a} \tag{12.33}$$

which is then applied to the prior matrix before using the ME2 model. In this way the structure of the prior matrix is preserved as much as possible. The estimation of τ above is only an approximation; for a more rigorous approach see Bell (1983).

A second limitation of ME2 is the fact that it considers the traffic counts as error-free observations on non-stochastic variables. In effect, the model gives complete credence to the traffic counts and uses the prior matrix only to compensate for the fact that they do not contain sufficient information for estimation purposes. However, this may not be very appropriate in practice. For a start, one must acknowledge that traffic counts are certainly not error free. Apart from counting errors there is the problem of time variations (hourly, seasonal, etc.). Traffic counts obtained on different days or at different times can hardly be considered to be observations on a non-stochastic variable.

Willumsen (1984) has suggested an approach to compensate for this second difficulty. It starts from the idea that functions of the type $\{X \log X/Y - X + Y\}$ can be seen as useful measures of the difference between X and Y. He then constructs a composite objective function to satisfy the following:

$$\text{Minimise } S_3 = \sum_{ij}(T_{ij} \log T_{ij}/t_{ij} - T_j + t_{ij}) + \sum_a \phi_a(V_a \log V_a/v_a - V_a + v_a) \quad (12.34)$$

where

V_a is now the 'true' value of the traffic count at a
v_a is the value of one observation of the flow made at a
ϕ_a is a weighting factor which depends on the confidence attached to the observation v_a.

The use of the Lagrangian method now leads to the solution:

$$T_{ij} = t_{ij} \prod_a X_a^{p_{ij}^a} \quad (12.22)$$

$$V_a = v_a X_a^{1/\phi_a} \quad (12.35)$$

Again this model can be solved using the multi-proportional algorithm but in this case we also need to correct the observations to obtain a better estimation of the true value of the link flows. Note that if ϕ_a is very large, i.e. we assign a high weight to the counts as we believe them to be very accurate, V_a tends to v_a; in the limit with $\phi_a = \infty$ we revert to the original model as $V_a = v_a$. On the other hand, the smaller the value of ϕ_a, the greater the credence given to the prior matrix $\mathbf{t}$.

One would expect that the weights ϕ_a depend on the variability of the observations. Brenninger-Gothe *et al*. (1989) have discussed this model in detail. They have shown that a very natural value for the weights ϕ_a is the variance (or standard deviation) associated to the observations. If these are not available they can be estimated using some assumption about the distribution of the error terms. These authors have further extended the model to consider weights attached to both the prior matrix (μ_{ij}) and the traffic counts (ϕ_a); thus the new objective function becomes:

Minimise $S_3 = \sum_j \mu_{ij}(T_{ij}\log T_{ij}/t_{ij} - T_{ij} + t_{ij}) + \sum_a \phi_a(V_a\log V_a/v_a - V_a + v_a)$ (12.36)

The main limitations of ME2 can therefore be reduced using reasonably simple methods. However, other authors have proposed alternative approaches to solve the matrix estimation problem, some of which start from a different basic framework.

12.4.7 Improved Matrix Estimation Models

Bell (1983) has formulated a model which tries to preserve the structure of the prior matrix, in the sense described in the previous section, adding a new constraint and thus modifying the mathematical programme as follows:

Minimise $-S_2$ subject to

$$\hat{V}_a - \sum_{ij} T_{ij}p_{ij}^a = 0 \text{ for each counted link } a \qquad (12.19)$$

$$\tau = \sum_{ij} T_{ij}\bigg/\sum_{ij} t_{ij} \qquad (12.37)$$

and

$$T_{ij} \geq 0$$

In addition to this, Bell suggests the use of a Newton–Raphson method to solve this model with an iterative estimation for τ. Alternatively, one may assume an initial value for τ, solve the standard model using a multi-proportional method and then check if it is consistent with equation (12.37). The cycle should be repeated until the value of τ converges.

The use of a Newton–Raphson algorithm has advantages in terms of computer time and is also useful in tracing the effect of errors in the traffic counts through to the estimated trip matrix (Bell 1983); this type of sensitivity analysis is an alternative to the treatment of errors in the traffic counts suggested above. However, the Newton–Raphson method requires more memory and is therefore restricted to small and medium-size networks.

A variant to the standard objective function (S_1) is either to linearise it using Taylor's expansion or to construct a generalised least squares formulation. In both cases we still try to minimise the difference between prior and estimated matrices subject to the same constraint (12.19). Bell (1984) suggested the Taylor series expansion solution whereas McNeil and Hendrickson (1985) and Cascetta (1984) have put forward versions involving generalised least squares approaches. One problem is that under certain circumstances these models may produce negative entries in the estimated trip matrix, in particular where the prior matrix originally had small values. This is not an uncommon occurrence and therefore this feature is undesirable.

Maher (1983) proposed the use of a Bayesian approach to the trip matrix estimation problem which results in functional forms equivalent to the generalised least squares method. A prior estimate of the trip matrix is updated in the light of a set of traffic counts; both are assumed to be multivariate Normally distributed variables with known covariances.

Spiess (1987) has put forward a maximum likelihood model to solve the problem. He considers a specific formulation where for each O–D pair t_{ij} is obtained by observing an independent Poisson process with mean $\Omega_{ij}T_{ij}$. This corresponds to the problem of taking a sample of an existing trip matrix with a sampling rate of $\Omega_{ij} < 1$. The probability of observing t_{ij} is:

$$\text{Prob}\,[\text{Poisson}\,(\Omega_{ij}T_{ij}) = t_{ij}] = (\Omega_{ij}T_{ij})^{t_{ij}}\exp{(-\Omega_{ij}T_{ij})}/t_{ij}! \qquad (12.38)$$

The joint probability of observing the sample matrix $\{t_{ij}\}$ is therefore:

$$\text{Prob}\,[\{t_{ij}\}] = \prod_{ij}\text{Prob}\,[t_{ij}] = \prod_{ij}(\Omega_{ij}T_{ij})^{t_{ij}}\exp{(-\Omega_{ij}T_{ij})}/t_{ij}! \qquad (12.39)$$

Applying the maximum likelihood estimation technique to this problem requires finding the matrix $\{T_{ij}^*\}$ which satisfies the constraints and yields the maximum probability (12.39) of observing $\{t_{ij}\}$. By taking logarithm of equation (12.39) and adopting the usual convention that $0\log 0 = 0$, we can formulate the maximum likelihood model as:

$$\text{Max}\sum_{ij}(t_{ij}\log{(\Omega_{ij}T_{ij})} - \Omega_{ij}T_{ij} - \log{t_{ij}!}) \qquad (12.40)$$

subject to the usual non-negativity constraints and to equation (12.19). Separating the logarithm into the sum and discarding constant terms one can rewrite (12.40) as:

$$\text{Min}\sum_{ij}(\Omega_{ij}T_{ij} - t_{ij}\log{T_{ij}}) \qquad (12.41)$$

This objective function is convex in T_{ij}; provided the set of constraints is consistent and the flows feasible, then the existence of an optimal solution is assured. The solution may be obtained by any standard solution method for convex programming problems. However, Spiess (1987) has developed an algorithm that exploits some of the specific properties of this problem.

For further comments on this problem and possibilities for extensions see Cascetta and Nguyen (1988) and Willumsen (1991).

12.4.8 Treatment of Non-proportional Assignment

The ME2 model discussed in the preceding sections is based on the assumption that it is possible to obtain the route choice proportions $\{p_{ij}^a\}$ independently from the matrix estimation process. Wherever congestion plays an important role in route choice this assumption becomes questionable as the route choice proportions and the trip matrix become interdependent. Because of its theoretical and practical advantages, equilibrium assignment is the natural framework for extending the ME2 model for the congested network case.

The main problem in incorporating Wardrop's equilibrium into trip matrix estimation is that now the route choice proportions and the trip matrix to be estimated are interdependent. One way of tackling this problem is to adopt an iterative approach: assume a set of route choice proportions $\{p_{ij}^a\}$, estimate a matrix **T**, load it onto the network and obtain a new set of route choice proportions; repeat the process until route choice proportions and estimated matrices are mutually consistent.

This general scheme can be implemented in different ways. For example, in SATURN (Hall *et al*. 1980) the route choice proportions are estimated using the value ϕ in the Frank–Wolfe algorithm (the optimum linear combination of accumulated and auxiliary flows; see section 11.2.2) at each iteration. It is recognised that in general the path flows under equilibrium conditions are not unique. However, this method assumes them to be unique.

An alternative approach requires restating the original problem in terms of a three-dimensional matrix (origin, destination, route) as follows:

$$\text{Maximise } S_4 = -\sum_{ijr} T_{ijr}(\log T_{ijr}/t_{ijr} - 1) \tag{12.42}$$

subject to

$$\sum_{ijr} T_{ijr}\delta_{ijr}^a - \hat{V}_a = 0 \tag{12.43}$$

and

$$T_{ijr} \geq 0$$

where the index r indicates the route or path chosen; δ_{ijr}^a is 1 if route r between i and j uses link a, and zero otherwise.

It is always possible, of course, to reconstruct the O–D matrix $\{T_{ij}\}$ by aggregating the path flow matrices $\{T_{ijr}\}$. Again the solution to this new program is:

$$T_{ijr} = t_{ijr}\prod_a X_a^{\delta_{ijr}^a} \tag{12.44}$$

and

$$T_{ij} = \sum_r T_{ijr} \tag{12.45}$$

The prior path flows may be calculated from the prior trip matrix as $t_{ijr} = t_{ij}/R_{ij}$, where R_{ij} is the number of paths between i and j. In this case, the path flows can take any value as they are not assumed unique. The Frank–Wolfe algorithm for equilibrium assignment is used to identify attractive paths (those selected at each all-or-nothing step) but not to define the strict proportions of the trip matrix using them. This is only a heuristic scheme and a suitable algorithm for its solution is as follows:

1. Assign, using equilibrium assignment methods, a base-year matrix $\{t_{ij}\}$ to the network and save the corresponding routes (trees). Set the cycle counter n to 1.
2. Estimate a trip matrix $\{T_{ij}\}^n$ for iteration n, using independent routes $\{\delta_{ijr}^a\}$ and observed flows $\{\hat{V}_a\}$.
3. Assign $\{T_{ij}\}^n$ to equilibrium, saving the routes (trees) used in the process.
4. Increment n by 1 and return to step 2 unless the changes in routes $\{\delta_{ijr}^a\}$ or estimated matrices have been sufficiently small.

For a test of this approach and a comparison with proportional assignment techniques in the case of a comprehensive data set for Reading in the UK, see Willumsen (1982).

A more general approach has been put forward by Fisk (1988), in which a maximum-entropy matrix estimation and user equilibrium assignment are combined as a single mathematical program. A similar approach has been proposed by Oh (1989). The computer requirements of such methods seem very high, thus reducing their practical value.

12.4.9 Estimation of Trip Matrix and Mode Choice

The idea of extending this type of approach to matrix and mode choice estimation is attractive. Let us consider a singly constrained destination/mode choice model of the following logit form:

$$T_{ij} = O_i \frac{S_j \Sigma_k \exp(\Sigma_p \theta_p X_{ijk}^p)}{\Sigma_d S_d \Sigma_k \exp(\Sigma_p \theta_p X_{idk}^p)} \tag{12.46}$$

where the mode choice component of the model is given by:

$$P_{ij}^k = \frac{\Sigma_p \exp(\Sigma_p \theta_p X_{ijk}^p)}{\Sigma_m \exp(\Sigma_p \theta_p X_{ijm}^p)} \tag{12.47}$$

T_{ij} are trips between zones i and j, O_i is the total number of trips originating at zone i, S_j is a measure of the attractiveness of zone j, P_{ij}^k is the proportion of trips using mode k between zones i and j, X_{ijk}^p is the pth explanatory variable for mode k (for example, in-vehicle travel time) and θ are model parameters.

Although the derivations we will present below are for the simpler multinomial logit case, they can easily be extended to consider the simultaneous estimation of more general nested logit forms (Ortúzar and Willumsen 1991).

12.4.9.1 Simple Unimodal Case

Let us consider first a single mode case with just one scale parameter μ, multiplying a 'generalised cost' variable X_{ij}, to be estimated. In this simple case (12.46) reduces to:

$$T_{ij} = O_i \frac{S_j \exp(\mu X_{ij})}{\Sigma_d S_d \exp(\mu X_{id})} \tag{12.48}$$

Now, assume we possess observations on a set of link flows $\hat{V}_a$, and also that we know, from an assignment model, the proportions p_{ij}^a for all links with observed flows. In such a case we can postulate that equation (12.19) holds and to estimate the value of μ we can, for example, seek to minimise the following normalised non-linear (generalised) least squares function:

$$S = \sum_a \left[\left(\hat{V}_a - \sum_{ij} T_{ij} p_{ij}^a \right) \Big/ \hat{V}_a \right]^2 \tag{12.49}$$

In order to find the minimum we usually require first and second derivatives of S with respect to μ. These are provided by Ortúzar and Willumsen (1991); unfortunately, even in this simple case the derivatives look rather intractable so a unique solution to the problem may be difficult to establish.

12.4.9.2 Updating with Aggregate Modal Shares

Let us consider the transference of model (12.46)–(12.47) with parameters $\boldsymbol{\theta}$ estimated in another context; we ignore the original mode-specific constants as they ensure reproduction of the aggregate market shares in that context. Define a transfer utility function as:

$$V_{ijk} = \mu \left(\sum_p \theta_p X_{ijk}^p \right) + M_k \tag{12.50}$$

where X_{ijk}^p are zonal values for the level-of-service and socioeconomic variables in the new context, μ and $\mathbf{M}$ are a scale parameter as before, and a set of $(K - 1)$ mode-specific constants to be estimated; K is the total number of modes.

In this case it is possible to find maximum likelihood estimators for μ and $\mathbf{M}$ but it is possible to guarantee a unique optimum only for fixed μ, i.e. when only the constants are updated.

12.4.9.3 Updating with Traffic Counts

The main problems arise in this case if we are interested in mixed-mode combinations but only have counts for the 'pure' modes. For example, consider the case of choice between car, bus, underground and combinations of the latter with the first two. It is obvious that even if we have separate counts for each pure mode, these include observations corresponding to the mixed-mode movements. If we settle for a mode aggregation and are interested in estimating the scale parameter μ and a set of constants for the pure modes, the problem can be solved using a generalised least squares formulation similar to (12.49), as shown by Ortúzar and Willumsen (1991).

12.4.9.4 *Updating with Combined Information*

Assume we wish to update μ and $\mathbf{M}$ of (12.50) and have available observed aggregate shares P_k and sets of observed passenger counts $\hat{\mathbf{V}}$ for each competing mode. The problem can be formulated either as a maximum likelihood or generalised least squares one.

In the first case we will get different functions to maximise and hence different first-order conditions and optima, depending on the assumptions made about the distribution of count errors. The favourite assumptions have been multinomial, independent Poisson and independent Normal (see Tamin and Willumsen 1988). As it can be assumed that data on counts are independent of data on aggregate shares, the log-likelihood function takes the form of a sum of two expressions. If it is assumed that the counts have no error, a final case of interest results which requires maximising a much simpler function subject to (12.19). Expressions for each of these cases are given by Ortúzar and Willumsen (1991); there is no guarantee, however, that either of them leads to a unique optimum.

The generalised least squares formulation has two advantages: the first is that no distributional assumptions are needed on the data set; the second is the possibility of incorporating explicitly differences in the accuracy of each data item prior to estimation. A need for normalising, which is also a feature of this approach, is very evident here given the different order of magnitude of the differences between observed and modelled values for both types of data. For example, the maximum difference in the case of aggregate shares is just 1, while differences in count data may easily run to figures in the hundreds or thousands.

The range of methodologies available in principle to solve this important problem is difficult to evaluate without recourse to experimentation; by the end of the 1980s such an exercise had not been conducted but work was in progress.

12.5 MARGINAL AND CORRIDOR MODELS

12.5.1 Introduction

We have seen how conventional modelling approaches often require large amounts of resources (especially computing time and technical expertise), have sometimes a slow response rate, may not be sensitive to some of the policy options needing analysis and may be based on weak theoretical frameworks (see for example, Supernak 1983). In previous chapters we have discussed how to avoid most of these common pitfalls; in this section we wish to explore some shortcuts which can be taken to speed up the response time of modelling exercises.

Having considered some of the simplified approaches in the preceding sections one must recognise that they would seldom satisfy, on their own, all the requirements of a large scale project or major policy change. The use of trip matrix estimation techniques from traffic counts may be acceptable for situations where a fixed-matrix assumption is reasonable, for example, the design of traffic managements schemes. However, the adoption of model estimation from traffic counts methods is still weak in terms of modal choice, an important element in most project assessments. Sketch planning methods offer quick response but at a high risk in terms of coarseness of the analysis. It is interesting to explore whether these approaches can be combined to utilise their strengths and avoid their weak points.

The basic idea is to adopt an approach which would use simpler models to provide a planning background and would selectively apply 'state-of-the-art' models to the most relevant decision elements of the problem. Technical journals devote little space to report systematically on the many shortcuts planners and consultants by necessity adopt in practice (Leamer 1978). Some conferences offer better illustrations of these; see for example Ashley *et al.* (1985) and Clancy *et al.* (1985).

The first element in the development of practical simplified approaches is to recognise that there is always some implicit or explicit planning context providing local experience and data. How to utilise these two effectively should always be the first step in this task. The production of sound advice to decision makers under severe time constraints should deal with questions like the following:

- how best to simplify or select models that will appropriately represent the impacts of the project to be analysed;
- how to make adequate use of existing data and local experience;
- how to take advantage of some of the special characteristics of the problem in hand; and
- how to deal with the inevitable biases introduced through the pragmatic answers adopted to the questions above.

12.5.2 Corridor Models

A typical opportunity for simplifying modelling tasks without compromising realism too much is provided in corridor studies. Corridors are strong, basically linear, transport facilities sometimes combining high-capacity and limited-access arterial roads with rail rapid transit or busway provisions. The linear nature of the facilities may help to simplify the modelling task; it may be sufficient to model the linear corridor and consider only the points of entry and exit to it as origins and destinations. There may be a major destination at one end of the corridor (the central business district for example) or they may be distributed throughout its length.

In any case, assignment problems will be minimal or non-existent and the modelling effort will be able to concentrate on issues such as mode and, in some cases, destination choice. The basic information needed will be the current flow levels by mode and section of the corridor, data on level of service variables for each mode and section, and the relevant characteristics of travellers. A good deal of these data are obtainable through choice-based interviewing either in-vehicle (train) or at the main destinations (workplace).

The extreme simplification of the network structure generates considerable savings in data collection and coding. The transfer of discrete choice models from other contexts may be undertaken using the techniques discussed in section 9.4. If necessary, trip generation transfer may be performed using the methods discussed in section 4.5; however, in most cases a fixed multimode trip matrix is assumed for these studies. If the study is to cover several years in the future, it may be necessary to use a matrix updating technique based on growth factors, as discussed in section 5.2.

Corridor modelling with severe capacity constraints require some care. Bottleneck effects in the corridor should be treated specifically and sometimes micro-assignment models may be applied to them. Direct demand models also appear as suitable choices for this type of problems.

12.5.3 Marginal Demand Models

Faced with problems which cannot be tackled through a full-scale transportation study due to limited resources and time availability, one would like to concentrate efforts on that part of the transport demand most likely to be affected by the project or policy in question. If the project is not corridor based this requires a little more care and attention. However, often the special characteristics of the problem may be utilised to simplify the task in hand. A generalised approach to this problem was proposed by De Cea *et al.* (1986). This approach is outlined below and shown in Figure 12.4.

1. Definition of the problem. The terms of reference of the study, if available, should facilitate an identification of the main elements of a problem, be it a particular investment project or the consideration of a new policy option. Terms of reference, however, do not exempt the analyst from identifying the wider implications of the alternatives to be considered.
2. Identification of the relevant population and the impact potential of the project. At this stage, one seeks to identify the most likely impacts of a project or policy option and the sections of the population most likely to be affected. In principle anything is likely to affect everything else, but one should try to identify first-order effects and those most likely to perceive (gain or lose) the costs and benefits of the project.
3. Identification of the technical resources available to analyse the main impacts of the project on the relevant population. The existence of data, perhaps not up to date, other studies and models, and in particular local expertise, can play a key role in providing sound and quick advice to decision makers. Updating data sets and adjusting existing models should normally require less resources than starting from a clean slate. Local knowledge may be crucial at this stage.

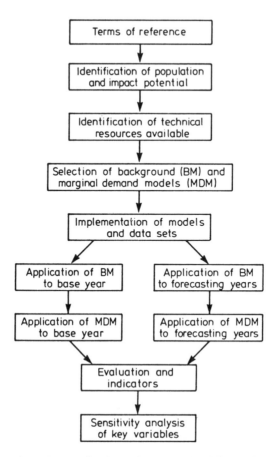

Figure 12.4   Steps in project evaluation using a marginal demand estimation approach

4. Selection of *background* and *marginal demand* models. A key element of this approach is the use of a coarser background model to estimate the general level of demand, and a finer marginal demand model to identify the specific impacts of the project on that general demand. The choice of background and marginal demand models depends on the nature of the problem and on the technical resources available. The marginal demand model is applied to the relevant population only and should, of course, be able to discriminate the impacts of the project and/or policy options on that population. In selecting these models the feasibility of their implementation and use within the time and resources framework of the study is paramount. The simplifying assumptions adopted at this stage should be properly documented.
5. Implementation of the models and data sets. Background and marginal demand models should then be mounted on a computer together with the data sets to be used and updated as part of the study. In many cases it will be necessary to write short programs to convert data sets to suitable formats and to perform the required tests and report production.
6. Application of the background and marginal demand models to the base year and

their validation. This may require some additional data collection, ideally on a small scale.

7. Application of the background and marginal demand models to forecast relevant future years. This will require first forecasting the values of the planning variables for those years and then applying the models with and without the project or under different policy options.

8. Evaluation. Model runs in the previous two steps should provide the indicators required for an evaluation of the options open to decision makers. Attention should be paid to frame this evaluation in terms of good local practice and to produce the indicators which decision makers consider most meaningful.

9. Sensitivity analysis. The simplifying assumptions adopted in previous stages and the uncertainty about the future make it necessary to test how sensitive the advice produced is to changes in the inputs and weights adopted in the study. Budget and time constraints will usually limit the amount of sensitivity tests that can be performed. It is often possible, however, to elicit preferences from decision makers on what they consider to be the most important elements to be examined in these tests. These may take the form of questions like:

> Would the project still be feasible if . . . oil prices double, discount rate is increased by 2% etc.?

These preferences could then be used to select sensitivity tests complementing those required by the simplifying assumptions adopted above. This is a pragmatic methodology whose virtues and limitations can only be assessed in practice.

De Cea *et al*. (1986) followed this approach to study a possible extension to the Santiago (Chile) underground network. In outline their approach involved:

- The identification of the population of interest as that in zones with walk access to the Metro before or after the potential extension, including mixed-mode journeys.
- The use of trip matrix estimation techniques based on traffic counts to provide background trip matrices for both cars and public transport; use was made of an extensive set of traffic counts supplemented by *ad hoc* surveys at bus stops and Metro stations.
- The transfer of a corridor-based disaggregate mode choice model to the study area through the recalibration of the mode-specific constants. The availability of the corridor model and suitable income data made this possible.
- Economic, financial and environmental evaluation of the project complemented by sensitivity analysis of key parameters.

This complete study was undertaken in four months. The cost-benefit analysis predicted a reasonable return on investment. The extension of the Metro has now been implemented and the results apparently confirm the accuracy of the study.

12.6 GAMING SIMULATION

Mathematical models do not solve any real-life transport problems: it is the interpretation of mathematical solutions which is useful to make decisions concerning transport problems. Simplified models may help in reducing the effort required to find a mathematical answer and in facilitating the subsequent interpretation of this solution in relation to the real problem. We use conceptual or mental models to understand, interpret and act in our professional life. Mental models are, in effect, a prerequisite for the development and application of mathematical ones run on a computer.

Despite their significance and because of their character, it is difficult to examine mental models and this often leads to quite unmanageable communication problems. Better and richer mental models in the minds of planners and decision makers are probably as important as the use of rigorous and sound behavioural models in the computer, if transport planning is to be improved. Given the key role played by mental models in the use and application of mathematical ones, it seems sensible to investigate techniques for improving the first in order to get better solutions through the second.

But how are mental models acquired, revised, rejected and enhanced? The main factors seem to be formal and informal education, discussions and, above all, practical experience. One of the main problems facing planning education and training is how to provide realistic experience. This is particularly acute in the transport field where the most important consequences of a policy measure or infrastructure project may follow only after considerable time. Besides, it is surprisingly easy to become too involved in the details of particular techniques and lose sight of the wider process where they must fit.

The need for methods of developing a general comprehension of a system rather than detailed information about its parts has been recognised in several fields, particularly in management and business training. Several educational techniques have been developed to this end: case studies, role playing and different types of exercises. Gaming simulation is a particularly attractive technique in this field. It was originally developed for military purposes in the form of war games but since computers became widely available it has spread successfully into management science, politics, sociology, and regional and transport planning.

Educational games are sequential decision-making exercises structured around an artificial environment acting as surrogate for the real world. This artificial environment may be just a set of instructions and graphical material or may involve an elaborate simulation exercise using computer programs, physical models and animated displays. As in real life, games usually have a competitive dimension. This feature can be incorporated in at least two forms: by dividing the players into teams with partially conflicting objectives (e.g. car owners, environmental protection officers, local residents, etc.) or, by facing each player with a computer model of a complex system plus a common set of initial conditions and final objective. Key indicators can then be used to assess the performance of each player in achieving these objectives. The first approach stresses the need for negotiation and compromise whilst the second emphasises efficiency in pursuing objectives. Both methods enhance understanding of complex systems and support the development of learning skills. In both cases the success of players depends on their ability to learn from the outcome of their own decisions, that of others and from the effect of unexpected events like a strike or fuel price increases. The final objective of any gaming-simulation exercise is augmenting the ability to learn through the enrichment of the conceptual model every player has of a system. For a good background on gaming-simulation design and experience the reader is directed to Greenblat and Duke (1975) or Taylor (1971), and in the transport field to Ortúzar and Willumsen (1978).

A number of gaming simulations have been developed specifically for the transport field. Some of these cover problems like negotiating the alignment for a new road or planning new public-transport services. Probably the most widely used game in the urban transport management field is GUTS (Willumsen and Ortúzar 1985). The original objectives for this computer-based game were:

- The game should treat the transport sector of an urban area as a system, i.e. it should highlight the interrelations between modes, traffic management and investment decisions, and financial constraints; therefore, the computer program contains relationships conveying these interactions.
- The game should be realistic but manageable; the most common types of investment and traffic management decisions should be included and key financial and resource constraints be simulated.
- The model should allow for a range of alternative and even conflicting objectives to be pursued, and consequently the program should produce not a single but multiple performance indicators; at the same time, the information available to players should not be too different from that commonly available to decision makers.
- The game should stress the importance of continually monitoring the performance of a transport system.
- The model should allow the representation of different types of urban areas in terms of residence, employment, car ownership, income distributions and growth rates, public-transport patronage and related indicators.

GUTS is available as an interactive program for micro computers with modest memory requirements. The model is based on a simplified, urban area with circular symmetry. Two modes of transport, car and buses, operate freely and in competition; the user can make decisions on public-transport fares and levels of service, the introduction of bus lanes, supplementary licence schemes, parking provision and charges, as well as major investment projects. The program checks these decisions and runs the model to represent one year of operation of the transport system. At the end of the run indicators on flow levels, speeds, modal split, travel time and expenditure by person type are produced, and the financial performance of the bus company is reported. Changes in accessibility levels and the impact of new investment are also simulated, as are unexpected events inducing changes to the cost structure of the transport modes operating in the city. The symmetry condition imposed on the city simplifies the model with advantages in terms of speeding up the learning curve of the user and enhancing running time in the computer.

Games like GUTS can enhance transport planning in a number of ways. First, in their normal training-tool mode, they can be used to educate new recruits to a team and to develop a common language throughout an office. Second, a model of this type may be seen as a simple 'sketch planning' tool valuable in discussing broad policy options and particular conceptions of decision makers. GUTS, and similar programs, are no substitute for full-scale models but may help bridge the gap between broad strategies and specific modelling studies. A third use of tools of this kind is in demonstrating the advantages and limitations of mathematical models. The extremes of total rejection of transport models or their blind acceptance are still present in most political and planning quarters. The evident simplicity of a gaming-simulation exercise combined with its capacity to represent interactions between modes and decisions, provide a good example of what the formal modelling approach can offer. The use and subsequent critique of the game by politicians and planners would help them to understand each other's activities and interests better.

The development and widespread distribution of games of this kind may well play a role in enhancing public awareness of transport problems, the role of alternative strategies to tackle them, their implications for resource allocation and quality of life. The Dutch Government has pioneered this approach through the development and widespread distribution of a 'Traffic Oracle' program for microcomputers to permit the simulation and discussion of different future scenarios and policies for transport planning and management in the Netherlands.

EXERCISES

12.1 The network in Figure 12.5 represents a small area with two origins A and B and two destinations Y and Z.

Traffic counts have been made of the car flows using the network with the following results:

Link	Flow
M–N	400
N–P	700
P–Q	500

(a) Use an entropy-maximising model to estimate a trip matrix from the information above. Assume a suitable prior matrix for this problem if necessary. A 3% error in the modelled flow is considered acceptable for this question.

(b) Repeat the calculations above but assuming the prior matrix is given by:

	Y	Z
A	100	50
B	80	200

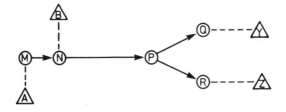

Figure 12.5 Simple network for Exercise 12.1

12.2 The network in Figure 12.6 represents links connecting two origins A and B to two destinations C and D in a developing country. The populations of the two origins are 10 000 and 20 000 inhabitants respectively and the markets held at C and D are equally attractive in terms of size and prices. The link distances (in km) are indicated in the figure.

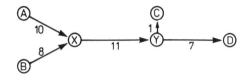

Figure 12.6 Simple network for Exercise 12.2

Person counts have been obtained for three links as follows:

Link	Persons/day
A–X	3 400
X–Y	11 900
Y–D	4 100

Calibrate a model of the type

$$T_{ij} = \frac{bP_iD_jd_{ij}^{-n}}{\sum_j D_j}$$

where P_i is the population of zone i, D_j is the attractiveness index for the market in zone j, and d_{ij} is the travel distance between i and j. Try at least two values for the power n, including $n = 2$ and $n = 2.5$.

12.3 Three villages, A, B and C, are connected by a navigable river in an underdeveloped country. Village A has a population of 1000 inhabitants; village B is 30 km downstream of A and has a population of 2000; village C is 10 km downstream of B and has a population of 300 inhabitants. The value of the goods exchanged in each village per day is 500, 600 and 600 pesos respectively.

Two observers have spent some time making directional counts of passengers travelling in boats along the river with the following results:

River section	Passengers per half day
A–B	45
B–A	60
B–C	360
C–B	560

(a) Calibrate a gravity model of the form suggested in Exercise 12.2, where D_j is replaced by the population of village j. Use $n = 2.0$.

(b) Calibrate a similar model but replace D_j by the value of the goods exchanged in each village per day.

(c) Which model do you think is best? Why?

12.4 The elasticity of the demand for buses to the fare is typically acknowledged to be in the region of -0.3. The average trip maker between zone A and the centre of town (CBD) currently faces a bus fare of $2 per trip; the bus share of all trips between A and the CBD is 60%, other trips use either car or underground.

If the total number of trips between both zones is 2,000 estimate the loss in patronage of the buses if the fare is raised to $3 per trip, all other things being equal, using the incremental logit method. Compare your result with the more crude elasticity calculation; discuss your findings (*Hint*: estimate the parameter θ_c from the data given the simple expression for the logit direct elasticity).

13 Other Important Topics

This chapter covers four important aspects of transport modelling. The first is the modelling of freight; section 13.1 highlights the main differences with passenger demand modelling and outlines suitable approaches for freight.

Section 13.2 is devoted to the forecasting of planning variables. These are variables like future population, employment, school places, shopping areas and income distribution, which are needed to make predictions with transport planning models. Sometimes these variables are provided externally to the study; in others they must be estimated as part of the planning exercise. In either case, they play a key role in determining the forecasting ability of the models discussed in this book.

One of the most important planning variables is car ownership and this is the subject of section 13.3. Both time-series and econometric models to forecast car ownership are discussed, together with some more recent approaches.

Finally, the issues surrounding the concept, estimation and application of the *value of time* are presented in section 13.4. The book would not have been complete without this discussion.

13.1 FREIGHT DEMAND MODELS

13.1.1 Importance

Most of this book has concentrated on demand modelling for passengers, with a strong emphasis on urban problems. However, freight movements, and in particular road haulage, are an important source of congestion and other traffic problems. The noise and nuisance generated by heavy lorries, the problems created by on-street loading and unloading of goods vehicles to serve shops and premises, and the usual complaint about lorries taking up a good deal of the capacity of inter-urban roads are only some of the problems associated with this type of traffic.

Unfortunately, in urban areas the policy options available to influence road haulage are very limited. They are mainly controls on loading/unloading, on the size of vehicles allowed in certain areas (lorry routeing), the provision of major freight interchanges, the encouragement of rear access to premises and improved layouts at new developments.

Freight demand modelling may play a particularly important role in developing countries where the efforts to increase exports and to gain access to underdeveloped areas are even more urgent. Facilitating the movement of freight in these cases is likely to have a major impact on economic development. Moreover, the competition between road and rail in some of these countries is a key issue in resource allocation for investment and maintenance.

In the case of inter-urban movements there is greater scope for policies to influence freight mode choice and to regulate competition between rail and road. Improved allocation of road user charges and targeting subsidies to key rail or road services, are also an important policy option. The design of these tools may require more refined modelling efforts than those used in urban studies.

Given these issues, it appears surprising that much less research has been undertaken on modelling this type of movement than the effort allocated to passenger demand. Why would this be the case? We believe there are several reasons for this:

- There are many aspects of freight demand that make it more difficult to model than passenger movements; some of these are discussed below.
- For some time urban congestion has been highest in the political agenda of most industrialised countries and in this field passenger movements play a more important role than freight.
- The movement of freight involves more actors than the movement of passengers; we have the industrial *firm* or firms sending and receiving the goods, the *shippers* organising the consignment and modes, the *carrier(s)* undertaking the movement and several others running transhipment, storage and custom facilities. In some cases two or more of these may coincide, for example in own-account operations, but there is always scope for conflicting objectives which are difficult to model in detail in practice.
- Recent trends in freight research have emphasised the role it plays in the overall production process, inventory control and management of stocks. These trends are a departure from more traditional passenger modelling techniques and share little in common.

This section will summarise approaches to freight demand modelling as reported in the literature. It starts with a discussion of the main difficulties associated with modelling freight movements. It then presents what is probably the most traditional approach to the problem, that is to adapt the conventional four-stage aggregate demand model to the case of commodities. Extensions of the disaggregate approach to freight demand are also outlined. The section closes with some practical considerations for the implementation of these ideas. The interested reader is directed to the very valuable book by Harker (1987) for further details.

13.1.2 Factors Affecting Goods Movements

As in the case of passenger demand, it is useful to consider first the factors that one would expect to influence freight movements. The following is not an exhaustive list but covers the most important ones.

- Locational factors; freight is always a derived demand and usually part of an industrial process. Therefore, the location of sources for raw materials and other inputs to a production process as well as the location of intermediate and final markets for their products, will determine the levels of freight movements involved as well as their origins and destinations;

- The range of products needed and produced is very high, much greater than even the most exaggerated or detailed segmentation of travel demand by person types and journey purposes. A given demand for bolts cannot be satisfied by providing cashew nuts. There will be very many commodity matrices in any study of freight demand.

- Physical factors. The characteristics and nature of raw materials and end products influences the way in which they can be transported: in bulk, packaged in light vans, in very secure vehicles if the products are of high value, in refrigerated containers if they are perishable. There is a greater variety, therefore, of vehicle types to match commodity classes than in the case of passenger transport.

- Operational factors. The size of the firm, its policy for distribution channels, its geographical dispersion and so on, strongly influence the possible use of different modes and shipping strategies.

- Geographical factors. The location and density of population may influence the distribution of end products;

- Dynamic factors. Seasonal variations in demand and changes in consumers' tastes play an important role in changing goods' movement patterns.

- Pricing factors. As opposed to the case of passenger demand, prices are not, in general, published material because they are much more flexible and subject to negotiations and bargaining power.

13.1.3 Pricing Freight Services

It is usually quite difficult for the analyst to obtain reliable data about freight charges. For example, in Europe both transport firms and users try to keep them confidential so as to strengthen their position when it comes to renegotiate them. The factors affecting charges or cost imputations, and therefore mode choice, are thought to be:

- The length of the supply contracts. A better price can be obtained if the shipper guarantees demand for one or more years rather than just for one single shipment. The existence of price adjustment clauses helps to extend the lengths of contracts.

- The extent of volume discounts. Following from the above, a contract guaranteeing steady high-volume shipments is likely to benefit from a lower price.

- The importance of terminal facilities. The availability of a rail terminal nearby, or even at the firm, would certainly reduce the cost of shipping by rail; its absence would increase the likelihood of using road transport all the way, without even considering rail or water transport.

- The use of own-account operations, especially road haulage. Some firms prefer this type of operation for reasons other than transport (image, reliability, integration). These firms will tend to extend the use of own-account operation for marginal products rather than consider a completely new mode.

- Some modes are more suited to transport particular commodities. For example, *pipelines* are ideal for bulk liquids and some suspensions and *merry-go-round* (non-stop) trains are very suited for movements from coal-mines to power stations. This closer fit of supply characteristics to demand would certainly influence the charges made for those products.
- Hierarchical transport systems. For example, in the case of petroleum products, use of large tankers to refineries, then small tankers and pipelines to major terminals, rail to other terminals, and lorries to petrol stations and final users. These structures are difficult to modify in the short run as they have evolved over a long period and are well established; thus, their pricing mechanisms may be very difficult to change.

13.1.4 Aggregate Freight Demand Modelling

The great majority of freight demand models applied in practice have been of the aggregate kind (see for example Van Es 1982; Friesz *et al.* 1983; Harker 1985). These applications follow the classic four-stage model with some adaptations specific to freight. A typical example of this approach is the work of Kim and Hinkle (1982), who used the American Urban Transport Planning Suite (UTPS) with some adaptations to model statewide freight movements. In outline this approach involves:

- Estimation of freight generations and attractions by zone.
- Distribution of generated volumes to satisfy 'trip-end' generation and attraction constraints. The usual methods for this task are linear programming or use of a gravity model.
- Assignment of origin–destination movements to modes and routes.

We shall look at these and other factors in some detail below.

13.1.4.1 Freight Generations and Attractions

The techniques used to obtain total trip ends depend on the level of aggregation originally envisaged and on the type of products considered:

- Direct survey of demand and supply may be undertaken for major flows for some homogeneous products: sugar, petroleum products, iron ore, coal, cement, fertilizers, grains, etc. These may be forecast using industry or sector studies. This approach is usable for inter-urban movements but is not recommended for urban problems.
- The use of macroeconomic models, for example of the input–output nature, based on regional rather than national data.
- Growth-factor methods, such as those discussed in Chapter 4, are often used in forecasting future trip ends.
- Zonal multiple linear regression is often used to obtain more aggregate measures of freight generations and attractions, in particular in urban areas.
- Demand may be associated with warehouse capacity or with total shopping area at each zone (urban studies) rather than with industrial development.

13.1.4.2 Distribution Models

Many urban studies simply apply growth-factor methods to observed goods movement matrices, as discussed in Chapter 5. However, many inter-urban freight transport studies have used synthetic aggregate models, even of the direct-demand type. The two aggregate techniques most used in this area are briefly discussed here: a gravity model and a linear programming approach.

In the case of the gravity model it is relatively simple to re-interpret its functional form as:

$$T_{ij}^k = A_i^k B_j^k O_i^k D_j^k \exp\left(-\beta^k C_{ij}^k\right) \qquad (13.1)$$

where

k is a commodity type index;
T_{ij}^k are tonnes of product k moved from i to j;
A_i^k, B_j^k are balancing factors with their usual interpretation;
O_i^k, D_j^k are supply and demand for product k at zone i (or j);
β^k are calibration parameters, one per product k; and
C_{ij}^k are generalised transport costs per tonne of product k between zones i and j.

The idea of using a generalised cost function formulation for freight demand is apparently due to Kresge and Roberts (1971). This can be interpreted as follows (omitting superscript k for simplicity):

$$C_{ij} = f_{ij} + b_1 s_{ij} + b_2 \sigma s_{ij} + b_3 w_{ij} + b_4 p_{ij} \qquad (13.2)$$

where

f_{ij} is the out-of-pocket charge for using a service from i to j;
s_{ij} is door-to-door travel time between i and j;
σs_{ij} is the variability of travel time s;
w_{ij} is the waiting time or delay from request for service to actual delivery—it may be a long time for maritime transport, for example;
p_{ij} is the probability of loss or damage to goods in transit.

All of these depend on the mode used and to some extent on the commodity being transported. The constants b_n are, in general, proportional to the value of the goods. For example, in the case of the probability of loss the cost is at least the goods value, but probably more, due to penalties for delays in delivery. In the case of delay, variability of delay and transit times, the values of b_n are at least proportional to those of the goods, essentially through increased inventory costs. Modern industrial production techniques, such as those emphasising 'just-in-time' deliveries, try to minimise these elements together with stocking costs. The minimum for b_1 to b_3 is the cost of the interest rate applied to the value of the goods during the time period considered.

In general terms, it is important to consider the relative contribution of transport (generalised) costs to the final cost of a commodity. For example, in the case of wheat, coal, cement and bricks, where transport costs are a main element in their final price; however, in the case of convenience foods, consumer goods, chocolates or electronics, transport costs have a low (direct) contribution to price.

A second approach to distribution modelling is *linear programming* (LP). This usually takes the form of a minimisation program: minimise total haulage costs (in money terms, very rarely in terms of generalised costs), subject to supply and demand constraints.

$$\text{Minimise } Z = \sum_{ij} T_{ij} C_{ij} \qquad (13.3)$$

subject to:

$$\sum_i T_{ij} = D_j \qquad (13.4)$$

$$\sum_j T_{ij} = O_i \qquad (13.5)$$

This is the well-known Hitchcock's transportation problem which can be solved efficiently in a very simple way. More advanced formulations may involve non-linear costs and perhaps more elaborate constraints involving a time element and minimum shipment sizes.

This minimisation problem makes some sense from the point of view of a large firm trying to satisfy its customers at a minimum cost. Alternatively, if an industry has several plants with different productions capacities and costs, the objective function may be to maximise profits or to minimise total cost at the market place. From the point of view of modelling, the LP approach has a better chance of being realistic when:

- the industry is concentrated in a few firms;
- there are low value goods and relatively high transport costs;
- there are few demand points (zones), perhaps a monopsony (a single buyer).

However, it must be recognised that although LP may be a good model for the behaviour of a single client or industrial firm, it cannot hope to represent aggregate behaviour for various commodities. The LP solution will tend to be too sparse, with particular destinations being served only by certain origins. On the other hand, the gravity model is quite flexible. By changing the value of β it is possible to vary the relative importance of cost compared with supply and demand constraints.

The formal relationship between LP and gravity models has been explored by Evans (1973). She has shown that in the limit, $\beta = 0$ in (13.1) will produce a matrix of movements where transport costs play no role (in fact this is Furness's solution to the growth-factor problem); whereas a very large value for β will generate a solution closer to an LP model, i.e. where transport costs are dominant (in the limit $\beta = \infty$ will reproduce the LP solution). Therefore, it is possible to use the gravity model formulation to represent the whole range of client behaviour for destination choice, from that almost indifferent to transport costs (electronics?) to the behaviour expected in the case of low-cost, high-bulk commodities like cement, sand, and so on, where transport costs are paramount.

13.1.4.3 Mode Choice

This is essentially a shipper's decision as to which carrier should be used to deliver the goods to their destination. When modelled at this very aggregate level, modal choice is often treated using a multinomial logit formulation based on generalised costs, as described above. This may turn out to be very approximate because the information can only capture those elements of mode choice incorporated in the generalised costs concept above.

These shippers' decisions are, of course, dependent on the rates charged by carriers, which in turn depend on the volumes they move between each O–D pair. As the size of many consignments is significant in terms of the impact on carriers' rates, there are interactions inside mode choice which go beyond that encountered between passengers and public-transport operators. This problem is often ignored at high levels of aggregation.

In the case of urban freight movements the problem of mode choice is trivial; the coverage provided by non-road modes is extremely limited.

13.1.4.4 Assignment

This is now a carrier's decision: the choice of the best route to take the goods from origin to destination. To some extent this is the least difficult of the problems. The use of capacity restraint is probably relevant to most urban situations. In the case of inter-urban movements, on the other hand, it may be sufficient to use a stochastic assignment model. However, it may be argued that different types of vehicles must be modelled in different ways; for example, light vans may be much less sensitive to the hilliness of routes than heavy lorries; also, vehicles carrying perishable goods might give greater priority to minimising time than those carrying, say, bulk coal. The use of multi-class assignment methods may then be warranted to cope with this variety of cost concepts.

13.1.4.5 Equilibrium

As in the case of passenger demand, the problem of system or market equilibrium pervades the whole modelling exercise but the techniques to achieve it are still under development. One of the early formulations of this problem is due to Friesz *et al.* (1983) who developed a freight network equilibrium model (FNEM). This model considers explicitly the decisions of both shippers and carriers for an inter-modal freight network with non-linear costs and delay functions that vary with commodity volumes.

FNEM treats shippers and carriers sequentially; shippers are assumed to be user optimisers trying to minimise the delivered price of the commodities they send, and therefore Wardrop's first principle is used to replicate their behaviour. This sub-model is an elastic transport demand model expressed as a mathematical programming problem solvable by the usual extension to the Frank–Wolfe algorithm, as discussed in Chapter 11. The assignment to carriers is performed through the use of a 'perceived' network including only the O–D pairs, transhipment nodes, and associated links considered by shippers in their decisions.

The carrier sub-model uses a full description of the actual transportation networks. Carriers are assumed to be individual operating-cost minimisers and are modelled using Wardrop's second principle. The flow patterns of individual carriers are aggregated to obtain global network flows.

A similar approach was formulated by Moavenzadeh *et al.* (1983) for planning intercity transport demand in Egypt. In this case the approach is based on the simultaneous transportation equilibrium model (STEM) (Safwat and Magnanti 1988), as discussed in Chapter 11.

At a higher level of analysis, it may well be that the macroeconomic models used to generate the total demand and supply levels, and in some cases the matrix of movements, use transport costs which are inconsistent with those generated by other parts of the model. Consequently, when such models are employed sequentially with a detailed freight network model, the two may well fail to converge to stable solutions.

Harker (1985) has formulated a model called the generalised spatial price equilibrium model (GSPEM) which ties together the concepts of spatial process and shipper-carrier equilibrium to simultaneously predict:

• the production and consumption of goods;
• the shippers' routeing of freight;
• the freight rates; and
• the shippers' routeing of the freight traffic.

A variant of the Frank–Wolfe algorithm has been developed to solve a particular implementation of this problem and this has been applied to a large-scale problem (with approximately 3560 nodes and 14 600 arcs) concerning the US coal economy.

13.1.5 Disaggregate Approaches

Since discrete choice models were developed and applied to model passenger demand, the idea of extending them to cover freight movements also gained currency; see for example Gray (1982) and Van Es (1982). In the case of freight, the demand for transport is seen as that for a number of individual consignments, each with its own characteristics, for which the individual shipper has to take a number of transport-related decisions. Every decision is seen as a choice made from a discrete set of alternatives. There is a number of related choices to be made in each case, e.g. to transport x tonnes at time t of commodity k by transport mode m from origin i to destination j. The carrier would then have to choose the route to perform this task.

The general flexibility of discrete choice modelling permits the construction of very general utility functions for these types of choices. They can include, for example:

• the characteristics of the transport services, such as tariffs, times, reliability, damage and loss, minimum consignment, and so on;

- the attributes of the goods to be transported, such as type of product, volume/weight ratio, value/weight ratio, perishability, inventory system and ownership;
- the characteristics of the market, such as its relative prices, firm size, availability of loading/unloading facilities, general infrastructure facilities;
- the attributes of the shipping firm, such as its production level, sales prices, plant location, available infrastructure facilities, storage policy, and so on.

This type of approach has found limited application on a national scale. The main reasons for this are the more limited understanding of all the elements involved in developing these utility functions and the very demanding data-collection efforts required to estimate this type of model.

However, its application to particular sub-markets or commodities may provide very valuable insights for policy formulation. For example, Ortúzar (1989) was able to use stated-preference data to examine the question of offering a new service (refrigerated containers) for international maritime cargo. This type of approach has also been used by Fowkes and Tweddle (1988). Future efforts in this direction are likely to prove fruitful from both research and practical viewpoints.

13.1.6 Some Practical Issues

Despite efforts in the 1980s, freight demand modelling is still less advanced than passenger demand modelling approaches. The leading edge of research and development seems to have been passenger demand forecasting, with freight following its footsteps trying to adapt models to its particular needs.

The problems of data collection may be compounded in the case of freight. For example, data collection for disaggregate approaches suffers from confidentiality and reliability problems. Even collecting data for aggregate modelling represents a much greater effort than that for passenger movements: great dispersion of firms, important daily and seasonal variations, and so on.

Opportunities for extensive roadside interviews are very limited, except at points where long delays are inevitable (waiting for a ferry, for example). In some cases, such as international travel, it may be advantageous to collate data from customs or a collection of waybills.

Example 13.1: Three types of aggregate models were estimated by Tamin and Willumsen (1988) for the island of Bali, Indonesia: a gravity (GR), an intervening opportunities (OP) and a combined gravity-opportunities (GO) model. All these models were estimated with five different types of commodities but using traffic counts alone. The resulting freight matrices were then compared with those observed in a major survey of the island. It was found that although the GO model performed slightly better than the pure gravity model, the gain in accuracy did not compensate the greater computational effort. The GR model calibrated in this way was capable of discriminating between the five groups of commodities obtaining a different β value for each. This model was far superior to the simple application of the Furness growth factor method. For more details see Tamin and Willumsen (1988).

Because simplified models use low-cost and regularly collected data (traffic counts), it may be possible to run them often enough to update forecasts and provide corrective measures for plans, i.e. they offer opportunities for implementing a continuous planning approach.

In the case of urban freight modelling very simple approaches are normally followed. They are usually based on models of vehicle movements disregarding the commodities shifted, the type of locations served and the underlying economic activities that originate this demand. It is often considered sufficient to obtain a commercial-vehicle matrix using roadside interviews (at cordon and screenline points) and then to gross it up to the planning horizon by means of growth-factor methods.

13.2 FORECASTING PLANNING VARIABLES

13.2.1 Introduction

As discussed in Chapter 1, modellers always distinguish between endogenous variables, i.e. those to be forecast as part of the modelling exercise like flows, and exogenous or independent variables. The latter are required to run the models but are supposed to originate externally to the models themselves. Typical examples in the transport field are population, employment, car ownership and income. Values for these variables should be provided for the base year and for each of the years for which forecasts are needed from the transport model.

The level of detail and disaggregation required for these variables depends on the type of model being used. In general terms an aggregate demand model makes fewer requirements than a disaggregate one in this sense. For example, at the trip generation level an aggregate, zonal-based, linear regression model may only require population, car ownership and average income by zone; a cross-classification or category analysis model, on the other hand, will need the number of households in each of the categories used, typically 108 per zone, as we saw in Chapter 4, when the model is stratified by income (6 levels), household structure (6 levels) and car ownership (3 levels).

The importance of these variables in influencing the accuracy of the whole modelling exercise is very high, as established by Mackinder and Evan (1981) in a study of 44 British urban transport studies. It was found that all the models overestimated key indicators of performance but that the most important element in explaining this overestimation was errors in the values used for the planning variables. Specification errors played a much lesser role in the overall inaccuracies. Interestingly, enough, the planning variables were often wrong because they followed official global forecasts which were also wrong in the first place.

The question then arises: how can we reduce as much as possible the errors in these planning variables? This is a difficult problem with no simple or single answer. A full discussion of the techniques available for forecasting these variables is outside the scope of this book; for practical methods the reader may consult England *et al.* (1985). However, we will discuss some of the ideas behind these techniques to appraise their strengths and weaknesses.

13.2.2 Use of Official Forecasts

The apparently simplest option in dealing with planning variables is to use official forecasts. In the UK, for example, there are estimates, at the District Council (and London Borough) level, of:

- population, households, employed residents and employment;
- number of households owning 0, 1 and 2 or more cars;
- private-vehicle trip ends by journey purposes.

The Department of Transport also produces forecasts, from time to time, of future demand expressed as expected vehicle kilometres for different types of vehicles. Other official institutions will provide other types of forecasts for planning variables, at least at a highly aggregate level.

Of course, these forecasts are seldom at a sufficient level of disaggregation to be directly usable in a detailed modelling exercise; however, they do reduce the amount of work needed to generate the required values for the planning variables at zonal level. Some of the techniques to achieve this are discussed in the next section.

To some extent the problem with using official forecasts is that they sometimes reflect the expected effect of economic and regional policies whose success may actually depend on other uncontrollable factors like international trade and cooperation. Mackinder and Evans (1981) found that errors in forecasting these global indicators were at the root of the problem of mistakes for the planning variables at the local level.

We shall come back to this problem again. How can we accurately forecast transport activity if there are significant errors in some of the key inputs used in our transport models?

13.2.3 Forecasting Population and Employment

Whenever forecasts of these planning variables are not provided for cities or districts, the planning team will need to develop methods for their estimation. There are several methods that can be used to this end, some more appropriate than others for each particular application.

13.2.3.1 Trend Extrapolation

The direct extrapolation of current trends is the simplest but least satisfactory procedure, even if it is only applied at the level of the whole study area. Trend extrapolation does not take into account decisions already made about the availability of land for future development; it does not value new regional development policies nor does it consider the expected growth in employment in the study area. In addition to this, it does not provide any information about the age structure of the population, an important element in trip generation modelling.

13.2.3.2 Cohort Survival

A more detailed technique considers deaths, births and immigration, in and out of a study area, to forecast future population:

$$P_{t_1} = P_{t_0} + B_{t_0t_1} - D_{t_0t_1} + NI_{t_0t_1} \qquad (13.6)$$

where

P_{t_1} is population at time t_1
P_{t_0} is population at time t_0
$B_{t_0t_1}$ are surviving births in the period t_0 to t_1
$D_{t_0t_1}$ are deaths in the same period
$NI_{t_0t_1}$ is net migration in the same period.

Used in this very aggregate fashion, equation (13.6) ignores the age structure of the population and may under- or over-estimate, for example, the corresponding fertility rates. For this reason the method is usually applied to subgroups of the population, or *cohorts*, and the method becomes a *cohort survival* approach. This involves the following stages:

1. The population is separated into cohorts; males are separated from females and each sex group divided into age strata (usually of five years) to give a population structure for the base year.
2. Fertility rates are then applied to females of child-bearing age.
3. The new-borns are added up and 'sexed' in known proportions.
4. The female and male babies make up the first cohort at the next round of calculations.
5. Survival rates are applied to females and males in all cohorts, starting from the youngest generation; survivors are then 'aged', that is moved forward to the next cohort.
6. The process is repeated, re-starting from stage 2 until the forecasting period has been reached.

If migration of population is to be treated in the forecasts, additional information regarding the sex and age structure of migrants is required. It is easy to see how the method may be adapted to include that new input.

The information demanded by this technique includes the initial number, age/sex structure of the population, and its associated survival, fertility and migration rates. The main source of uncertainty lies in the prediction of the rates, in particular fertility and migration rates.

13.2.3.3 Transitional Probabilities

An interesting alternative approach to cohort survival methods is to follow *family cycles* and use *transitional probabilities* reflecting the chances of moving from one stage in the cycle to another, for example, from married couple with no children to married couple with one child under school age, and from there to married couple with two children, and so on. A whole matrix of transitional probabilities is then built and processed to obtain the population in households at different stages in the family cycle in the forecast years. This approach certainly offers the potential of providing a very detailed account of population growth, very much at the level required for trip generation modelling. However, the uncertainty in the estimation and stability of the transitional probabilities is likely to be greater than that associated with fertility and migration rates in cohort survival methods.

Both cohort survival and transitional probability approaches can be usefully adapted to a continuous planning framework, where periodically collected data about fertility, migration and survival rates, and/or probabilities of changing family cycle status, permit the updating of previous estimates of population in the future and hence the changing of trip generation rates, and so on.

When forecasting employment change we are faced with similar problems. General trends in employment depend on economic policy, international trade and regional incentives. At a more local level aspects like the availability of land and qualified labour force in the study area, play an important role as well as the type of economic activity prevailing there. Moreover, the type and levels of employment also play a key role in determining the levels of income available to the households in the study area, which in turn influence car ownership and trip making behaviour.

13.2.3.4 Economic Base

A useful distinction in employment forecasting is that of *basic* and *non-basic* activities. Non-basic activities are those which are created in response to local demands where as basic activities are those which require an external stimulus of some kind. Basic activities produce goods or services which are exported to other areas and regions. Non-basic activities produce goods and services to attend the needs of the local population. It is believed that the growth of basic activities creates additional non-basic ones (shops, banks, services, and so on) to satisfy the needs of additional population. The basic activities of a region constitute its *economic base* and strengthening it would result in economic, employment and population growth.

13.2.3.5 Input–Output Analysis

Finally, in forecasting the growth of a particular activity one should also follow the concomitant growth it generates in other industries providing inputs to it. Some of these will be based outside the study area whilst others may be located inside it. The use of an input–output matrix is the traditional method of following these linkages at national or regional levels. Such a matrix depicts how much input from other sectors of the economy is needed to increase output from one particular activity. The availability of such matrices at local level is questionable; the lowest level of disaggregation seems to be a regional one.

13.2.4 The Spatial Location of Population and Employment

Having estimated population and employment (in different subgroups) for the study area, it becomes necessary to allocate them to specific zones in order to apply our transport models. This work is usually carried out in conjunction with local planning authorities who have established plans for future development and re-allocation of land uses to zones in the study area. The use of age or life-cycle specific forecasts is helpful in this process as different types of housing development are more likely to attract different types of families.

The location of employment depends on its nature; for example, industrial development, commercial services, consumer services, and so on. Major changes in the location of economic activities should probably be discussed with those involved in carrying them out. Industrial development may require special sites, good availability of water services and access to major roads and railway/port terminals. In the absence of restrictive planning controls, office employment tends to be located close to good communication facilities and as close as possible to other office developments.

These two examples show that in the final analysis the location of population and employment is not independent of the transport system. Changes in accessibility are likely to affect the potential for development of different parts of a study area. This can be taken into account in the discussions with planning authorities, or more formally, in a more comprehensive model, as outlined in the next section.

In summary, the allocation of population and employment to zones usually requires a combination of formal models and discussions with planning authorities. The practical ways in which these tasks are carried out owes a good deal to heuristic approaches and context-dependent choices. It seems difficult to eliminate current uncertainties about national, regional and local forecasts for these planning variables and this has important implications for the whole planning process.

13.2.5 Land-use and Transport Modelling

One interesting approach to forecasting population and employment and allocating them to zones is to internalise these exogenous planning variables in an integrated model of land use and transport. This has been an active area of research since the early 1960s; see for example McLoughlin (1969), Wilson *et al.* (1977) and Foot (1981). After an initial period of optimistic claims about the success of such models, researchers have now become more modest in their aspirations (see Mackett 1985).

The importance of the interaction between transport and land use is two fold. First, if transport strategies significantly change accessibility this will change demand for land and generate new development in some areas; these will in turn affect the pattern of trips (trip matrices) and therefore have an impact on the performance of the transport system. Second, changes in the attractiveness of some areas will affect the price of land there; this can be interpreted as the capitalisation of user benefits into land prices and implies a transfer of benefits to land owners. This capitalisation issue raises the question of who gains and who loses as a result of a transport scheme and how can a local authority recover from land owners some of the increase in land prices.

13.2.5.1 The Lowry Model

Many practical applications in the past have followed the lines put forward by Lowry (1965) in the 1960s. His model considers the spatial characteristics of an urban area in terms of three broad sectors of activity: employment in basic industries, employment in population-serving industries, and the household or population sector.

The Lowry model starts by allocating exogenously specified basic employment to zones and then the spatial distribution of households and non-basic employment are assigned using endogenous relationships. In addition, there are constraints on the maximum number of households for each zone (according to local regulations) and on the service employment thresholds for any zone; different types of service employment are assumed to have different minimum thresholds for their viability in any one zone.

The basic equations of the Lowry model can be written as:

$$\mathbf{P} = \mathbf{EA} \qquad (13.7)$$

$$\mathbf{E}^s = \mathbf{PB} \qquad (13.8)$$

$$\mathbf{E} = \mathbf{E}^b + \mathbf{E}^s \qquad (13.9)$$

where

$\mathbf{P}$ is a vector of population in each zone i
$\mathbf{E}$ is a row vector for total employment in each zone i,
$\mathbf{E}^b$ and $\mathbf{E}^s$ are row vectors for basic and non-basic (service) employment in each zone i,
$\mathbf{A}$ and $\mathbf{B}$ are zone-to-zone matrices of workplace-to-household and household-to-service-centre accessibilities.

The accessibility variables have two components, one corresponding to the participation rate in each zone (households per employee for $\mathbf{A}$ and service employment per household for $\mathbf{B}$) and a second corresponding to proper accessibility indices. These are normally calculated as:

$$A'_{ij} = \frac{E_j \exp\left(-\beta C_{ij}\right)}{\sum_{ij} E_j \exp\left(-\beta C_{ij}\right)} \qquad (13.10)$$

$$B'_{ij} = E^s_j \exp\left(-\alpha C_{ij}\right) \sum_{ij} E^s_j \exp\left(-\alpha C_{ij}\right) \qquad (13.11)$$

which are accessibility indices derived directly from the gravity model; see Chapter 5.

Lowry (1965) proposed a sequential solution to this problem including the constraints and thresholds mentioned above. More recent research efforts have emphasised the simultaneous solution of the same model and its extensions. Most of the latter have to do with additional disaggregation into different person and household types and their treatment over space. For example, certain types of person may be more willing (or capable) to pay for increased accessibility than others, thus influencing land prices and the type of development to be undertaken in different zones.

The integrated land-use and transport model has been implemented in a number of computer suites. In order to keep the problem tractable, some compromise in the level of detail of the transport part of the model is necessary; the hope is that what is lost in richness in the representation of the transport sector is more than compensated by gains in the forecasting of employment, population and household development in the study area. An important problem of these models, however, is that they may suffer greatly from convergence problems in their extremely complex equilibration mechanisms. For a comparison of different implementations and extensions to this approach the reader should consult Webster *et al.* (1988).

13.2.5.2 The Bid–Choice Model

A more contemporary approach has been put forward by Martínez (1991) following two modelling streams. The first one, originally proposed by Alonso (1964), is a *bid* model: land is assigned to the highest bidder. The proportion $P_{h/i}$ of customers type h making a successful bid for a given location i depends on whether h's willingness to pay WP_{hi} is the highest among the bidders $g \in \mathbf{H}$. The assumption that WP_{hi} is a function of the attributes of the plot and the bidder plus an IID Gumbel distributed disturbance (error) term, leads to a MNL expression:

$$P_{h/i} = \frac{H_h \exp(\mu WP_{hi})}{\sum_g H_g \exp(\mu WP_{gi})} \tag{13.12}$$

where μ is the usual scaling parameter of the distribution of disturbances. The expected market price p_i is equal to the expected maximum bid from potential buyers, given by

$$p_i = (1/\mu) \log \left\{ \sum_g H_g \exp(\mu WP_{gi}) \right\} \tag{13.13}$$

The second modelling stream is a maximum consumer surplus model or *choice* model derived from maximum utility following Anas (1982). The consumer surplus CS_{hi} of individual h from choosing location i is given by the difference between its willingness to pay and the price of the lot:

$$CS_{hi} = WP_{hi} - p_i$$

Under some simplified assumptions the proportion $P_{h/i}$ of consumers h choosing location i is given by:

$$P_{h/i} = \frac{S_i \exp\left[\mu(\mathrm{WP}_{hi} - p_i)\right]}{\sum_j S_j \exp\left[\mu(\mathrm{WP}_{hj} - p_i)\right]} \tag{13.14}$$

Martínez (1991) then proves that the distribution of households and firms obtained from the bid model in equations (13.12) and (13.13) is identical to that obtained from the choice version in equation (13.14). His bid–choice model is then summarised in these equations. These in turn can be simplified further when used at an aggregate level.

The transport system is represented in his model by suitable accessibility (to destinations) and attractiveness (with respect to origins) functions. The next task is to specify the WP functions; this must be done more or less on a case by case basis as the best function will depend on the availability of data.

These are powerful and flexible models. They respond to an urgent need to look closer into the issues of transport and land-use interaction, recovery of surpluses and distribution of benefits, in addition to changes in trip patterns. The widespread availability of general-purpose model estimation software has made possible the development of these models and their increasing application to practical problems.

It has been argued that this type of model is likely to work better where there are fewer constraints to the land market and type of development permitted by local authorities. This is probably the case in many developing countries, as reported by Chadwick (1987). However, as we have seen in this section, forecasting of planning variables is far from being accurate and its internalisation into an integrated land-use and transport model is unlikely to make it more reliable or robust. Our degree of understanding in this area is probably even more limited than in the transport sector alone. This problem highlights again the advantages of a continuous planning approach where regular updating of forecasts and plans reduces the risk of inaccurate predictions.

13.3 CAR-OWNERSHIP FORECASTING

13.3.1 Background

Although the total number of passenger cars active on the road in industrialised countries almost doubled between 1970 and 1986 (see for example de Jong 1989), the rate of growth was dramatically higher in developing countries. For example, the fall in import duties for small cars of less than 850 cc in Chile (from 120 to only 10%) in 1977, meant that average car ownership in Santiago went up by more than 100% in only five years (see Fernández *et al.* 1983). Even if the annual kilometrage per vehicle had remained constant in this period, it must be noted that the total increase in passenger-car kilometres represented a high cost to society in terms of accidents, fuel, pollution, increased traffic congestion and additional road construction and maintenance costs.

One problem faced by planners of vastly different nations is that forecasts of the number of cars and/or vehicle kilometres for, say, the year 2010, imply that these adverse effects may assume catastrophic proportions. In fact, by the end of the 1980s there were already cities like Athens, Los Angeles, Mexico, Santiago, Seoul and Tokyo which had become notorious for their congestion and pollution problems.

Models to predict changes in car ownership, an essential input to transport planning, have been under development since the early 1940s. It can be said in general that these efforts have been made with the following three different purposes in mind:

- Market research studies for vehicle manufacturers and petrol companies which are of no direct interest to transport modellers, as they are more concerned with vehicle attributes like size, engine capacity, and so on.
- Government-sponsored studies seeking to determine the need for new infrastructure (basically highways) at a national level; until the end of the 1970s simple time-series models were used for this task.
- Local studies, which are usually part of strategic transport studies, and which have made use of more advanced econometric methods with either cross-sectional and/or longitudinal data.

We will not attempt to cover all aspects of the car-ownership forecasting problem here, as whole books have been devoted to the subject (see for example Mogridge 1983; Train 1986). Here we briefly discuss the two following basic methods:

- Time-series extrapolations using aggregate data at a national or regional level (basically the work of John Tanner at the British Transport and Road Research Laboratory).
- Econometric methods using disaggregate data at the household level, as it has been argued that the decision to acquire a car cannot be modelled correctly at the strictly individual level nor at the zonal level (see for example Bates *et al.* 1978).

Modern methods sometimes incorporate features of both approaches and extend estimates to car usage as well. Critical reviews of these and other methods have been given by Button *et al.* (1982) and de Jong (1989).

13.3.2 Time-series Extrapolations

It seems clear that car ownership rates (e.g. cars/head of population) should not increase indefinitely in time (i.e. in general people with a driving licence are not going to indulge in several cars each); for this reason the increment curves which are usually put forward to model this phenomenon are S-shaped. If the number of cars/person in the USA and in the UK are plotted against time, one can find approximately the shapes depicted in Figure 13.1.

One curve which has proved popular in this field is the logistic, pioneered by Tanner (1978). The following three parameters are needed to adjust it:

C_0, the car-ownership rate in the base year (cars/person)

g_0, the rate of increase of the car-ownership rate in the base year given by $\dfrac{1}{C}\dfrac{dc}{dt}$

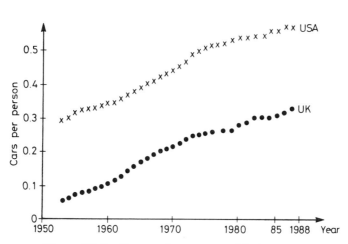

Figure 13.1   Shape of car-ownership increase

evaluated at $t = 0$

S, the saturation level of car ownership.

In logistic curves we have that:

$$\frac{dC}{dt} = aC_t(S - C_t) \tag{13.15}$$

where a is a constant. Solving this differential equation yields:

$$C_t = \frac{S}{1 + b \exp(-aSt)} \tag{13.16}$$

where b is an integration constant. To find the values of a and b we can resort to the boundary conditions at $t = 0$; from (13.15) and (13.16) we obtain respectively:

$$g_0 = a(S - C_0) \quad \text{and} \quad C_0 = \frac{S}{1 + b}$$

and replacing these values in (13.16) we finally get:

$$C_t = \frac{S}{1 + [(S - C_0)/C_0]\exp[-g_0St/(S - C_0)]} \tag{13.17}$$

Therefore, a knowledge of C_0 and g_0 for one year taken as a base permits us to extrapolate C_t for any future year if S is known; however, S is not known but must be estimated. Tanner's method consists of fitting the following regression line (see Figure 13.2):

$$g = \alpha + \beta C_t$$

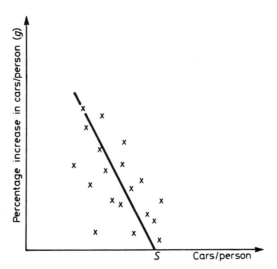

Figure 13.2   Determining the saturation level

Saturation corresponds by definition to that instant when the rate of change in the number of cars per capita (g) is zero; in this case we get $S = -\alpha/\beta$, and as we would expect α to be positive and β less than zero, we can deduce that $S > 0$.

Unfortunately constructing the graph of Figure 13.2 with data for the USA and the UK yields Figure 13.3; this implies that the method could work in the latter case but it is much more doubtful in the former. The method has been heavily criticised by Button *et al.* (1982).

With the above data, Tanner (1974) estimated S as 0.45 for Great Britain. Table 13.1 compares predictions for 1975 made at different years with the observed figure of 0.25 cars/head in that year. As can be seen, the method is not very reliable.

In summary, the main objections to the logistic extrapolation method are as follows:

1. The model is not sensitive to policy variables. It is impossible to study the effects on car ownership of changes in car prices, road tax and import duties, fuel costs,

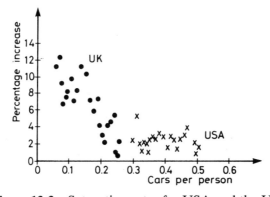

Figure 13.3   Saturation rates for USA and the UK

Table 13.1   Errors in prediction using extrapolation

Base year	Cars per person		Predicted growth
	In base year	Predicted for 1975	Actual growth
1960	0.11	0.28	1.14
1964	0.16	0.32	1.57
1966	0.18	0.31	1.67
1968	0.20	0.30	1.84
1969	0.21	0.28	1.66
1971	0.22	0.27	1.62
1972	0.23	0.26	1.48

and so on. Neither does it consider the influence of economic variables; therefore if the correlation among these variables changes in time, perverse results may be obtained (i.e. consider the effect in car-ownership increase brought about by the petrol crisis in 1973, or the aforementioned effect of the reduction of import duties in Chile in 1977).

2. S is assumed to be a constant; however, this may not be true in practice as attitudes tend to change with time.

3. The model does not yield information about different types of cars or, more importantly for planning purposes, the proportion of people belonging to households with 0, 1 and 2 or more cars.

13.3.3 Econometric Methods

These attempt to explain consumer behaviour directly rather than looking at general trends, and normally employ cross-sectional data. We will consider only two methods out of several which have been proposed; for a more comprehensive review see de Jong (1989).

13.3.3.1 The Method of Quarmby and Bates (1970)

This method uses just two independent variables, income and residential density, although it recognises the existence of several other factors of interest, such as household size and vehicle price. The basic relations of the model are:

$$\frac{P_0}{1 - P_0} = \alpha_0 I^{-b_0} D^{c_0} \tag{13.18}$$

$$\frac{P_2}{P_1} = a_1 \exp(b_1 I) D^{-c_1} \tag{13.19}$$

$$P_0 + P_1 + P_2 = 1 \tag{13.20}$$

where I is annual family income (thousands of $), D is the number of residents per acre and P_i is the probability of owning 0, 1 and 2 or more cars; a_i, b_i and c_i are parameters to be estimated.

Substituting P_1 from (13.20) into (13.19) and taking logarithms we get:

$$\log\{P_2/(1 - P_0 - P_2)\} = \log(a_1) + b_1 I - c_1 \log(D) \qquad (13.21)$$

Now, because D is a discrete variable for any given segment it may be considered a constant and (13.21) reduces to:

$$\log\{P_2/(1 - P_0 - P_2)\} = b_1 I + \text{constant}$$

It is instructive to consider that as income (I) increases, so does the left-hand side term of equation (13.21); therefore one can deduce that $(1 - P_0 - P_2)$ tends to zero or, what comes out to be the same, P_2 tends to $(1 - P_0)$. However, as P_0 is nearly zero for high incomes, that would mean that P_2 would tend to 1 and this is obviously incorrect as one would expect a lower limit for it. This upper bound, or saturation level (S) of P_2, may be incorporated to the model by adjusting (13.21), yielding:

$$\log\{P_2/[S(1 - P_0) - P_2]\} = \log(a_1) + b_1 I - c_1 \log(D) \qquad (13.22)$$

where S must be determined empirically; now, as this may be difficult in practice, the usual procedure involves trying different values in a sensitivity analysis. The type of curves obtained by this method are illustrated in Figure 13.4.

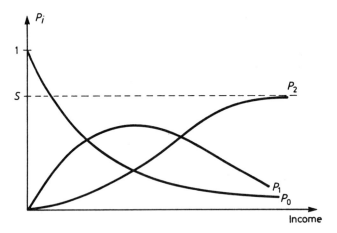

Figure 13.4   Car ownership versus income

Example 13.2: Consider the data in the table overleaf and assume a value of $S = 0.78$; the problem is to estimate the parameters of the Quarmby and Bates's model for a fixed residential density value.

If we take the logarithm of (13.18) for fixed D (i.e. c_0 is of no interest) we get:

$$\log\{P_0/(1 - P_0)\} = \log(a_0) - b_0 \log(I)$$

Income	P_0	P_1	P_2
1	0.61	0.34	0.05
2	0.35	0.47	0.18
3	0.22	0.44	0.34
4	0.16	0.37	0.47
5	0.10	0.30	0.60
6	0.08	0.24	0.68

and fitting a regression line to the data we obtain $a_0 = 1.74$ and $b_0 = 1.60$. On the other hand, if we replace the value of S in equation (13.22) for constant D, we get:

$$\log \{ P_2/[0.78(1 - P_0) - P_2]\} = \log (a_1) + b_1 I$$

and fitting another regression line to the data we finally obtain $a_1 = 0.10$ and $b_1 = 0.84$.

13.3.3.2 The Regional Highway Transport Model (RHTM) Method (Bates et al. 1978)

This combines the best features of the previous two approaches. First, it is necessary to define the following variables:

$P(1+)$ = percentage of households with one or more cars, with a saturation level of $S(1+)$

$P(2+)$ = percentage of households with two or more cars, with a saturation level of $S(2+)$

Therefore the previous method's values can be derived as:

$$P_0 = 1 - P(1+)$$

$$P_1 = P(1+) - P(2+)$$

$$P_2 = P(2+)$$

but it must be noted that the saturation levels are different from those of Tanner. The model takes the following form:

$$P_t(1+) = \frac{S(1+)}{1 + \exp \{-a_1(I_t/p_t)^{-b_1}\}} \tag{13.23}$$

$$P_t(2+) = \frac{S(2+)}{1 + \exp \{-a_2 - b_2(I_t/p_t)\}} \tag{13.24}$$

where (I_t/p_t) is annual family income ($/week) deflated by a car price index. The model was calibrated using British data for the period 1969–75, yielding the following parameter values (income in £):

$$a_1 = -7.76 \quad b_1 = 2.26 \quad S(1+) = 0.95$$
$$a_2 = -3.76 \quad b_2 = 0.04 \quad S(2+) = 0.60$$

To forecast it is necessary to assume a certain distribution of income (for example, one of the Gamma type); also, to convert the modelled results to cars/person (C_p) it is necessary to use census data. For example, Bates *et al.* (1978) postulated the following conversion rule:

$$C_p = P(1+) + 2.17P(2+)$$

To obtain cars/household we finally require information about the future average number of persons per household.

13.3.3.3 Models of Car Ownership and Use

Khan and Willumsen (1986) argued that in developing countries, growth in car ownership (and use) commits future resources to additional investment in roads and road maintenance. They insisted that car ownership should be considered as a policy variable rather than an exogenous factor; in order to support these ideas, they developed policy-sensitive models of car ownership and used and calibrated them using data from different countries and time periods. They studied a number of functional forms, one of the most useful models being:

$$\log C_{1000} = -361 + 70.5 \log \text{GNPH} - 0.373 \log \text{PURTAX} - 2.58 \log \text{OWNTAX}$$
$$-0.682 \log \text{IMPDUTY} - 29.4 \log \text{FUELPR} - 2.04 \log \text{POPDEN}$$
$$R^2 = 0.86$$

where C_{1000} is the number of cars per 1000 inhabitants, GNPH is the gross national product per capita, PURTAX is the purchase tax associated with cars, OWNTAX is the associated ownership tax (road licence), IMPDUTY is the import duty for cars, FUELPR is the price per litre of fuel and POPDEN is the population density.

A second model was developed to estimate annual kilometrage per car, KM/C. One such model was:

$$\log \text{KM/C} = 5.76 - 0.434 \log \text{GNHP} - 0.368 \log \text{FUELPR} - 0.67 \log \text{ROADPOP}$$

where ROADPOP is the paved road length per head of population.

Finally, Khan and Willumsen (1986) developed an 'analysis' model where the total number of cars, car-km, fuel consumption, tax revenues, and road maintenance and investment costs are calculated for one or more years in the future. Alternative policies regarding taxation, import duties and road construction can then be compared in terms of their implied costs to the country. The general structure of these models is shown in Figure 13.5.

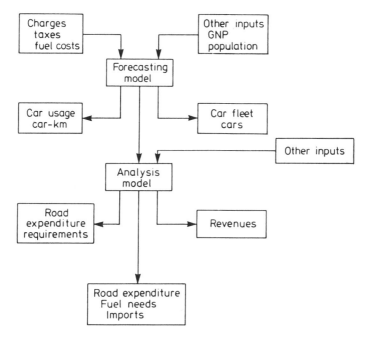

Figure 13.5   Khan and Willumsen's 'analysis' model

These models were developed essentially as a research and policy analysis tool; more work is needed in this direction to apply this approach to specific countries. Although the models were written in FORTRAN they are simple enough to permit their use in a good microcomputer spreadsheet program.

13.4 THE VALUE OF TRAVEL TIME

13.4.1 Introduction

The question 'has time a value?' is answered in the affirmative by most people. A more serious problem is 'what value?' and under what circumstances it can or must be measured.

This theme has generated an enormous debate in the literature (see for example Bruzelius 1979) simply because time savings are the single most important benefit of transport improvement projects all over the world. However, and in spite of its importance, a consensus has not been reached about the size and nature of the values to be used in project evaluation. We will not attempt to review the subject in great detail here, but refer the reader to Gunn (1985a) for a deeper discussion.

For example, in Great Britain (and other countries, such as Chile) values of time corresponding to a fixed proportion of the average hourly rate are recommended. In the USA, on the other hand, increasing values for three ranges of time savings:

0–5 min, 5–15 min and 15 or more minutes have been recommended (AASHTO 1977). Clearly the use of linear or non-linear valuation functions should lead to different benefits and hence to different investment priorities. For example, the British norm tends to favour schemes generating small time savings while the American norm favours schemes generating more substantive time savings.

Some studies distinguish between *subjective* (or behavioural) and *evaluation* values of time. The first corresponds to, for example, the value of the parameter associated with in-vehicle travel time in the generalised cost functions we studied in Chapter 5 and which should have been derived by estimating a demand model with empirical data. The evaluation value is that used, as the name implies, to compare alternative schemes which produce different levels of time and other resource savings. It is argued, therefore, that the behavioural value of time reflects mostly the ability of the traveller to pay and not the intrinsic value of a particular time saving. This is why very often the value of time used for evaluation purposes is an *equity* value, taken as being the same for all travellers, independently from their age or socioeconomic group, as we will see below.

On the other hand, it may be argued that the use of different 'values of time' for evaluation and demand modelling purposes introduces inconsistencies of approach at different stages of the same exercise. There is little dispute, however, that the subjective values of time are heavily dependent on model specification and data (see Gaudry *et al.* 1989); this is an undesirable property because consistent evaluation of projects is sought over a wide range of models and areas.

Heggie (1983) has argued that the debate has tended to be more empirical than theoretical. The enormous practical difficulties associated with measuring the value of time directly have encouraged the use of indirect methods, generally of the disaggregate demand modelling type. However, indirect methods generate problems, such as:

- how to choose an appropriate sample, i.e. one which basically contains people with a real choice among clearly defined alternatives in terms of time and cost of travel;
- how to measure the travel attributes, i.e. avoiding aggregation, perception and other sources of bias;
- which demand function to use that is consistent with the situation under study.

All the above suggests that values derived from models estimated with revealed-preference data (the large majority of cases) may be suspect.

Perhaps the most complete recent study about the value of travel time savings was performed between 1981 and 1986 by a consortium of consultants and academic experts in Britain, using a series of models estimated with revealed- and stated-preference data for various choice situations in several areas of Great Britain (Bates and Roberts 1986). Its principal recommendations (Department of Transport 1987) were:

1. The value of working time (i.e. trips made during or as part of work) is equal to the gross hourly income of the traveller, including all additional costs to the employer.

2. The trips for all other purposes, including trips to work, increased their valuation from 27 to 43% of the average hourly income of full-time employed adults (this is an increment of 58%).
3. For the majority of cases a single *equity* value of time should be used; however, in cases where the proportion of children, pensioners or employed adults is judged to differ significantly from the national average, an *ad hoc* equity value of time should be estimated using the individual values for each of these groups.
4. To update these values, information about real hourly incomes on each year should be used; to forecast, these incomes should be estimated as a function of the domestic per capita product.
5. The values of waiting and walking time should be taken as twice the value of in-vehicle travel time; bicycle users should be treated as pedestrians in this sense.
6. Small time savings should be valued equally as more significative savings.

13.4.2 Methods of Analysis

13.4.2.1 The Revealed-Preference Approach

In Chapter 8 we discussed alternative specifications for the utility functions of discrete choice models (Train and McFadden 1978; Jara-Díaz and Farah 1987; Gaudry and Wills 1978). A useful way to calculate the value of time (VOT) is to find the rate of substitution between perceived times and costs of travelling at constant utility (Gaudry *et al.* 1989). In general this approach yields the following expression:

$$\text{VOT} = -\frac{dC_i}{dt_i}\bigg|_v = \frac{\partial V_i/\partial t_i}{\partial V_i/\partial c_i} \tag{13.25}$$

For the linear-in-parameters wage rate (w) specification, the above simply yields:

$$\text{VOT} = \frac{w\theta_t}{\theta_c} \tag{13.26}$$

where θ_t is the coefficient of t_i (in-vehicle, waiting or walking time) and θ_c the coefficient of the term c_i/w. From (13.26) one can easily see that the ratio θ_t/θ_c represents VOT as a percentage of income.

For the linear-in-parameters expenditure rate (g) specification, where g is given by (8.7), equation (13.25) yields:

$$\text{VOT} = \frac{g\theta_t}{\theta_c} \tag{13.27}$$

Finally, for the Box–Cox case (8.3) we get:

$$\frac{\text{VOT}}{w} = \frac{\theta_t t_i^{(\tau_t-1)}}{\theta_c(C_i/w)^{(\tau_c-1)}} \tag{13.28}$$

which will clearly vary across alternatives if τ_k is not equal to 1. This latter formula implies that if both τ's are equal and are less than one, the model will necessarily yield higher value of time estimates for modes which are more expensive per minute; however, this may not be the case if the τ's differ (see Gaudry *et al.* 1989).

Because the estimated θ_i are random variables, the VOT values (13.26)–(13.28) are also randomly distributed. For this reason it is necessary to find a way of judging their significance. Jara-Díaz *et al.* (1988) show that by making a first-order expansion of a Taylor series for the random variable θ_t/θ_c around its mean value (the ratio of the estimated coefficients), the following, t-ratio may be constructed:

$$t_{tc} = \left(\sqrt{\frac{\sigma_t^2}{\theta_t^2} + \frac{\sigma_c^2}{\theta_c^2} - \frac{2\,\mathrm{Cov}\,(\theta_t, \theta_c)}{\theta_t\theta_c}} \right)^{-1} \tag{13.29}$$

It is important to mention that values of time estimated using different model structures (i.e. multinomial or nested logit) and/or different specification differ significantly for the same database (Gaudry *et al.* 1989). The problem is that with cross-sectional data it is not easy to clearly reject any reasonable model form (see the discussion in Jara-Díaz and Ortúzar 1989).

13.4.2.2 The Transfer Price Approach

In the context of travel demand analysis, *transfer price* is understood as the amount by which the cost of one option would have to be varied to equalise its overall attractiveness with that of another predefined option (see Bonsall 1983).

A typical application of the method involves asking individuals, for example, by how much should the fare of their currently preferred option increase to persuade them to switch to another alternative. It is clear that an important problem of the technique (in common with other forms of stated-preference analyses) has to do with the reliability that the analyst can associate with such a data set. On the other hand, a strong advantage of the method, if it works, is that it makes it possible to know not only the direction of individual preferences but also the difference (in preference terms) among the various available options. Thus in theory, and in common with other SP studies, less data than for an RP study are required to obtain a model of similar accuracy. We will not attempt to discuss the method in detail here, but interested readers are referred to Gunn (1984) for a good discussion of its advantages and problems, in particular its general inconsistency with conventional random utility theory.

Example 13.3: Consider a random utility model such as (7.2) in a binary-choice situation and assume that the transfer price (TP) corresponds to the difference between the utility of the chosen alternative (U_c) and the other (U_r), i.e. it represents the increment in the cost of the chosen one that would made the traveller indifferent to both options. Thus we have:

$$TP = U_c - U_r$$

However, the expected value of $(U_c - U_r)$ is precisely the difference in representative utilities $(V_c - V_r)$; so assuming these to be linear in the parameters, as usual, we can form the following linear regression system:

$$\text{TP(observed)} = \theta_1(X_{1c} - X_{1r}) + \theta_2(X_{2c} - X_{2r}) + \ldots$$

which should allow us to estimate the unknown parameters θ knowing the attributes **X** for both options. Furthermore, different values of time for *time savers* and *cost savers* may be calculated with this method (see Lee and Dalvi 1969).

One important problem, first noted by Hensher (1976), concerns the treatment of habit in transfer price models. Gunn (1984) shows that specifications which use TP as a dependent variable but restrict its sign (i.e. by modelling the options separately or by switching the observable characteristics to reflect the difference between chosen and rejected option) cannot easily be made consistent with conventional random utility theory (see also the discussion in Chapter 8).

EXERCISES

13.1 Consider the following simple econometric model to determine car ownership as a function of income:

$$P_0/(1 - P_0) = \alpha I^\beta$$
$$P_2/[0.8(1 - P_0) - P_2] = 0.09 \exp(0.751)$$
$$P_0 + P_1 + P_2 = 1$$

(a) Calibrate the model using the data in the table below (*Hint*: do it graphically)
(b) Indicate what proportions with 0, 1 and 2 or more cars would the model predict for an annual income of 6 monetary units.

I	P_0	P_1	P_2
1	0.60	0.35	0.05
2	0.40	0.50	0.10
3	0.25	0.55	0.20
4	0.20	0.45	0.35
5	0.15	0.35	0.50

13.2 The following table presents the results of a transfer price survey made on a sample of 8 individuals in Exercise 9.3; TP indicates the reported increment in the monetary cost (expressed in time units after deflating by income) of the currently chosen mode that would leave each individual indifferent to both alternatives. The study assumed that only time (t) and cost/income (c) were relevant variables.

Individual	Chosen option	TP	t_1 (min)	t_2 (min)	c_1 (min)	c_2 (min)
1	1	8.0	47.5	83.2	14.8	7.0
2	1	6.5	30.2	45.0	10.4	5.0
3	1	2.5	22.0	30.4	12.6	4.0
4	2	0.5	45.0	50.6	8.2	5.0
5	2	1.5	15.3	20.5	50.0	17.0
6	1	8.5	34.8	50.2	55.0	35.0
7	2	130.0	65.5	100.5	200.3	53.5
8	2	6.0	12.0	14.0	44.6	17.0

(a) Use the data to estimate the individuals' subjective value of time. Discuss the role, size and sign of the intercept of the transfer price linear regression equation (*Hint*: if you do not have available a calculator with a linear regression facility, do it graphically assuming the coefficient of time, θ_t is known and equal to -0.03).

(b) If the revealed preferences parameter for the time variable is indeed -0.03 and the mode specific constant of option 1 is 1.35, estimate the subjective value of time using another method. Compare your results and discuss.

References

AASHTO (1977) *A Manual of User Benefit Analysis of Highway and Bus Transport Improvements*. American Association of State Highway and Transportation Officials, Washington, DC.

Abraham, H., Shaw, N. and Willumsen, L. (1992) A micro-based incremental four stage transportation model for London. *Proceedings 20th PTRC Summer Annual Meeting*, University of Manchester Institute of Science and Technology, September 1992, England.

Allaman, P.M., Tardiff, T.J. and Dunbar, F.C. (1982) New approaches to understanding travel behaviour. *NCHRP Report 250*, National Cooperative Highway Research Program, Transportation Research Board, Washington, DC.

Alonso, W. (1964) *Location and Land Use.* Harvard University Press, Cambridge.

Alonso, W. (1968) Predicting best with imperfect data. *Journal of the American Institute of Planners*, **34**(3), 248–255.

Allsop, R.E. and Charlesworth (1977) Traffic in a signal-controlled road network: an example of different signal timings inducing different routeing, *Traffic Engineering and Control* **18**(5), 262–264.

Ampt, E. (1981) Recent advances in large-scale travel surveys. *Proceedings 9th PTRC Summer Annual Meeting*, University of Warwick, July 1981, England.

Anas, A. (1982) *Residential Location Markets and Urban Transportation*. Academic Press, London.

Arezki, Y. (1986) Comparison of some algorithms for equilibrium traffic assignment with fixed demand. *Proceedings 14th PTRC Summer Annual Meeting*, University of Sussex, July 1986, England.

Arezki, Y., Chadwick, N. and Willumsen, L. (1991) Congestion, evaluation and equilibrium: some empirical results. *Proceedings 19th PTRC Summer Annual Meeting,* University of Sussex, September 1991, England.

Ashley, D.J. (1978) The Regional Highway Traffic Model: the home based trip end model. *Proceedings 6th PTRC Summer Annual Meeting*, University of Warwick, July 1978, England.

Ashley, D.J., Lowe, S., Mundy, R., Stanley, R. and Baanders, A. (1985) The long distance travel model for The Netherlands: its specification and application. *Proceedings 13th PTRC Summer Annual Meeting*, University of Sussex, July 1985, England.

Atherton, T.J. and Ben Akiva, M.E. (1976) Transferability and updating of disaggregate travel demand models. *Transportation Research Record 610*, 12–18.

Babin, A., Florian, M., James-Lefebre, L. and Spiess, H. (1982) EMME/2: interactive graphic method for road and transit planning. *Transportation Research Record 866*, 1–9.

Bacharach, M. (1970) *Biproportional Matrices and Input Output Change*. Cambridge University Press, New York.

Bates, J.J. (1988a) Econometric issues in stated preference analysis. *Journal of Transport Economics and Policy*, **XXII**(1), 59–69.

Bates, J.J. (guest ed.) (1988b) Stated preference methods in transport research. *Journal of Transport Economics and Policy*, **XXII**(1), 1–137.

Bates, J.J., Ashley, D.J. and Hyman, G. (1987) The nested incremental logit model: theory and application to modal choice. *Proceedings 15th PTRC Summer Annual Meeting*, University of Bath, September 1987, England.

Bates, J.J., Gunn, H.F. and Roberts, M. (1978) A model of household car ownership. *Traffic Engineering and Control*, **19**(11/12), 486–491, 562–566.

Bates, J.J. and Roberts, M. (1983) Recent experience with models fitted to stated preference data. *Proceedings 11th PTRC Summer Annual Meeting*, University of Sussex, July 1983, England.

Bates, J.J. and Roberts, M. (1986) Value of time research: summary of methodology and findings. *Proceedings 14th PTRC Summer Annual Meeting*, University of Sussex, July 1986, England.

Beckman, M.J., McGuire, C.B. and Winsten, C.B. (1956) *Studies in the Economics of Transportation*. Yale University Press, New Haven.

Bell, M.G.H. (1983) The estimation of an origin destination matrix from traffic counts. *Transportation Science*, **17**(2), 198–217.

Bell, M.G.H. (1984) Log-linear models for the estimation of origin-destination matrices from traffic counts: an approximation. In J. Volmüller and R. Hamerslag (eds), *Proceedings of the Ninth International Symposium on Transportation and Traffic Theory*. VNU Science Press, Utrecht.

Ben Akiva, M.E. (1977) Choice models with simple choice set generating processes. *Working Paper*, Centre for Transportation Studies. M.I.T.

Ben Akiva, M.E. and Bolduc, D. (1987) Approaches to model transferability and updating: the combined transfer estimator. *Transportation Research Record 1139*, 1–7.

Ben Akiva, M.E. and Lerman, S.R. (1985) *Discrete Choice Analysis: Theory and Application to Travel Demand*. The MIT Press, Cambridge, Mass.

Ben-Akiva, M. and Morikawa, T. (1990) Estimation of travel demand models from multiple data sources. *Proceedings 11th International Symposium on Transportation and Traffic Theory*, Yokohama, July 1990, Japan.

Ben Akiva, M., Morikawa, T. and Shiroishi, F. (1992) Analysis of the reliability of preference ranking data. *Journal of Business Research* **24**(2), 149–164.

Ben Akiva, M.E. and Watanatada, T. (1980) Application of a continuous spatial choice logit model. In C.F. Manski and D. McFadden (eds), *Structural Analysis of Discrete Data: With Econometric Applications*. The MIT Press, Cambridge, Mass.

Blase, J.H. (1979) Hysteresis and catastrophe theory: a demonstration of habit and threshold effects in travel behaviour. *Proceedings 7th PTRC Summer Annual Meeting*, University of Warwick, July 1979, England.

Bonsall, P.W. (1983) Transfer price data: its use and abuse. *Proceedings 11th PTRC Summer Annual Meeting*, University of Sussex, July 1983, England.

Boyce, D.E., Day, N.D. and McDonald, C. (1970) *Metropolitan Plan Making*. Monograph Series No. 4, Regional Science Research Institute, Philadelphia.

Bradley, M. (1988) Realism and adaptation in designing hypothetical travel choice concepts. *Journal of Transport Economics and Policy*, **XXII**(1), 121–37.

Bradley, M.A. and Daly, A.J. (1991) Estimation of logit choice models using mixed stated preference and revealed preference information. *Preprints 6th International Conference on Travel Behaviour*, Quebec, May 1991, Canada.

Bradley, M.A. and Daly, A.J. (1992) Uses of the logit scaling approach in stated preference analysis. *Proceedings 6th World Conference on Transport Research*, Lyon, July 1992, France.

Bradley, M.A. and Kroes, E. (1990) Forecasting issues in stated preference survey research. *69th TRB Annual Meeting*, Washington DC, January 1990, USA.

Branston, D. (1976) Link capacity functions: a review. *Transportation Research*, **10**(4), 223–36.

Brenninger-Gothe, M., Jornsten K. and Lundgren, J. (1989) Estimation of origin-destination matrices from traffic counts using multiobjective programming formulations. *Transportation Research* **23**B(4), 257–269.

Brög, W. and Ampt, E. (1982) State of the art in the collection of travel behaviour data. In *Travel Behaviour for the 1980's*, Special Report 201, National Research Council, Washington, DC.

Brög, W. and Erl, E. (1982) Application of correction and weighting factors to obtain a representative data base. *Proceedings 10th PTRC Summer Annual Meeting*, University of Warwick, July 1982, England.

Brög, W., Erl, E., Meyburg, A.H. and Wermuth, M.J. (1982) Problems of non-reported trips in surveys of non-home activity patterns. *Transportation Research Record 891*, 1–5.

Brög, W. and Meyburg, A.H. (1980) The non-response problem in travel surveys: an empirical investigation. *Transportation Research Record 775*, 34–8.

Bruton, M.J. (1985) *Introduction to Transportation Planning*. Hutchinson, London.

Bruzelius, N. (1979) *The Value of Travel Time*. Croom Helm, London.

Bureau of Public Roads (1964) *Traffic Assignment Manual*. Urban Planning Division, US Department of Commerce, Washington DC.

Burrell, J.E. (1968) Multiple route assignment and its application to capacity restraint. In W. Leutzbach and P. Baron (eds), *Beiträge zur Theorie des Verkehrsflusses*. Strassenbau und Strassenverkehrstechnik Heft, Karlsruhe.

Button, K.J., Pearman, A.D. and Fowkes, A.S. (1982) *Car Ownership Modelling and Forecasting*. Gower, Aldershot.

Caldwell, L.C. and Demetski, M.J. (1980) Transferability of trip generation models. *Transportation Research Record 751*, 56–62.

Casey, H.J. (1955) Applications to traffic engineering of the law of retail gravitation. *Traffic Quarterly*, **IX**(1), 23–35.

Cascetta, E. (1984) Estimation of trip matrices from traffic counts and survey data: a generalised least squares approach estimator. *Transportation Research*, **18B**(4/5), 289–99.

Cascetta, E. and Nguyen, S (1988) A unified framework for estimating or updating origin/destination matrices from traffic counts. *Transportation Research*, **22B**(6), 437–455.

Chadwick, G. (1987) *Models of Urban and Regional Systems in Developing Countries*. Pergamon Press, Oxford.

Chapman, R.G. and Staelin, R. (1982) Exploiting rank ordered choice set data within the stochastic utility model. *Journal of Marketing Research*, **XIX**(3), 288–301.

Clancy, M., Dawson, J., Catling, I., Turner, J. and Harrison, W. (1985) Electronic road pricing in Hong Kong. *Proceedings 13th PTRC Summer Annual Meeting*, University of Sussex, July 1985, England.

Copley, G. and Lowe, S.R. (1981) The temporal stability of trip rates: some findings and implications. *Proceedings 9th PTRC Summer Annual Meeting*, University of Warwick, July 1981, England.

Coslett, S.R. (1981) Efficient estimation of discrete choice models. In C.F. Manski and D. McFadden (eds), *Structural Analysis of Discrete Data: With Econometric Applications*. The MIT Press, Cambridge, Mass.

Crow, R.T., Young, K.H. and Cooley, T. (1973) Alternative demand functions for abstract transportation modes. *Transportation Research*, 7(4), 335–54.

Daganzo, C.F. (1979) *Multinomial Probit: The Theory and its Applications to Demand Forecasting*. Academic Press, New York.

Daganzo, C.F. (1980) Optimal sampling strategies for statistical models with discrete dependent variables. *Transportation Science*, **14**(4), 324–45.

Daganzo, C.F. and Sheffi, Y. (1979) Estimation of choice models from panel data. *26th Annual Meeting of the Regional Science Association*, Los Angeles, November 1979, USA.

Daly, A.J. (1982a) Estimating choice models containing attraction variables. *Transportation Research*, **16B**(1), 5–15.

Daly, A.J. (1982b) Applicability of disaggregate models of behaviour: a question of methodology. *Transportation Research*, **16A**(5/6), 363–70.

Daly, A.J. (1987) Estimating 'tree' logit models. *Transportation Research*, **21B**(4), 251–68.

Daly, A.J. (1992) *ALOGIT 3.2 User's Guide*. Hague Consulting Group, The Hague.

Daly, A.J. and Gunn, H.F. (1986) Cost effective methods for national level demand forecasting. In A. Ruhl (ed.), *Behavioural Research for Transport Policy*. VNU Science Press, Utrecht.

Daly, A.J. and Ortúzar, J. de D. (1990) Forecasting and data aggregation: theory and practice. *Traffic Engineering and Control*, **31**(12), 632–643.

Daly, A.J., van der Valk, J. and van Zwam, H.P.H. (1983) Application of disaggregate models for a regional transportation study in the Netherlands. In P. Baron and H. Nuppnau (eds), *Research for Transport Policies in a Changing World*. SNV Studiengesellschaft Nahverkehr, Hamburg.

Daly, A.J. and Zachary, S. (1978) Improved multiple choice models. In D.A. Hensher and M.Q. Dalvi (eds), *Determinants of Travel Choice*. Saxon House, Westmead.

Daor, E. (1981) The transferability of independent variables in trip generation models. *Proceedings 9th PTRC Summer Annual Meeting*, University of Warwick, July 1981, England.

De Cea, J. and Cruz, G. (1986) ESMATUC: un modelo de estimación de matrices de viajes en transporte urbano colectivo. *Apuntes de Ingeniería*, **24**, 109–25.

De Cea, J. and Fernández, J.E. (1989) Transit assignment to minimal routes: an efficient new algorithm. *Traffic Engineering and Control*, **30**(10), 491–494.

De Cea, J., Ortúzar, J. de D. and Willumsen, L.G. (1986) Evaluating marginal improvements to a transport network: an application to the Santiago underground. *Transportation*, **13**(3), 211–33.

De Jong, G.C. (1989) *Some Joint Models of Car Ownership and Use*. Ph.D. Thesis, Faculteit der Economishe, Universiteit van Amsterdam.

Dehghani, Y. and Talvitie, A.P. (1980) Model specification, model aggregation and market segmentation in mode choice models: some empirical evidence. *Transportation Research Record 775*, 28–34.

Dehghani, Y. and Talvitie, A.P. (1983) Forecasting accuracy, transferability and updating of modal constants in disaggregate mode choice models with simple and complex specifications. *Proceedings 11th PTRC Summer Annual Meeting*, University of Sussex, July 1983, England.

Department of Transport (1985) *Traffic Appraisal Manual (TAM)*. HMSO, London.

Department of Transport (1987) *Values for Journey Time Savings and Accident Prevention*. HMSO, London.

Dial, R.B. (1971) A probabilistic multipath traffic assignment model which obviates path enumeration. *Transportation Research*, **5**(2), 83–11.

DICTUC (1978) *Encuesta origen y destino de viajes para el Gran Santiago*. Informe Final al Ministerio de Obras Públicas, Departamento de Ingeniería de Transporte, Universidad Católica de Chile, Santiago.

Dijkstra, E.W. (1959) Note on two problems in connection with graphs (spanning tree, shortest path). *Numerical Mathematics*, **1**(3), 269–71.

Domencich, T.A., Kraft, G. and Valette, J.P. (1968) Estimation of urban passenger travel behaviour: an economic demand model. *Highway Research Record 238*, 64–78.

Domencich, T. and McFadden, D. (1975) *Urban Travel Demand: A Behavioural Analysis*. North-Holland, Amsterdam.

Douglas, A.A. and Lewis, R.J. (1970) Trip generation techniques: (1) Introduction; (2) Zonal least squares regression analysis. *Traffic Engineering and Control*, **12**(7 and 8), 362–5, 428–31.

Douglas, A.A. and Lewis, R.J. (1971) Trip generation techniques: (3) Household least squares regression analysis; (4) Category analysis and sumary of trip generation techniques. *Traffic Engineering and Control*, **12**(9 and 10), 477–9, 532–5.

Downes, J.D. and Emmerson, P. (1983) Urban transport modelling with fixed travel budgets; an evaluation of the UMOT process. *TRRL Supplementary Report SR 799*, Transport and Road Research Laboratory, Crowthorne.

Downes, J.D. and Gyenes, L. (1976) Temporal stability and forecasting ability of trip generation models in Reading. *TRRL Report LR 726*, Transport and Road Research Laboratory, Crowthorne.

Duncan, G.J., Juster, F.T. and Morgan, J.N. (1987) The role of panel studies in research on economic behaviour. *Transportation Research*, **21**A(4/5), 249–63.

Dunphy, R.T. (1979) Workplace interviews as an efficient source of travel survey data. *Transportation Research Record, 701*, 26–9.

Eilon, S. (1972) Goals and constraints in decision making. *Operations Research Quarterly*, **23**(1), 3–15.

England, J., Hudson, K., Masters, R., Powell, K. and Shortridge, J. (eds) (1985) *Information Systems for Policy Planning in Local Government*. Longman, Harlon.

ESTRAUS (1989) *Estudio estratégico de transporte del Gran Santiago*. Informe Final a la Secretaría Ejecutiva de la Comisión de Transporte Urbano, Consorcio SIGDO-KOPPERS/CIS, Santiago.

Evans, S.P. (1973) A relationship between the gravity model for trip distribution and the transportation problem in linear programming. *Transportation Research*, **7**(1) 39–61.

Evans, S.P. (1976) Derivation and analysis of some models for combining trip distribution and assignment. *Transportation Research*, **10**(1), 37–57.

Evans, S.P. and Kirby, H.R. (1974). A three dimensional Furness procedure for calibrating gravity models. *Transportation Research*, **8**(2), 105–22.

Fernández, J.E., Coeymans, J.E. and Ortúzar, J. de D. (1983) Evaluating extensions to the Santiago underground system. *Proceedings 11th PTRC Summer Annual Meeting*, University of Sussex, July 1983, England.

Fernández, J.E. and Friesz, T.L. (1983) Equilibrium predictions in transportation markets: the state of the art. *Transportation Research*, **17**B(2), 155–72.

FHWA (1967) *Guidelines for Trip Generation Analysis*. Federal Highway Administration, US Department of Transportation, Washington, DC.

Fisk, C.S. (1988) On combining maximum entropy trip matrix estimation with user optimal assignment. *Transportation Research*, **22**B(1), 69–73.

Florian, M. and Nguyen, S. (1978) A combined trip distribution modal split and trip assignment model. *Transportation Research*, **12**(4), 241–6.

Florian, M., Gaudry, M. and Lardinois, C. (1988) A two-dimensional framework for the understanding of transportation planning models. *Transportation Research*, **22**B(6), 411–19.

Florian, M. and Spiess, H. (1982) The convergence of diagonalization algorithms for asymmetric network equilibrium problems. *Transportation Research*, **16**B(6), 477–84.

Foerster, J.F. (1979) Mode choice decision process models: a comparison of compensatory and non-compensatory structures. *Transportation Research*, **13**A(1), 17–28.

Foerster, J.F. (1981) Nonlinear and non-compensatory perceptual functions of evaluations and choice. In P.R. Stopher, A.H. Meyburg and W. Brög (eds), *New Horizons in Travel Behaviour Research*. D.C. Heath and Co., Lexington, Mass.

Foot, D. (1981) *Operational Urban Models*. Methuen, London.

Fowkes, A.S. and Tweddle, G. (1988) A computer guided stated preference experiment for freight mode choice. *Proceedings 16th PTRC Summer Annual Meeting*, University of Bath, September 1988, England.

Fowkes, A.S. and Wardman, M. (1988) The design of stated preference travel choice experiments, with special reference to interpersonal taste variations. *Journal of Transport Economics and Policy*, **XXII**(1), 27–44.

Friesz, T.L., Tobin, R. and Harker, P. (1983) Predictive intercity freight network models: the state of the art. *Transportation Research*, **17**A(6), 409–17.

Furness, K.P. (1965) Time function iteration. *Traffic Engineering and Control*, **7**(7), 458–60.

Galbraith, R.A. and Hensher, D.A. (1982) Intra-metropolitan transferability of mode choice models. *Journal of Transport Economics and Policy*, **XVI**(1), 7–29.

Gárate, C. (1988) Utilización y actualización de la información de la Encuesta Origen-Destino del Gran Valparaíso. *Actas del V Congreso Panamericano de Ingeniería de Tránsito y*

Transporte, Universidad de Puerto Rico en Mayagüez, July 1988, Puerto Rico.

Garrido, R.A. (1992) The influence of the semantic scale on the estimation of values of time from linear regression models fitted to stated preference data. *Working Paper 62*, Department of Transport Engineering, Pontificia Universidad Católica de Chile, Santiago.

Gaudry, M.J.I., Jara-Díaz, S.R. and Ortúzar, J. de D. (1989) Value of time sensitivity to model specification. *Transportation Research*, **23**B(2), 151–8.

Gaudry, M.J.I. and Wills, M.I. (1978) Estimating the functional form of travel demand models. *Transportation Research*, **12**(4), 257–89.

Gibson, J., Baeza, I. and Willumsen, L.G. (1989) Congestion, bus stops and congested bus stops. *Traffic Engineering and Control*, **30**(6), 291–6.

Golob, T.F. and Richardson, A.J. (1981) Non-compensatory and discontinuous constructs in travel behaviour models. In P.R. Stopher, A.H. Meyburg and W. Brög (eds), *New Horizons in Travel Behaviour Research*. D.C. Heath and Co., Lexington, Mass.

Goodwin, P. (1977) Habit and hysteresis in mode choice. *Urban Studies*, **14**(1), 95–8.

Gray, R. (1982) Behavioural approaches in freight transport modal choice. *Transport Reviews*, **2**(2), 161–84.

Greenblat, C. and Duke, R. (1975) *Gaming-Simulation: Rationale, Design and Applications*. John Wiley & Sons, New York.

Gunn, H.F. (1984) An analysis of transfer price data. *Proceedings 12th PTRC Summer Annual Meeting*, University of Sussex, July 1984, England.

Gunn, H.F. (1985a) Value of time for evaluation purposes: the state of the art. *Report No. 421–01*, Hague Consulting Group, The Hague.

Gunn, H.F. (1985b) Artificial sample applications for spatial interaction models. *Colloquium Vervoersplanologisch Speurwerk*, The Hague, November 1985, Holland.

Gunn, H.F. and Bates, J.J. (1982) Statistical aspects of travel demand modelling. *Transportation Research*, **16**A(5/6), 371–82.

Gunn, H.F., Ben Akiva, M.E. and Bradley, M.A. (1985) Tests of the scaling approach to transferring disaggregate travel demand models. *Transportation Research Record 1037*, 21–30.

Gunn, H.F., Fisher, P., Daly, A.J. and Pol, H. (1982) Synthetic samples as a basis for enumerating disaggregate models. *Proceedings 10th PTRC Summer Annual Meeting*, University of Warwick, July 1982, England.

Gunn, H.F., Kirby, H.R., Murchland, J.D. and Whittaker, J.C. (1980) The RHTM trip distribution investigation. *Proceedings 8th PTRC Summer Annual Meeting*, University of Warwick, July 1980, England.

Gunn, H.F. and Pol, H. (1986) Model transferability: the potential for increasing cost-effectiveness. In A. Ruhl (ed.), *Behavioural Research for Transport Policy*. VNU Science Press, Utrecht.

Gur, Y.J. (1982) Recalibration of disaggregate mode choice models based on on-board survey data. *Proceedings 10th PTRC Summer Annual Meeting*, University of Warwick, July 1982, England.

Hall, M.D., Daly, A.J., Davies, R.F. and Russell, C.H. (1987) Modelling for an expanding city. *Proceedings 15th PTRC Summer Annual Meeting*, University of Bath, September 1987, England.

Hall, M.D., Van Vliet, D. and Willumsen, L.G. (1980) SATURN-a simulation assignment model for the evaluation of traffic management schemes. *Traffic Engineering and Control*, **21**(4), 168–76.

Harker, P.T. (1985) The state of the art in the predictive analysis of freight transport systems. *Transport Reviews*, **5**(2), 143–64.

Harker, P.T. (1987) *Predicting Intercity Freight Flows*. VNU Science Press, Utrecht.

Hartley, T.M. and Ortúzar, J. de D. (1980) Aggregate modal split models: is current U.K. practice warranted? *Traffic Engineering and Control*, **21**(1), 7–13.

Hauser, J.R. (1978) Testing the accuracy, usefulness and significance of probabilistic choice models: an information theoretic approach. *Operations Research*, **26**(4), 406–21.

Heckman, J.J. (1981) Statistical models for discrete panel data. In C.F. Manski and D. McFadden (eds), *Structural Analysis of Discrete Data: With Econometric Applications*. The MIT Press, Cambridge, Mass.

Heggie, I.G. (1983) Valueing savings in non working time: the empirical dilemma. *Transportation Research*, **17A**(1), 13–23.

Hensher, D.A. (1976) Valuations of commuter travel time savings: an alternative procedure. In I.G. Heggie (ed.), *Modal Choice and the Value of Time*. Clarendon Press, Oxford.

Hensher, D.A. (1986) Sequential and full information maximum likelihood estimation of a nested logit model. *The Review of Economics and Statistics*, **68**(4), 657–67.

Hensher, D.A. (1987) Issues in the pre-analysis of panel data. *Transportation Research*, **21A**(4/5), 265–85.

Hensher, D.A. (1991) Hierarchical stated response designs and estimation in the context of bus use preferences. *Logistics and Transportation Reviews*, **26**(4), 299–323.

Hensher, D.A. (1993a) Stated preference analysis of travel choices: the state of practice. *Transportation* (in press).

Hensher, D.A. (guest ed.) (1993b) Practice of stated preference methods. *Transportation* (in press).

Hensher, D.A. and Johnson, L.W. (1981) *Applied Discrete Choice Modelling*. Croom Helm, London.

Högberg, P. (1976) Estimation of parameters in models for traffic prediction: a non-linear-regression approach. *Transportation Research*, **10**(4), 263–5.

Holden, D., Fowkes, A.S. and Wardman, M. (1992) Automatic stated preference design algorithms. *Proceedings 20th PTRC Summer Annual Meeting*. University of Manchester Institute of Science and Technology, September 1992, England.

Holm, J., Jensen, T., Nielsen, S., Christensen, A. Johnsen, B. and Ronby, G. (1976) Calibrating traffic models on traffic census results only. *Traffic Engineering and Control*, **17**(4), 137–40.

Horowitz, J.L. (1981a) Sources of error and uncertainty in behavioural travel demand models. In P.R. Stopher, A.H. Meyburg and W. Brög (eds), *New Horizons in Travel Behaviour Research*. D.C. Heath and Co., Lexington, Mass.

Horowitz, J.L. (1981b) Sampling error, specification and data errors in probabilistic discrete choice models. Appendix C of D.A. Hensher and L.W. Johnson, *Applied Discrete Choice Modelling*. Croom Helm, London.

Horowitz, J.L. (1982) Specification tests for probabilistic choice models. *Transportation Research*, **16**A(5/6), 383–94.

Huber, J. and Hanson, D. (1986) Testing the impact of dimensional complexity and affective differences of paired concepts in adaptive conjoint analysis. *Working Paper*, School of Business, Duke University.

Hyman, G.M. (1969) The calibration of trip distribution models. *Environment and Planning* **1**(3), 105–112.

Jansen, G.R.M. and Bovy, P.H.L. (1982) The effect of zone size and network detail on all-or-nothing and equilibrium assignment outcomes. *Traffic Engineering and Control*, **23**(6), 311–17.

Jara-Díaz, S.R. and Farah, M. (1987) Transport demand and user's benefits with fixed income: the goods/leisure trade-off revisited. *Transportation Research*, **21**B(2), 165–70.

Jara-Díaz, S.R. and Ortúzar, J. de D. (1989) Introducing the expenditure rate in the estimation of mode choice models. *Journal of Transport Economics and Policy*, **XXIII**(3), 293–308.

Jara-Díaz, S.R. Ortúzar, J. de D. and Parra, R. (1988) Valor subjetivo del tiempo considerando efecto ingreso en la partición modal. *Actas del V Congreso Panamericano de Ingeniería de Tránsito y Transporte*, Universidad de Puerto Rico en Mayagüez, July 1988, Puerto Rico.

Johnson, L.W. (1990) Discrete choice analysis with ordered alternatives. In M.M. Fisher, P. Nijkamp and Y.Y. Papageorgiou (eds.), *Spatial Choices and Processes*. North Holland, Amsterdam.

Johnson, L.W. and Hensher, D.A. (1982) Application of multinomial probit to a two-period panel data set. *Transportation Research*, **16**A(5/6), 457–64.

Kannel, E.J. and Heathington, K.W. (1973) Temporal stability of trip generation relations. *Highway Research Record 472*, 17–27.

Khan, A. and Willumsen, L.G. (1986) Modelling car ownership and use in developing countries. *Traffic Engineering and Control*, **27**(11), 554–60.

Kim, T.J. and Hinkle, J. (1982) Model for statewide freight transportation planning. *Transportation Research Record 889*, 15–19.

Kimber, R.M. and Hollis, E.M. (1979) Traffic queues and delays at road junctions. *TRRL Report LR 909*, Transport and Road Research Laboratory, Crowthorne.

Kirby, H.R. (1979) Partial matrix techniques. *Traffic Engineering and Control*, **20**(8/9), 422–8.

Kitamura, R. and Bovy, P.H.L. (1987) Analysis of attrition biases and trip reporting errors for panel data. *Transportation Research*, **21**A(4/5), 287–302.

Kocur, G.T., Adler, T., Hyman, W. and Aunet, B. (1982) Guide to forecasting travel demand with direct utility assessment. *Report No. UMTA-NH-11-0001-82*, Urban Mass Transportation Administration, US Department of Transportation, Washington, DC.

Koppelman, F.S. (1976) Guidelines for aggregate travel prediction using disaggregate choice models. *Transportation Research Record 610*, 19–24.

Koppelman, F.S., Kuah, G-K and Rose G. (1985a) Transfer model updating with aggregate data. *64th Annual TRB Meeting*, Washington, DC, January 1985, USA.

Koppelman, F.S., Kuah, G-K and Wilmot, C.G. (1985b) Transfer model updating using disaggregate data. *Transportation Research Record 1037*, 102–7.

Koppelman, F.S. and Wilmot, C.G. (1982) Transferability analysis of disaggregate choice models. *Transportation Research Record 895*, 18–24.

Kraft, G. (1968) Demand for intercity passenger travel in the Washington–Boston corridor. North-East Corridor Project Report, *Systems Analysis and Research Corporation*, Boston, Mass.

Kresge, D.T. and Roberts, P.O. (1971) *Techniques of Transport Planning: Systems Analysis and Simulation Models*. Brookings Institution, Washington, DC.

Kruithof, J. (1937) Calculation of telephone traffic. *Der Ingenieur*, **52**(8), E15–E25.

Kruskal, J.B. (1965) Analysis of factorial experiments by estimating monotone transformations of the data. *Journal of the Royal Statistical Society*, **27**B(**4**), 251–263.

Kumar, A. (1980) Use of incremental form of logit models in demand analysis. *Transportation Research Record 775*, 21–7.

Lamb, G.M. and Havers, G.E. (1970) Introduction to transportation planning: treatment of networks. *Traffic Engineering and Control*, **11**(10), 486–9.

Lamond, B. and Stewart, N.F. (1981) Bregman's balancing method. *Transportation Research*, **15**B(4), 239–48.

Lancaster, K.J. (1966) A new approach to consumer theory. *Journal of Political Economy*, **14**(2), 132–57.

Langdon, M.G. (1976) Modal split models for more than two modes. *Proceedings 4th PTRC Summer Annual Meeting*, University of Warwick, July 1976, England.

Langdon, M.G. (1984) Methods of determining choice probability in utility maximising multiple alternative models. *Transportation Research*, **18**B(3), 209–34.

Larson, R.C. and Odoni, A.R. (1981) *Urban Operations Research*. Prentice-Hall, Englewood Cliffs, N.J.

Leamer, E. (1978) *Specification Searches: Ad-Hoc Inference with Nonexperimental Data*. John Wiley & Sons, New York.

Lee, N. and Dalvi, M.Q. (1969) Variations on the value of travel time. *Manchester School*, **37**(3), 213–36.

Leonard, D.R. and Gower, P. (1982) User guide to CONTRAM Version 4. *TRRL Supplementary Report 735*, Transport and Road Research Laboratory, Crowthorne.

Lerman, S.R. (1984) Recent advances in disaggregate demand modelling. In M. Florian (ed.), *Transportation Planning Models*. North-Holland, Amsterdam.

Lerman, S.R. and Louviere, J.J. (1978) The use of functional measurement to identify the form of utility functions in travel demand models. *Transportation Research Record 673*, 78–85.

Lerman, S.R. and Manski, C.F. (1976) Alternative sampling procedures for calibrating disaggregate choice models. *Transportation Research Record 592*, 24–8.

Lerman, S.R. and Manski, C.F. (1979) Sample design for discrete choice analysis of travel behaviour: the state of the art. *Transportation Research*, **13**B(1), 29–44.

Lerman, S.R., Manski, C.F. and Atherton, T.J. (1976) *Non-random sampling in the calibration of disaggregate choice models*. Final Report to the Urban Planning Division, Federal Highway Administration, US Department of Transportation, Washington, DC.

Lieberman, E. (1981) Enhanced NETSIM program. *Transportation Research Board Special Report 194*, 32–5.

Liem, T.C. and Gaudry, M.J.I. (1987) P-2: A program for the Box–Cox logit model with disaggregate data. *Publication 525*, Centre de Recherche sur les Transports, Université de Montréal.

Louviere, J.J. (1988) Conjoint analysis modelling of stated preferences: a review of theory, methods, recent developments and external validity. *Journal of Transport Economics and Policy*, **XXII**(1), 93–119.

Louviere, J.J. (guest ed.) (1992) Special issue on experimental choice analysis. *Journal of Business Research* **24**(2), 89–189.

Low, D.E. (1972) A new approach to transportation systems modelling. *Traffic Quarterly*, **26**(3), 391–404.

Lowry, I.S. (1965) A model of a metropolis. *Technical Memorandum RM-4035-RC,* The Rand Corporation, California.

Luce, R.D. and Suppes, P. (1965) Preference, utility and subjective probability. In R.D. Luce, R.R. Bush and E. Galanter (eds), *Handbook of Mathematical Psychology*. John Wiley & Sons, New York.

McDonald, K.G. and Stopher, P.R. (1983) Some contrary indications for the use of household structure in trip generation analysis. *Transportation Research Record 944*, 92–100.

McFadden, D. (1978) Modelling the choice of residential location. In A. Karlquist, L. Lundquist, F. Snickars and J.W. Weibull (eds), *Spatial Interaction Theory and Planning Models*. North-Holland, Amsterdam.

McFadden, D. and Reid, F.A. (1975) Aggregate travel demand forecasting from disaggregate behavioural models. *Transportation Research Record 534*, 24–37.

McKelvey, R.D. and Zavoina W. (1975) A statistical model for the analysis of ordinal level dependent variables. *Journal of Mathematical Sociology* **4**(2), 103–120.

McLeod, W.T. and Hanks, P. (eds.) (1986) *The New Collins Concise Dictionary of the English Language*. William Collins, Sons & Co., Glasgow.

McLoughlin, J. (1969) *Urban and Regional Planning: A Systems Approach*. Faber & Faber, London.

McLynn, J.M. and Woronka, T. (1969) Passenger demand and modal split models. *Report 230, Northeast Corridor Transportation Project*, U.S. Department of Transport, Washington D.C.

McNeil, S. and Hendrickson, C. (1985) A regression formulation of the matrix estimation problem. *Transportation Science*, **19**(3), 278–92.

Mackett, R.L. (1985) Integrated land use-transport models. *Transport Reviews*, **5**(4), 325–43.

Mackinder, I.H., and Evans, S.E. (1981) The predictive accuracy of British transport studies in urban areas. TRRL *Supplementary Report SR 699*, Transport and Road Research Laboratory, Crowthorne.

Maher, M.J. (1983) Inferences on trip matrices from observations on link volumes: a Bayesian statistical approach. *Transportation Research*, **17**B(6), 435–47.

Mahmassani, H.S. and Sinha, K.C. (1981) Bayesian updating of trip generation parameters. *Transportation Engineering Journal*, **107**(TE5), 581–9.

Manheim, M.L. (1979) *Fundamentals of Transportation Systems Analysis*. The MIT Press, Cambridge, Mass.

Manski, C.F. and Lerman, S.R. (1977) The estimation of choice probabilities from choice based samples. *Econometrica*, **45**(8), 1977–88.

Manski, C.F. and McFadden, D. (1981) Alternative estimators and sample designs for discrete choice analysis. In C.F. Manski and D. McFadden (eds), *Structural Analysis of Discrete Data: With Econometric Applications*. The MIT Press, Cambridge, Mass.

Martínez, F.J. (1987) La forma incremental del modelo logit: aplicaciones. *Actas del III Congreso Chileno de Ingeniería de Transporte,* Universidad de Concepción, November 1987, Chile.

Martínez, F.J. (1992) The bid–choice land-use model: an integrated economic framework. *Environment and Planning*, **24**A(6), 871–875.

Mayberry, J.P. (1973) Structural requirements for abstract-mode models of passenger transportation. In R.E. Quandt (ed.), *The Demand for Travel: Theory and Measurement*. D.C. Heath and Co., Lexington, Mass.

Meyer, R.J., Levin, I.P. and Louviere, J.J. (1978) Functional analysis of mode choice. *Transportation Research Record 673*, 1–7.

Moavenzadeh, F., Markow, M., Brademeyer, B. and Safwat, K. (1983) A methodology for intercity transportation planning in Egypt. *Transportation Research*, **17**A(6), 481–91.

Mogridge, M.J.H. (1983) *The Car Market*. Pion, London.

Moore, E.F. (1957) The shortest path through a maze. *Proceedings International Symposium on the Theory of Switching*. Harvard University Press, Cambridge, Mass.

Morikawa, T., Ben-Akiva, M. and Yamada, K. (1992) Estimation of mode choice models with serially correlated RP and SP data. *Proceedings 6th World Conference on Transport Research*, Lyon, June 1992, France.

Morley, R. (1972) *Mathematics for Modern Economics*. Fontana, London.

Moser, C.A. and Kalton, G.K. (1971) *Survey Methods in Social Investigation*. Heinemann Educational Books, London.

Murchland, J.D. (1977) The multiproportional problem. *TSG Note JDM-263*, Transport Studies Group, University College London.

MVA Systematica (1982) *TRIPS highway assignment model*. MVA House, Woking.

Newell, G.F. (1980) *Traffic Flow on Transportation Networks*. The MIT Press, Cambridge, Mass.

Nutt, P.C. (1981) Some guides to the selection of a decision making strategy. *Technological Forecasting and Social Change*, **19**(2), 133–45.

OECD (1974) *Urban traffic models: possibilities for simplification*. OECD Road Research Group, Paris.

Oh, J-H. (1989) Estimation of trip matrices in networks with equilibrium link flows. *Proceedings 17th PTRC Summer Annual Meeting*, University of Sussex, September 1989, England.

Oi, K.I.Y. and Shuldiner, P.W. (1962) *An Analysis of Urban Travel Demands*. Northwestern University Press, Evanston.

Ortúzar, J. de D. (1980a) Mixed-mode demand forecasting techniques. *Transportation Planning and Technology*, **6**(2), 81–95.

Ortúzar, J. de D. (1980b) Modelling park'n'ride and kiss'n'ride as submodal choices: a comment. *Transportation*, **11**(4), 383–5.

Ortúzar, J. de D. (1982) Fundamentals of discrete multimodal choice modelling. *Transport Reviews*, **2**(1), 47–78.

Ortúzar, J. de D. (1983) Nested logit models for mixed-mode travel in urban corridors. *Transportation Research*, **17**A(4), 283–99.

Ortúzar, J. de D. (1986) The cultural and temporal transferability of discrete choice disaggregate modal split models. In T.D. Heaver (ed), *Research for Tomorrow's Transport Requirements*, University of British Columbia, Vancouver.

Ortúzar, J. de D. (1989) Determining the preferences for frozen cargo exports. In World Conference on Transport Research (eds), *Transport Policy, Management and Technology Towards 2001*, Western Periodicals Co, Ventura, Ca.

Ortúzar, J. de D. Editor (1992) *Simplified Transport Demand Modelling*. PTRC Education and Research Services Ltd., London.

Ortúzar, J. de D., Achondo, F.J. and Espinosa, A. (1986) On the stability of logit mode choice models. *Proceedings 14th PTRC Summer Annual Meeting*, University of Sussex, July 1986, England.

Ortúzar, J. de D., Achondo, F.J. and Ivelic, A.M. (1987) Sequential and full information estimation of hierarchical logit models: some new evidence. *Proceedings 11th Triennial Conference on Operations Research*, Buenos Aires, August 1987, Argentina.

Ortúzar, J. de D. and Donoso, P.C.F. (1983) Survey design, implementation, data coding and evaluation for the estimation of disaggregate choice models in Santiago, Chile. *2nd International Conference on New Survey Methods in Transport*, Sydney, September 1983, Australia.

Ortúzar, J. de. D., Donoso, P.C.F. and Hutt, G.A. (1983) The effects of measurement techniques, variable definition and model specification on demand model functions. *Proceedings 11th PTRC Summer Annual Meeting*, University of Sussex, July 1983, England.

Ortúzar J. de D. and Garrido R.A. (1991) Rank, rate or choice? an evaluation of SP methods in Santiago. *Proceedings 19th PTRC Summer Annual Meeting*, University of Manchester Institute of Science and Technology, September 1991, England.

Ortúzar, J. de D. and Garrido, R.A. (1993a) On the semantic scale problem in stated preference rating experiments. *Transportation* (in press).

Ortúzar, J. de D. and Garrido, R.A. (1993b) Estimation of discrete choice models with mixed SP and RP data. *XIII World Conference on Operation Research (IFORS '93)*, Lisbon, July 1993, Portugal.

Ortúzar, J. de D. and Hutt, G.A. (1988) Travel diaries in Chile: the state of the art. *Proceedings 16th PTRC Summer Annual Meeting*, University of Bath, September 1988, England.

Ortúzar, J. de D. and Ivelic, A.M. (1987) Effects of using more accurately measured level-of-service variables on the specification and stability of mode choice models. *Proceedings 15th PTRC Summer Annual Meeting*, University of Bath, September 1987, England.

Ortúzar, J. de D. and Ivelic, A.M. (1988) Influencia del nivel de agregación de los datos en la estimación de modelos logit de elección discreta. *Actas del V Congreso Panamericano de Ingeniería de Tránsito y Transporte*, Universidad de Puerto Rico en Mayagüez, July 1988, Puerto Rico.

Ortúzar, J. de D. and Williams, H.C.W.L. (1982) Una interpretactión geométrica de los modelos de elección entre alternativas discretas basados en la teoría de la utilidad aleatoria. *Apuntes de Ingeniería 7*, 25–50.

Ortúzar, J. de D. and Willumsen, L.G. (1978) Learning to manage transport systems. *Traffic Engineering and Control*, **19**(5), 239–239.

Ortúzar, J. de D. and Willumsen, L.G. (1991) Flexible long range planning using low cost information. *Transportation*, **18**(2), 151–173.

Outram, V.E. and Thompson, E. (1978) Driver route choice–behavioural and motivational studies. *Proceedings 5th PTRC Summer Annual Meeting*, University of Warwick, July 1977, England.

Overgaard, K.R. (1967) Urban transportation planning: traffic estimation. *Traffic Quarterly*, **XXI**(2), 197–218.

Papageorgiou, M. (ed) (1991) *Concise Encyclopedia of Traffic & Transportation Systems*. Pergamon Press, Oxford.

Pape, U. (1974) Implementation and efficiency of Moore algorithms for the shortest route problem. *Mathematical Programming*, **7**(3), 212–22.

Pearmain, D. and Swanson, J. (1990) The use of stated preference techniques in the quantitative analysis of travel behaviour. *Proceedings of the IMA Conference on Mathematics in Transport*, University of Cardiff, September 1990, Wales.

Pearmain, D., Swanson, J., Kroes, E. and Bradley, M. (1991) *Stated preference techniques: a guide to practice.* Steer Davies Gleave and Hague Consulting Group, London.

Quandt, R. and Baumol, W. (1966) The demand for abstract transport modes: theory and measurement. *Journal of Regional Science,* **6**(2), 13–26.

Quarmby, D.A. and Bates, J.J. (1970) An econometric method of car ownership forecasting in discrete areas. *MAU Note 219.* Department of the Environment, London.

Richardson, A.J. (1982) Search models and choice set generation. *Transportation Research,* **16**A(5/6), 403–19.

Roberts, F.S. (1975) Weighted di-graph models for the assessment of energy use and air pollution in transportation systems. *Environment and Planning,* **7**A(6), 703–24.

Robertson, D. (1969) TRANSYT: a traffic network study tool. *TRRL Report LR 253,* Transport and Road Research Laboratory, Crowthorne.

Robillard, P. (1975) Estimating the O–D matric from observed link volumes. *Transportation Research,* **9**(2/3), 123–8.

Rose, G. and Koppelman, F.S. (1984) Transferability of disaggregate trip generation models. In J. Volmüller and R. Hamerslag (eds). *Proceedings of the Ninth International Symposium on Transportation and Traffic Theory.* VNU Science Press, Utrecht.

Ruijgrok, C.J. (1979) Disaggregate choice models: an evaluation. In G.R.M. Jansen, P.H.L. Bovy, J.P.J.M. van Est and F. Le Clercq (eds), *New Developments in Modelling Travel Demand and Urban Systems.* Saxon House, Westmead.

Safwat, K.N.A. and Magnanti, T. (1988) A combined trip generation, trip distribution, modal split and trip assignment model. *Transportation Science,* **22**(1), 14–30.

Schneider, M. (1959) Gravity models and trip distribution theory. *Papers and Proceedings of the Regional Science Association,* **V**(1), 51–6.

Sheffi, Y. (1985) *Urban Transportation Networks.* Prentice-Hall, Englewood Cliffs, N.J.

Sheffi, Y., Hall, R. and Daganzo, C.F. (1982) On the estimation of the multinomial probit model. *Transportation Research,* **16**A(5/6), 447–56.

Sikdar, P.K. and Hutchinson, B.G. (1981) Empirical studies of work trip distribution models. *Transportation Research,* **15**A(3), 233–43.

Skelton, N. (1982) Determining appropriate sample sizes when two means are to be compared. *Traffic Engineering and Control,* **23**(1), 29–37.

Small, K.A. and Brownstone, D. (1982) Efficient estimation of nested logit models: an application to trip timing. *Research Memorandum 296,* Economic Research Program, Princeton University.

Smeed, R.J. (1968) Traffic studies and urban congestion. *Journal of Transport Economics and Policy,* **II**(1), 2–38.

Smith, M.E. (1979) Design of small sample home interview travel surveys. *Transportation Research Record 701,* 29–35.

Smith, M.J. (1979) Traffic control and route choice: a simple example. *Transportation Research,* **13**B(4), 289–94.

Smith, M.J. (1981) Properties of a traffic control policy which ensures the existence of a traffic equilibrium consistent with the policy. *Transportation Research,* **15**B(6), 453–62.

Smith, R.L. and Cleveland, D.E. (1976) Time stability analysis of trip generation and predistribution modal choice models. *Transportation Research Record 569,* 76–86.

Smit-Kroes, N. and Nijpels, E.H.T.M. (1988) *Tweede Struktuurschema Verker en Vervoer.* State Publisher, The Hague.

Smock, R.J. (1962) An iterative assignment approach to capacity restraint on arterial networks. *Highway Research Board Bulletin 156,* 1–13.

Sobel, K.L. (1980) Travel demand forecasting by using the nested multinomial logit model. *Transportation Research Record 775,* 48–55.

Sosslau, A.B., Hassam, A., Carter, M. and Wickstrom, G. (1978) Quick response urban travel estimation techniques and transferable parameters. *NCHRP Report 817,* National Cooperative Highway Research Program, Transportation Research Board, Washington D.C..

Spear, B.D. (1977) *Applications of new travel demand forecasting techniques to transportation: a study of individual choice models*. Final Report to the Office of Highway Planning, Federal Highway Administration, US Department of Transportation, Washington, DC.

Spielberg, F., Weiner, E. and Ernst, U. (1981) The shape of the 1980's: demographic, economic and travel characteristics. *Transportation Research Record 807*, 27–34.

Spiess, H. (1983) On optimal route choice strategies in transit networks. *Publication 286, Centre de Recherche sur les Transports*, Université de Montréal, Canada.

Spiess, H. (1987) A maximum likelihood model for estimating origin-destination matrices. *Transportation Research*, **21**B(5), 395–412.

Spiess, H. and Florian, M. (1989) Optimal strategies: a new assignment model for transit networks. *Transportation Research*, **23**B(2), 82–102.

Steenbrink, P.A. (1974) *Optimisation of Transport Networks*. John Wiley & Sons, New York.

Steer, J. and Willumsen, L.G. (1983) An investigation of passenger preference structures. In S. Carpenter and P.M. Jones (eds), *Recent Advances in Travel Demand Analysis*. Gower, Aldershot.

Stone, R. (1966) *Mathematics in the Social Sciences*. Chapman and Hall, London.

Stopher, P.R. (1975) Goodness-of-fit measures for probabilistic travel demand models. *Transportation*, **4**(1), 67–83.

Stopher, P.R. and McDonald, K.G. (1983) Trip generation by cross-classification: an alternative methodology. *Transportation Research Record 944*, 84–91.

Stopher, P.R. and Meyburg, A.H. (1979) *Survey Sampling and Multivariate Analysis for Social Scientists and Engineers*. D.C. Heath and Co., Lexington, Mass.

Stouffer, A. (1940) Intervening opportunities: a theory relating mobility and distance *American Sociological Review*, **5**(6), 845–67.

Suh, S., Park, C. and Kim, T.J. (1990) A highway capacity function in Korea: measurement and calibration. *Transportation Research*, **24**A(3), 177–186.

Supernak, J. (1979) A behavioural approach to trip generation modelling. *Proceedings 7th PTRC Summer Annual Meeting*, University of Warwick, July 1979, England.

Supernak, J. (1971) Transferability of the person category trip generation model. *Proceedings 9th PTRC Summer Annual Meeting*, University of Warwick, July 1981, England.

Supernak, J. (1983) Transportation modelling: lessons from the past and tasks for the future. *Transportation*, **12**(1), 79–90.

Supernak, J., Talvitie, A.P. and DeJohn, A. (1983) Person category trip generation modelling. *Transportation Research Record 944*, 74–83.

Swanson, J., Pearmain, D. and Loughead, K. (1992) Stated preference sample sizes. *Proceedings 20th PTRC Summer Annual Meeting*. University of Manchester Institute of Science and Technology, September 1992, England.

Tamin, O.Z. and Willumsen, L.G. (1988) Freight demand model estimation from traffic counts. *Proceedings 16th PTRC Summer Annual Meeting*, University of Bath, September 1988, England.

Tamin, O.Z. and Willumsen, L.G. (1989) Transport demand model estimation from traffic counts. *Transportation*, **16**(1), 3–26.

Tanner, J.C. (1974) Forecasts of vehicles and traffic in Great Britain: 1974 revision. *TRRL Report LR 650*, Transport and Road Research Laboratory, Crowthorne.

Tanner, J.C. (1978) Long term forecasting of vehicle ownership and road traffic. *Journal of the Royal Statistical Society*, **141**A(1), 14–63.

Tardiff, T.J. (1976) A note on goodness-of-fit statistics for probit and logit models. *Transportation*, **5**(4), 377–88.

Tardiff, T.J. (1979) Specification analysis for quantal choice models. *Transportation Science*, **13**(3), 179–90.

Taylor, T.L. (1971) *Instructional Planning Systems: A Gaming-Simulation Approach to Urban Problems*. Cambridge University Press, New York.

Terza, J.V. (1985) Ordinal probit: a generalization. *Communications in Statistics–Theory and Method*, **14**(1), 1–11.

Timberlake, R.S. (1988) Traffic modelling techniques for the developing world. *67th Annual TRB Meeting*, Washington DC, January 1988, USA.

Train, K.E. (1977) Valuations of modal attributes in urban travel: questions of non-linearity, non-genericity and taste variations. *Working Paper*, Cambridge Systematics Inc. West, San Francisco.

Train, K.E. (1986) *Qualitative Choice Analysis: Theory, Econometrics and an Application to Automobile Demand*. The MIT Press, Cambridge, Mass.

Train, K.E. and McFadden, D. (1978) The goods/leisure trade-off and disaggregate work trip mode choice models. *Transportation Research*, **12**(5), 349–53.

Traugott, M.W. and Katosh, J.P. (1979) Response validity in surveys of voting behaviour. *Public Opinion Quarterly*, **42**(4), 359–77.

Tverski, A. (1972) Elimination by aspects: a theory of choice. *Psychological Review*, **79**(4), 281–99.

Van Es, J.V. (1982) Freight transport, an evaluation. *ECMT Round Table 58*, European Conference of Ministers of Transport, Paris.

Van Vliet, D. (1977) D'Esopo: a forgotten tree-building algorithm. *Traffic Engineering and Control*, **18**(7/8), 372–5.

Van Vliet, D. (1978) Improved shortest path algorithms for transport networks. *Transportation Research*, **12**(1), 7–20.

Van Vliet, D. (1982) SATURN: a modern assignment model. *Traffic Engineering and Control*, **23**(12), 578–81.

Van Vliet, D. Bergman, T. and Scheltes, W. (1987) Equilibrium assignment with multiple user classes. *Proceedings 15th PTRC Summer Annual Meeting*, University of Sussex, July 1987, England.

Van Vliet, D. and Dow, P. (1979) Capacity restrained road assignment. *Traffic Engineering and Control*, **20**(6), 296–305.

Van Zuylen, H. and Willumsen, L.G. (1980) The most likely trip matrix estimated from traffic counts. *Transportation Research*, **14**B(3), 281–93.

Wardrop, J.G. (1952) Some theoretical aspects of road traffic research. *Proceedings of the Institution of Cival Engineers, Part II*, **1**(36), 325–62.

Wardrop, J.G. (1968) Journey speed and flow in central urban areas. *Traffic Engineering and Control*, **9**(11), 528–32, 539.

Warner, S.L. (1962) *Strategic Choice of Mode in Urban Travel: A Study of Binary Choice*. Northwestern University Press, Evanston.

Webster, F.V., Bly, P.H. and Paulley, N.J. (eds) (1988) *Urban Land-Use and Transport Interaction: Policies and Models*. Gower, Aldershot.

Wermuth, M.J. (1981) Effects of survey methods and measurement techniques on the accuracy of household travel-behaviour surveys. In P.R. Stopher, A.H. Meyburg and W. Brög (eds), *New Horizons in Travel Behaviour Research*. D.C. Heath and Co., Lexington, Mass.

Williams, H.C.W.L. (1977) On the formation of travel demand models and economic evaluation measures of user benefit. *Environment and Planning*, **9**A(3), 285–344.

Williams, H.C.W.L. (1981) Travel demand forecasting: an overview of theoretical developments. In D.J. Banister and P.G. Hall (eds), *Transport and Public Policy Planning*. Mansell, London.

Williams, H.C.W.L. and Ortúzar, J. de D. (1982a) Behavioural theories of dispersion and the mis-specification of travel demand models. *Transportation Research*, **16**B(3), 167–219.

Williams, H.C.W.L. and Ortúzar, J. de D. (1982b) Travel demand and response analysis — some integrating themes. *Transportation Research*, **16**A(5/6), 345–62.

Williams, H.C.W.L. and Senior, M.L. (1977) Model based transport policy assessment: (2) Removing fundamental inconsistencies from the models. *Traffic Engineering and Control*, **18**(10), 464–9.

Williams, I. (1976) A comparison of some calibration techniques for doubly constrained models with an exponential cost function. *Transportation Research*, **10**(2), 91–104.

Wills, M.J. (1986) A flexible gravity-opportunities model for trip distribution. *Transportation Research*, **20**B(2), 89–111.

Willumsen, L.G. (1978) Estimation of an O–D matrix from traffic counts: a review. *Working Paper 99*, Institute for Transport Studies, University of Leeds.

Willumsen, L.G. (1981) Simplified transport demand models based on traffic counts. *Transportation*, **10**(3), 257–78.

Willumsen, L.G. (1982) Estimation of trip matrices from volume counts; validation of a model under congested conditions. *Proceedings 10th PTRC Summer Annual Meeting*, University of Warwick, July 1982, England.

Willumsen, L.G. (1984) Estimating time-dependent trip matrices from traffic counts. In J. Volmüller and R. Hamerslag (eds), *Proceedings of the Ninth International Symposium on Transportation and Traffic Theory*. VNU Science Press. Utrecht.

Willumsen, L.G. (1991) Origin–destination matrix: static estimation. In Papageorgiou, M. (ed), *Concise Encyclopedia of Traffic & Transportation Systems*. 315–322. Pergamon Press, Oxford.

Willumsen, L., Bolland, J., Arezki, Y. and Hall, M. (1993) Multi-modal modelling in congested networks: SATURN + SATCHMO. *Traffic Engineering and Control*, **34**(6), 294–301.

Willumsen, L.G. and Ortúzar, J. de D. (1985) Intuition and models in transport management. *Transportation Research*, **19**A(1), 51–8.

Willumsen, L.G. and Radovanać M. (1988) Testing the practical value of the UMOT model. *International Journal of Transport Economics*, **15**(2), 203–23.

Wilson, A.G. (1970) *Entropy in Urban and Regional Modelling*. Pion, London.

Wilson, A.G. (1974) *Urban and Regional Models in Geography and Planning*. John Wiley & Sons, London.

Wilson, A.G. and Kirby, M.J. (1980) *Mathematics for Geographers and Planners*. Clarendon Press, Oxford.

Wilson, A.G., Rees, P.H. and Leigh, C.M. (eds) (1977) *Models of Cities and Regions: Theoretical and Empirical Developments*. John Wiley & Sons, London.

Wonnacott, T.H. and Wonnacott, R.J. (1977) *Introductory Statistics for Business and Economics*. John Wiley & Sons, New York.

Wootton Jeffreys and Partners (1980) *JAM User Manual*. Brookwood.

Wootton, H.J., Ness, M.P. and Burton, R.S. (1981) Improved direction signs and the benefits for road users. *Traffic Engineering and Control*, **22**(5), 264–8.

Wootton, H.J. and Pick, G.W. (1967) A model for trips generated by households. *Journal of Transport Economics and Policy*, **I**(2), 137–53.

Young, W. and Richardson, A.J. (1980) Residential location preference models: compensatory and non-compensatory approaches. *Proceedings 8th PTRC Summer Annual Meeting*, University of Warwick, July 1980, England.

Zahavi, Y. (1979) The UMOT project. *Report No. DoT-RSPA-DPB-20–79–3*, US Department of Transportation, Washington DC.

Index